Windows 10 for the
Internet of Things

Charles Bell

Windows 10 for the Internet of Things

Charles Bell
Warsaw, Virginia, USA

ISBN-13 (pbk): 978-1-4842-2107-5   ISBN-13 (electronic): 978-1-4842-2108-2
DOI 10.1007/978-1-4842-2108-2

Library of Congress Control Number: 2016956879

Managing Director: Welmoed Spahr
Lead Editor: Jonathan Gennick
Technical Reviewer: Reggie Burnett
Editorial Board: Steve Anglin, Pramila Balan, Laura Berendson, Aaron Black, Louise Corrigan,
    Jonathan Gennick, Todd Green, Robert Hutchinson, Celestin Suresh John, Nikhil Karkal,
    James Markham, Susan McDermott, Matthew Moodie, Natalie Pao, Gwenan Spearing
Coordinating Editor: Jill Balzano
Copy Editor: Kim Burton-Weisman
Compositor: SPi Global
Indexer: SPi Global
Artist: SPi Global

Distributed to the book trade worldwide by Springer Science+Business Media New York, 233 Spring Street, 6th Floor, New York, NY 10013. Phone 1-800-SPRINGER, fax (201) 348-4505, e-mail orders-ny@springer-sbm.com, or visit www.springer.com. Apress Media, LLC is a California LLC and the sole member (owner) is Springer Science + Business Media Finance Inc (SSBM Finance Inc). SSBM Finance Inc is a **Delaware** corporation.

For information on translations, please e-mail rights@apress.com, or visit www.apress.com.

Apress and friends of ED books may be purchased in bulk for academic, corporate, or promotional use. eBook versions and licenses are also available for most titles. For more information, reference our Special Bulk Sales–eBook Licensing web page at www.apress.com/bulk-sales.

Any source code or other supplementary materials referenced by the author in this text are available to readers at www.apress.com. For detailed information about how to locate your book's source code, go to www.apress.com/source-code/. Readers can also access source code at SpringerLink in the Supplementary Material section for each chapter.

Printed on acid-free paper

Contents at a Glance

Contents

About the Author

Dr.Charles Bell conducts research in emerging technologies. He is a member of the Oracle MySQL Development team as a senior developer working on a variety of database administration and high-availability projects. He lives in a small town in rural Virginia with his loving wife. He received his doctorate of philosophy in engineering from Virginia Commonwealth University in 2005. His research interests include database systems, software engineering, sensor networks, and 3D printing. He spends his limited free time as a practicing maker, focusing on microcontroller and 3D printers and printing projects.

About the Technical Reviewer

Reggie Burnett is currently employed as senior software development manager for Oracle Corp., where he is in charge of development projects spanning many different platforms and architectures. Specializing in Windows and .NET technologies, Reggie has written articles for publications such as the *.NET Developers Journal*.

Reggie is married and has four children. He lives in central Tennessee where he plays golf and pool and works on his next geeky project.

Acknowledgments

I would like to thank all of the many talented and energetic professionals at Apress. I appreciate the understanding and patience of my editor, Jonathan Gennick, and managing editor, Jill Balzano. They were instrumental in the success of this project. I would also like to thank the army of publishing professionals at Apress for making me look so good in print. Thank you all very much!

I'd like to especially thank the technical reviewer, Reggie Burnett, for his often-profound insights, constructive criticism, and encouragement. I'd also like to thank my friends for their encouragement and suggestions for things to include in the book.

Most importantly, I want to thank my wife, Annette, for her unending patience and understanding while I spent so much time with my laptop.

Introduction

Internet of Things (IoT) solutions are not nearly as complicated as the name may seem to indicate. Indeed, the IoT is largely another name for what we have already been doing. You may have heard of connected devices or Internet-ready or even cloud-enabled. All of these refer to the same thing—be it a single device such as a toaster or a plant monitor or a complex, multidevice product like home automation solutions. They all share one thing in common: they can be accessed via the Internet to either display data or interact with the devices directly. The trick is applying knowledge of technologies to leverage them to the best advantages for your IoT solution. Until the release of Windows 10 IoT Core, those who use Windows wanting to experiment with IoT solutions and in particular hardware like the Raspberry Pi had to learn a new operating system in order to get started. That is no longer true! In this book, we explore how to leverage Windows 10 in your IoT solutions.

Intended Audience

I wrote this book to share my passion for IoT solutions and Windows 10. I especially wanted to show how anyone could use Windows 10 along with a low-cost computing board to create cool IoT projects—all without having to learn a new operating system!

The intended audience therefore includes anyone interested in learning how to use Windows 10 for IoT projects, such as hobbyists and enthusiasts, and even designers and engineers building commercial Windows 10-based IoT solutions.

How This Book Is Structured

The book was written to guide the reader from a general knowledge of IoT to expertise in developing Windows 10 solutions for the IoT. The first several chapters cover general topics, which includes a short introduction to the Internet of Things, the Windows 10 IoT Core technologies, and some of the available hardware for IoT. Additional chapters are primers on how to write IoT solutions in a variety of programming languages. Rather than focusing on a single language, which often forces readers unfamiliar with the language to learn new skills just to read the book, I've included tutorials in a number of languages to make the book usable by more readers. Throughout the book are examples of how to implement IoT solutions in the various languages. As you will see, some languages are better suited for certain projects. The book contains six detailed and increasingly complex projects for you to explore and enjoy as you develop IoT solutions with Windows 10. There is even a chapter that shows you how to work with Arduino-compatibe microcontroller boards. The book concludes with a look at how to grow beyond the material presented. An appendix listing the hardware components for each chapter is included for your convenience. The following is a brief overview of each chapter in this book.

- *Chapter 1: What Is the Internet of Things?* This chapter answers general questions about the IoT and how IoT solutions are constructed. You are introduced to some terminology describing the architecture of IoT solutions and you are provided examples of well-known IoT solutions. The chapter concludes with a brief introduction to Windows 10.

- *Chapter 2: Introducing the Windows 10 IoT Core.* This chapter presents a version of Windows 10 called the Windows 10 IoT Core that runs on low-cost computers, such as the Raspberry Pi. You discover the basic features of Windows 10, including how to prepare your PC and get started with Windows 10 on your device. You will also see how to boot up the Raspberry Pi with Windows 10!

- *Chapter 3: Introducing the Raspberry Pi.* This chapter explores the Raspberry Pi and how to set up and configure it using the Linux operating system in order to understand the platform and supporting technologies. You'll also discover a few key concepts of how to work with Linux and get a brief look at writing Python scripts, which are used to write Windows 10 IoT applications in later chapters.

- *Chapter 4: Developing IoT Solutions with Windows 10.* This chapter presents a demonstration on how to get started using Visual Studio 2015. The chapter introduces several Windows 10 IoT Core–compatible hardware boards, including the layout of the GPIO headers. The chapter demonstrates how to build, deploy, and test your first Windows 10 IoT Core application.

- *Chapter 5: Windows 10 IoT Development with C++.* This chapter provides a crash course on the basics of C++ programming in Visual Studio, including an explanation of some of the most commonly used language features. As such, this chapter provides you with the skills that you need to understand the growing number of IoT project examples available on the Internet. The chapter concludes by walking through a C++ example project that shows you how to interact with hardware.

- *Chapter 6: Windows 10 IoT Development with C#.* This chapter offers a crash course on the basics of C# programming in Visual Studio, including an explanation of some of the most commonly used language features. As such, this chapter provides you with the skills that you need to understand the growing number of IoT project examples available on the Internet. The chapter concludes by walking through a C# example project that shows you how to interact with hardware.

- *Chapter 7: Windows 10 IoT Development with Python.* This chapter is a crash course on the basics of Python programming in Visual Studio, including an explanation of some of the most commonly used language features. As such, this chapter provides you with the skills that you need to understand the growing number of IoT project examples available on the Internet. The chapter concludes by walking through a Python example project that shows you how to interact with hardware.

- *Chapter 8: Electronics for Beginners.* This chapter presents an overview of electronics for those who want to work with the types of electronic components commonly found in IoT projects. The chapter includes an overview of some of the basics, descriptions of common components, and a look at sensors. If you are new to electronics, this chapter gives you the extra boost that you need to understand the components used in the projects in this book.

- *Chapter 9: The Adafruit Microsoft IoT Pack for Raspberry Pi.* This chapter explores the Adafruit Microsoft IoT Pack for Raspberry Pi 3 and demonstrates a small project that uses the components in the kit (well, mostly) to read data from a simple sensor.

- *Chapter 10: Project 1: Building an LED Power Meter.* This chapter walks through a project using LEDs to display power (volts). You see how to use a potentiometer as a variable input device, read from an analog to digital converter (ADC), learn how to set up and use a serial peripheral interface (SPI), discover a powerful debugging technique, and learn how to create a class to encapsulate functionality.

- *Chapter 11: Project 2: Measuring Light.* This chapter explores a solution that demonstrates how to measure light using a sensor. The project measures the ambient light in the room and then calculates how much power to send to the LED using a technique called *pulse-width modulation* (PWM).

- *Chapter 12: Project 3: Using Weather Sensors.* This chapter demonstrates a very common type of IoT solution—a weather station. In this case, the project uses sensors from the Adafruit kit and implements the code by mixing C# and C++ in the same solution, reusing existing code, and combining it with new code in another language.

- *Chapter 13: Project 4: Using MySQL to Store Data.* This chapter revisits the project from Chapter 12 and modifies it to store the IoT data collected in a MySQL database. Thus, you see an example of how to complete the data storage element of your IoT solutions.

- *Chapter 14: Project 5: Using a Web Server to Control Hardware.* This chapter presents one method for building IoT solutions that control hardware remotely using a web page.

- *Chapter 15: Project 6: Windows IoT and Arduino.* In this chapter, you explore the Arduino platform along with the three Arduino technologies from Microsoft. You begin with a short tutorial on the Arduino and an in-depth look at using the Arduino Wiring libraries. This project combines many of the techniques and components from the previous chapters.

- *Chapter 16: Azure IoT Solutions: Cloud Services.* This chapter presents a few of the newer concepts and features of Microsoft Azure at a high level and in context of a sample project. You can therefore consider this chapter a bonus project chapter.

- *Chapter 17: Where to Go from Here?* This chapter explores what you can do to continue your craft of building IoT solutions. Most people want to simply continue to develop projects for themselves, either for fun or to solve problems around the home or office. However, some want to take their skills to the next level. This chapter shows you how to do just that.

- *Appendix.* Contains a list of the required hardware components for each chapter.

How to Use This Book

This book is designed to guide you through learning more about what the Internet of Things is, discovering the power of Windows 10 IoT Core, and seeing how to build your IoT solutions using the best language suited for the task.

If you are familiar with some of the topics early in the book, I recommend you skim them so that you are familiar with the context presented so that the later chapters—especially the examples—are easy to understand and implement on your own. You may also want to read some of the chapters out of order so that you can get your project moving, but I recommend going back to the chapters you skip to ensure that you get all of the data presented.

If you are just getting started with Windows 10 or are not well versed in using Visual Studio, I recommend reading Chapters 1-9 in their entirety before developing your own IoT solution or jumping to the example projects. That said, many of the examples permit you to build small examples that you can use to help learn the concepts.

Downloading the Code

The code for the examples shown in this book is available on the Apress web site, `www.apress.com`. You can find a link on the book's information page on the Source Code/Downloads tab. This tab is located in the Related Titles section of the page.

Contacting the Author

Should you have any questions or comments—or even spot a mistake you think I should know about—you can contact me, the author, at `drcharlesbell@gmail.com`.

CHAPTER 1

▨ ▨ ▨

What Is the Internet of Things?

Much has been written about the Internet of Things.[1] Some texts expand on the science and technology to implement and manage the Internet of Things, while other texts concentrate on the future or the inevitable evolution of our society as we become more connected to the world around us each and every day. However, you need not dive into such tomes or be able to recite rhetoric to get started with the Internet of Things. In fact, through the efforts of many companies, including Microsoft, you can explore the Internet of Things without intensive training or expensive hardware and software.

This book is intended to be a guide to help you understand the Internet of Things and to begin building solutions that you can use to learn more about it. We will explore many aspects—from understanding what the Internet of Things is, to basic knowledge of electronics, and even how to write custom software for building solutions for the Internet of Things. Best of all, we do so using one of the most popular platforms for personal computers: Windows 10.

So what is this Internet of Things, hence IoT?[2] I'll begin by explaining what it isn't. The IoT is not a new device or proprietary software, or some new piece of hardware. It is not a new marketing scheme to sell you more of what you already have by renaming it and pronouncing it "new and improved."[3] While it is true that the IoT employs technology and techniques that already exist, the way they are employed, coupled with the ability to access the solution from anywhere in the world, makes the IoT an exciting concept to explore. Now let's discuss what the IoT is.

The essence of the IoT is simply interconnected devices that generate and exchange data from observations, facts, and other data, making it available to anyone. While there seems to be some marketing efforts attempting to make anything connected to the Internet an IoT solution or device (not unlike the shameless labeling of everything "cloud"), IoT solutions are designed to make our knowledge of the world around us more timely and relevant by making it possible to get data about anything from anywhere at any time.

As you can imagine, if we were to connect every device around us to the Internet and make sensory data available for those devices, it is clear there is potential for the number of IoT devices to exceed the human population of the planet and for the data generated to rapidly exceed the capabilities of all but the most sophisticated database systems. These concepts are commonly known as *addressability* and *big data*, which are two of the most active and debated topics in IoT.

However, the IoT is all about understanding the world around us. That is, we can leverage the data to make our world and our understanding of it better.

Electronic supplementary material The online version of this chapter (doi:10.1007/978-1-4842-2108-2_1) contains supplementary material, which is available to authorized users.

---

[1] And here's some more!
[2] https://en.wikipedia.org/wiki/Internet_of_Things
[3] For example, everything seems to be cloud-this, cloud-that when in reality nothing was changed.

© Charles Bell 2016
C. Bell, *Windows 10 for the Internet of Things*, DOI 10.1007/978-1-4842-2108-2_1

The Internet of Things and You

How do we observe the world around us? The human body is a marvel of ingenious sensory apparatus that allow us to see, hear, taste, and even feel through touch anything we come into contact with or get near. Even our brains are capable of storing visual and auditory events recalling them at will. IoT solutions mimic many of these sensory capabilities and therefore can become an extension of our own abilities.

While that may sound a bit grandiose (and it is), IoT solutions can record observations in the form of data from one or more sensors. Sensors are devices that produce either analog or digital values. We can then use the data collected to draw conclusions about the subject matter.

This could be as simple as a sensor to detect when a mailbox is opened. In this case, the knowledge we gain from a simple switch opening or closing (depending on how it is implemented and interpreted) may be used to predict when incoming mail has arrived or when outgoing mail has been picked up. I use the term *predict* because the sensor (switch) only tells us the door was opened or closed, not that anything was placed in or removed from the mailbox itself—that would require additional sensors.

A more sophisticated example is using a series of sensors to record atmospheric data such as temperature, humidity, barometric pressure, wind speed, ambient light, rain fall, and so forth, to monitor the weather and perform analysis on the data to predict trends in weather. That is, we can predict within a reasonable certainty that precipitation is in the area.

Now, add the ability to see this data not only in real time (as it occurs), but also remotely from anywhere in the world and the solution becomes more than a simple weather station. It becomes a way to observe the weather about one particular place from anywhere in the world.

This example may be a bit commonplace since you can tune into any number of television, Web, and radio broadcasts to hear the weather from anywhere in the world. But consider the implications of building such a solution in your home. Now you can see data about the weather at your own home from anywhere!

In the same way, but perhaps on a smaller scale, we can build solutions to monitor plants to help us understand how often they need water and other nutrients. Or perhaps we can monitor our pets while we are away at work. Further, we can record data about wildlife in our area to better understand our effect on nature.

IoT Is More Than Just Connected to the Internet

So if a device is connected to the Internet, does that make it an IoT solution? That depends on whom you ask. Some believe the answer is yes. However, others (like me) contend that the answer is not unless there is some benefit from doing so.

For example, if you could connect your toaster to the Internet, what would be the benefit of doing so? It would be pointless (or at least extremely eccentric) to get a text on your phone from your toaster stating that your toast is ready. So in this case, the answer is no. However, if you have someone—such as a child or perhaps an older adult—whom you would like to monitor, it may be helpful to be able to check to see how often and when they use the toaster. That is, you can use the data to help you make decisions about their care and safety.

Allow me to illustrate with another example. I was fortunate to participate in a design workshop held on the Microsoft campus in the late 1990s. During our tour of the campus, we were introduced to the world's first Internet-enabled refrigerator (also called a *smart refrigerator*). There were sensors in the shelves to detect the weight of food. It was suggested that, with a little ingenuity, you could use the sensors to notify your grocer when your milk supply ran low, which would enable people to have their grocery shopping not only online but also automatic. This would have been great if you lived in a location where your grocer delivers, but not very helpful for those of us who live in rural areas. While it wasn't touted an IoT device (the term was coined later), many felt the device illustrated what could be possible if devices were connected to the Internet.

Thus, being connected to the Internet doesn't make something IoT. Rather, IoT solutions must be those things that provide some meaning—however small that has benefit is to someone or some other device or service. More importantly, IoT solutions allow us to sense the world around us and learn from those observations. The real tricky part is in how the data is collected, stored, and presented. We will see all of these in practice through examples in later chapters.

IoT solutions can also take advantage of companies that provide services that can help enhance or provide features that you can use in your IoT solutions. These features are commonly called *IoT services* and range from storage and presentation to infrastructure services, such as hosting.

IoT Services

Sadly, there are companies that tout having IoT products and services that are nothing more than marketing hype—much like what some companies have done by prepending "cloud" or appending "for the cloud" to the name. Fortunately, there are some really good products and services being built especially for IoT. These range from data storage and hosting to specialized hardware.

Indeed, businesses are adding IoT services to their product offerings and it isn't the usual suspects, such as the Internet giants. I have seen IoT solutions and services being offered by Cisco, AT&T, HP, and countless start-ups and smaller businesses. I use the term *IoT vendor* to describe those businesses that provide services for IoT solutions.

You may be wondering what these services and products are and why someone would consider using them. That is, what is an IoT service and why would you decide to buy it? The biggest concerns in the decision to buy a service are cost and time to market.

If your developers do not have the resources or expertise, and obtaining them will require more than the cost of the service, it may be more economical to purchase the service. However, you should also consider any additional software or hardware changes (sometimes called *retooling*) necessary in the decision. I once encountered a well-meaning and well-documented contracted service that permitted a product to go to market sooner than projected at a massive savings. Sadly, while the champions of that contract won awards for technical achievement, they failed to consider the fact that the systems had to be retooled to use the new service. More specifically, it took longer to adopt the new service than it would to write one from scratch. So instead of saving money, the organization spent nearly twice the original budget and were late to market. Clearly, you must consider all factors.

Similarly, if your time is short or you have hard deadlines to meet to make your solution production-ready, it may be quicker to purchase an IoT service rather than create or adapt your own. This may require spending a bit more, but in this case, the motivation is time and not (necessarily) cost. Of course, in reality project planning is a balance of cost, time, and features.

So what are some of the IoT services available? The following lists a few that have emerged in the last few years. It is likely more will be offered as IoT solutions and services mature.

- *Enterprise IoT data hosting and presentation*. Services that allow your users to develop enterprise IoT solutions from connecting to, managing, and customizing data presentation in a friendly form, such as graphs, charts, and so forth. Example: Xively (`https://xively.com`)

- *IoT data storage*. Services that permit you to store your IoT data and get simple reports. Example: SparkFun's IoT Data service (`https://data.sparkfun.com`)

- *Networking*. Services that provide networking and similar communication protocols or platforms for IoT. Most specialize in machine-to-machine (M2M) services Example: AT&T's cellular global SIM service (`business.att.com/enterprise/Family/mobility-services/internet-of-things`)

- *IoT hardware platforms*. Vendors that permit you to rapidly develop and prototype IoT devices using a hardware platform and a host of supported modules and tools for building devices ranging from a simple component to a complete device. Example: Intel's IoT gateway development kits (`www-ssl.intel.com/content/www/us/en/embedded/solutions/iot-gateway/overview.html`)

Now that you know more about what IoT is, let's look at a few examples of IoT solutions to get a better idea of what IoT solutions can do and how they are employed.

A Brief Look at IoT Solutions

An IoT solution is simply a set of devices designed to produce, consume, or present data about some event or series of events or observations. This can include devices that generate data, such as a sensor, devices that combine data to deduce something, devices or services designed to tabulate and store the data, and devices or systems designed to present the data. Any or all of these may be connected to the Internet.

IoT solutions may include one or all of these qualities, whether it is combined into a single device such as a web camera; use a sensor package and monitoring unit, such as a weather station; or use a complex system of dedicated sensors, aggregators, data storage, and presentation, such as complete home automation system. Figure 1-1 shows a futuristic picture of all devices—everywhere—connected to the Internet through databases, data collectors or integrators, display services, or other devices.

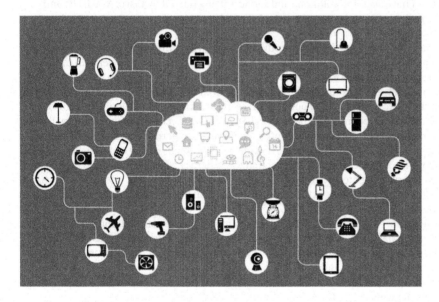

Figure 1-1. *The future of IoT – all devices, everywhere*[4]

Let's take a look at some example IoT solutions. The IoT solutions described in this section are a mix of solutions that should give you an idea of the ranges of sizes and complexities of IoT solutions. I also point out how some of these solutions leverage services from IoT vendors.

Sensor Networks

Sensor networks are one of the most common forms of IoT solutions. Simply stated, sensor networks allow you to observe the world around you and make sense of it. Sensor networks could take the form of a pond monitoring system that alerts you to water level, water purity (contamination), or water temperature; or detects predators; or even turns on features automatically, such as lighting or fish feeders.

---

[4]https://pixabay.com/en/network-iot-internet-of-things-782707/

If you, or someone you know, have spent any time in a medical facility, it's likely that a sensor network was employed to monitor body functions, such as temperature, cardiac and respiratory rates, and even movement. Modern automobiles also contain sensor networks dedicated to monitoring the engine, climate, and even in some cars, road conditions. For example, the lane-warning feature uses sensors (typically a camera, microprocessor, and software) to detect when you drift too far toward lane or road demarcations.

Thus, sensor networks employ one or more sensors that take measurements (observations) about an event or state, and communicate that data to another component or node in the network, which is then presented, in some form or another, for analysis. Let's take a look at an example of an important medical IoT solution.

Medical Applications

Medical applications—including health monitoring and fitness—are gaining a lot of attention as consumer products. These solutions cover a wide range of capabilities, such as the fitness features built into the Apple Watch to fitness bands that keep track of your workout, and even medical applications that help you control life-threatening conditions. For example, there are solutions that can help you manage diabetes.

Diabetes is a disease that affects millions of people worldwide (www.diabetes.org). There are several forms: the most serious being type 1 (www.diabetes.org/diabetes-basics/type-1/?loc=db-slabnav). Those afflicted with type 1 diabetes do not produce enough (or any) insulin due to genetic deficiencies, birth defects, or injuries to the pancreas. Insulin is a hormone that the body uses to extract a simple sugar called *glucose*, which is created from sugars and starches, from blood for use in cells.

Thus, type 1 diabetics must monitor their blood glucose to ensure that they are using their medications (primarily insulin) properly and balanced with a healthy lifestyle and diet. If their blood glucose levels become too low or too high, they can suffer from a host of symptoms. Worse, extremely low blood glucose levels are very dangerous and can be fatal.

One of the newest versions of a blood glucose tester consists of a small sensor that is inserted in the body along with a monitor that connects to the sensor via Bluetooth. You wear the monitor on your body (or keep it within 20 feet at all times). The solution is marketed by Dexcom (dexcom.com) and is called a *continuous glucose monitor* (CGM) that permits the patient to share their data to others via their phone. Thus, the patient pairs their CGM with their phone and then shares the data over the Internet to others. This could be loved ones, those that help with their care, or even medical professionals. Figure 1-2 shows an example of the Dexcom CGM monitor and sensor. The monitor is on the left and the sensor and transmitter are on the right. The sensor is the size of a small syringe needle and remains inserted in the body for up to a week.

Figure 1-2. Dexcom continuous glucose monitor with sensor

WHAT ABOUT BLOOD GLUCOSE TESTERS (GLUCOMETERS)?

Until solutions like the Dexcom CGM came about, diabetics had to use a manual tester. Traditional blood glucose testers are single-use events that require the patient to prick their finger or arm and draw a small amount of blood onto a test strip. While this device has been used for many years, it is only recently that manufacturers have started making blood glucose testers with memory features and even connectivity to other devices, such as laptops or phones. The ultimate evolution of these devices is a solution like Dexcom, which has become a medical IoT device that improves the quality of life for diabetics.

Dexcom also provides a free Windows application called Dexcom Studio (http://dexcom.com/dexcom-studio) to allow patients to see the data collected and generate a host of reports they can use to see their glucose levels over time. Reports include averages, patterns, daily trends, and more. They can even share their data with their doctor. Figure 1-3 shows an example of the Dexcom Studio with typical data loaded.

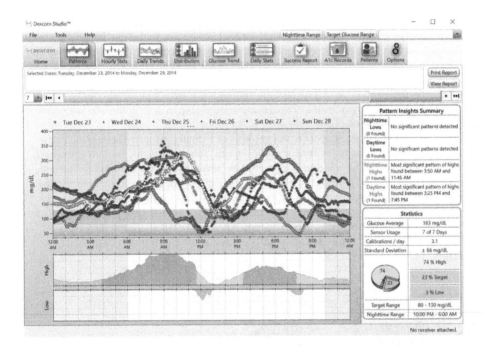

Figure 1-3. Dexcom Studio

A feature called Dexcom Share permits the patient to make their data available to others via an app on their phone. That is, the patient's phone transmits data to the Dexcom cloud servers, which is then sent to anyone who has the Dexcom Share app and has been given permission to see the data. Figure 1-4 shows an example of the Dexcom Share CGM report from the Dexcom Share iOS app, which allows you to easily and quickly check the blood glucose of a friend or loved one.

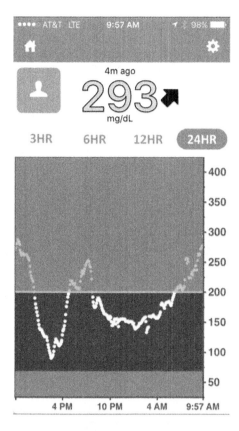

Figure 1-4. *Dexcom Share app report*

Not only does the app allow the visualization of the data, it can also relay alerts for low or high blood glucose levels, which has profound implications for patients who suffer from additional ailments or complications from diabetes. For example, if the patient's blood glucose level drops while they are alone, incapacitated, or unable to get treatment, loved ones with the Dexcom Share app can respond by checking on the patient and potentially avoiding a critical diabetic event.

While this solution is a single sensor connected to the Internet via a proprietary application, it is an excellent example of a medical IoT device that can enhance the lives of not only the patient but everyone who cares for them.

Combined with the programmable alerts, you and your loved ones can help manage the effects of diabetes. If you have a loved one who suffers with diabetes, a CGM is worth every penny for peace of mind alone. This is the true power of IoT materialized in a potentially life-saving solution.

Automotive IoT Solutions

Another personal IoT solution is the use of Internet-connected automotive features. One of the oldest products is called OnStar (`onstar.com`), which is available on most late-model and new General Motors (GM) vehicles. While OnStar predates the IoT evolution, it is a satellite-based service that has several levels and many fee-based options, it incorporates the Internet to permit communication with vehicle owners. Indeed, the newest GM vehicles come with a Wi-Fi access point built into the car! Better still, there are some basic features that are free to GM owners that, in my opinion, are very valuable.

The free, basic features include regular maintenance reports sent to you via e-mail, and the ability to use an app on your phone to remotely unlock, lock, and start the car—all the features on your key fob. This is a really cool feature if you have every locked your keys in your car! Figure 1-5 shows an example of the remote key fob app on iOS. Of course, there are even more features available for a fee, including navigation, telephone, Wi-Fi, and on-call support.

Figure 1-5. OnStar app key fob feature

The OnStar app works by connecting to the OnStar services in the cloud, requesting the feature (e.g., unlock) that is sent to the vehicle via the OnStar satellite network. So it is an excellent example of how IoT solutions use multiple communication protocols.

The feature I like most is the maintenance reports. You will receive an e-mail with an overview of the maintenance status of your vehicle. The report includes such things as oil life, tire pressure, engine and transmission warnings, emissions, air bag, and more. Figure 1-6 shows an excerpt of a typical e-mail that you receive.

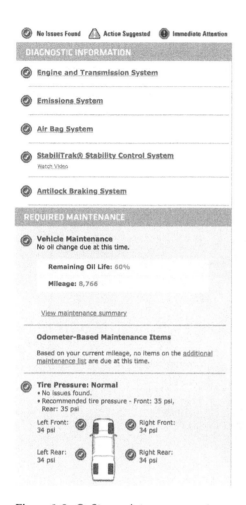

Figure 1-6. *OnStar maintenance report*

Notice the information displayed. This is no mere idiot light! Actual data is transmitted to OnStar from your vehicle. For example, the odometer reading and tire pressure data is taken directly from the vehicle's onboard data storage. That is, data from the sensors is read, interpreted, and the report generated for you. This feature demonstrates how automatic compilation of data in an IoT solution can help us keep our vehicles in good mechanical condition with early warning of needed maintenance. This serves us best by helping us keep our vehicles in prime condition and thus in a state of high resell value.

I should note that GM isn't the only automotive manufacturer offering such services. Many others are working on their own solutions, ranging from an OnStar-like feature set to solutions that focus on entertainment and connectivity.

Fleet Management

Another example of an IoT solution is a *fleet management system*.[5] While developed and deployed well before the coining of the phrase Interest of Things, fleet management systems allow businesses to monitor their cars, trucks, ships—just about any mobile unit—to not only track their current location but also to use the location data (GPS coordinates taken over time) to plan more efficient routes, thereby reducing the cost of shipment.

Fleet management systems aren't just for routing. Indeed, fleet management systems also allow businesses to monitor each unit to conduct diagnostics. For example, it is possible to know the amount of fuel in each truck, when its last maintenance was performed—or more importantly, when the next maintenance is due, and much more. The combination of vehicle geographic tracking and diagnostics is called *telematics*. Figure 1-7 shows a drawing of a fleet management system.

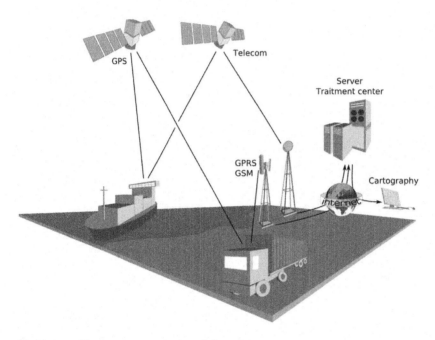

Figure 1-7. *Fleet management example[6]*

In Figure 1-7 you see the application of GPS systems to track location as well as satellite communication to transmit additional data, such as diagnostics, payload states, and more. All of these ultimately traverse the Internet and the data becomes accessible by the business analysts.

You may think fleet management systems are only for large shipping companies, but with the proliferation of GPS modules and even the microcontroller market, anyone can create a fleet management system. That is, they don't cost millions of dollars to develop.

For example, if you owned a bicycle delivery company, you could easily incorporate GPS modules with either cellular or wireless connectivity on each delivery person to track their location, average travel time, and more. More specifically, you can use such a solution to minimize delivery times by allowing packages to be handed off from one delivery person to another, rather than having them return to the depot each time they complete a set of deliveries.

---

[5]https://en.wikipedia.org/wiki/Fleet_management
[6]Éric Chassaing - via CC BY-SA 3.0 (http://creativecommons.org/licenses/by-sa/3.0/).

CAMERA DRONES AND THE IOT

One possible use of the IoT is making data that drones generate available over the Internet. Some people feel that drones are an invasion of privacy. I agree in situations where they are misused or established laws are violated. Fortunately, the vast majority of drone owners obey local laws, regulations, and property owners' wishes.[7]

However, there are many legitimate uses of drones, be they land-, air-, or sea-based. For example, I can imagine home monitoring solutions where you can check on your home remotely by viewing data from fixed cameras, as well as data from mobile drones. I for one would love to see a solution that allowed me to program a predetermined sentry flight path to monitor my properties with a flying camera drone.

While some vendors have Wi-Fi-enabled drones, there aren't many consumer-grade options available that stream data in real time over the Internet. However, it is just a matter of time before we see solutions that include drones. Of course, the current controversy and the movement of the US government to register and track drones, along with increasing restrictions on their use, may limit the expansion of drones and IoT solutions that include drone-acquired data.

IoT and Security

The recent rash of massive data breaches proves that basic security simply isn't good enough. We've seen everything from outright theft to exploitation of the data stolen from very well-known businesses, like Target (over 40 million credit card numbers may have been compromised), and government agencies, like the United States Office of Personnel Management (over 20 million Social Security numbers compromised).

IoT solutions are not immune to security threats. Indeed, as IoT solutions become more and more integrated into our lives, so too will our personal data. Thus, security must be taken extremely seriously and built into the solution from the start.

This includes solutions that we develop ourselves. More specifically, if you design a weather station for your own use, you should take reasonable steps to ensure that the data is protected from both accidental and deliberate exploitation. You may think weather data isn't a high risk but consider the case where you include GPS coordinates for your sensors (a reasonable feature) so that people can see where this weather is being observed. If someone could see that information and determine the solution uses an Internet connection, it is possible they could gain physical access to the Internet device and possibly use it to further penetrate and exploit your systems. Thus, security isn't just about the data; it should encompass all aspects of the solution—from data to software, to hardware to physical access.

There are four areas where you may want to consider spending extra care ensuring that your IoT solution is protected with good security. As you will see, this includes a number of things you should consider for your existing infrastructure, computers, and even safe computing habits. By leveraging all of these areas, you will be building a layered approach to security; often called a *defense-in-depth method*.

Security Begins at Home

Before introducing an IoT solution to your home network, you should consider taking precautions to ensure that the machines on your home network are protected. Some of the best practices for securing your home networking include the following.

---

[7]As of 21 December 2015, drones in the US that weigh more than 0.55 lbs. must be registered before flying. See https://registermyuas.faa.gov/.

- *Passwords.* This may seem like a simple thing, but always make sure that you use passwords on all of your computers and devices. Also, adopt good password habits, such as requiring longer strings, mixed case, numbers, and symbols to ensure that the passwords are not easily guessed.[8]

- *Secure your Wi-Fi.* If you have a Wi-Fi network, make sure that you add a password and use the latest security protocols, such as WPA2, or even better, the built-in secure setup features of some wireless routers.

- *Use a firewall.* You should also use a firewall to block all unused ports (TCP or UDP). For example, lock down all ports except those your solution uses, such as port 80 for HTML.

- *Restrict physical access.* Lock your doors! Just because your network has a great password and your computers use super world-espionage spy-encrypted biometric access, these things are meaningless if someone can gain access to your networking hardware directly. For IoT solutions, this means any external components should be installed in tamper-proof enclosures or locked away so that they cannot be discovered. This also includes any network wiring.

Secure Your Devices

As I mentioned, your IoT devices also need to be secured. The following are some practices to consider.

- *Use passwords.* Always add passwords to the user accounts you use on your IoT devices. This includes making sure that you rename any default passwords. For example, you may be tempted to consider a wee Raspberry Pi or BeagleBone Black too small of a device to be a security concern, but if you consider that these devices run one of the most powerful operating systems available (forms of Linux), a Raspberry Pi can be a very powerful hacking tool.

- *Keep your software up-to-date.* You should try to use the latest versions of any software that you use. This includes the operating system as well as any firmware that you may be running. Newer versions often have improved security or fewer security vulnerabilities.

- *If your software offers security features, use them.* If you have servers or services running on your devices, and they offer features such as automatic lockout for missed passwords, turn them on. Not all software has these features, but if they are available, they can be a great way to defeat repeated attacks.

Use Encryption

This is one area that is often overlooked. You can further protect yourself and your data if you encrypt the data as it is stored and the communication mechanism as it is transmitted. As long as you encrypt your data, it cannot be easily deciphered, even if someone were to gain physical access to the storage device. Use the same care with your encryption keys and passcodes as you do your computer passwords.

---

[8]You also need to balance complexity of passwords with your ability to remember them. If you have to write it down, you've just defeated your own security!

Security Doesn't End at the Cloud

There are many considerations for connecting IoT devices to cloud services. Indeed, Microsoft has made it very easy to use cloud services with your IoT solutions. However, there are two important considerations for security and your IoT data.

- *Do you need the cloud?* The first thing you should consider is whether you need to put any of your data in the cloud. It is often the case that cloud services make it very easy to store and view your data, but is it really necessary to do so? For example, you may be eager to view logistical data for where your dog spends his time while you are at work, but who else would really care to view this data? In this case, storing the data in the cloud to make it available to everyone is not really necessary.

- *Don't relax!* Many people seem to let their guard down when working with cloud services. For whatever reason, they consider the cloud more secure. The fact is—it isn't! In fact, you must apply the very same security best practices when working in the cloud that you do for your own network, computers, and security policies. Indeed, if anything, you need to be even more vigilant because cloud services are not in your control with respect to protecting against physical access (however remote and unlikely) nor are you guaranteed your data isn't on the same devices as tens, hundreds, or even thousands of other users' data.

Now that you have an idea of how you should include security in your projects, let's look at how Windows 10 has evolved into a modern platform that supports not only the usual productivity and gaming tasks but also help us build IoT solutions.

Introducing Windows 10

Microsoft has not been idle in recent years. In fact, the latest release of the Windows operating system, Windows 10, has shown Microsoft listens to its customers and delivers features that people want. More than any release in the past, Windows 10 is both familiar and capable on desktop, laptop, tablets, and even phones.

Sadly, the road to where we are now was not without its bumps and detours. Windows Vista, Windows 7, and later, Windows 8, made some dramatic changes that resulted in a lot of dissatisfied users. For example, making the Start menu a panel instead of a menu both alienated and confused many users. The avalanche of doubt continued to slide as the PC itself evolved from a bland box on the floor to a powerful machine that you can hold in your hands. But this evolution hasn't been without its problems and detours.

I've worked with and discussed the merits of various operating systems with many people. Most had loyalties for a particular platform that stemmed from familiarity, if not emotional attachment, more than technical merit. The rest were open-minded enough to choose to use a wide variety of platforms. Some people, like me, use several platforms every day at work with proficiency in all the major platforms: Linux, Mac, and Windows.

However, even some of the Windows faithful reluctantly agreed that they missed key features of other operating systems. So while they loved and used Windows exclusively—even choosing phones and tablets that run Windows rather than the more popular iOS and Android devices—they both loved and hated what Windows had become. Fortunately, Windows 10 makes up for many of these doubts and has redeemed itself quite well among the PC community.

As a long-term platform-independent user, I've had my favorites over the years, but some versions of Windows have not been high on the list and at times not on the list at all. This was mostly due to the changing face of the PC from beige boxes[9] to personal, tactile, sensitive devices through the proliferation of tablets and other smart devices.

---

[9]Yes, I was using PCs when IBM put the PC in personal computer. My first PC had an Intel 8088 processor running at 8MHz with a modest 512KB of memory. Most phones exceed these capabilities by orders of magnitude.

However, with the release of Windows 10, I count my Surface Pro 3 tablet not only a great laptop replacement (and the best PC tablet available) but also a platform that I am very comfortable using and wouldn't hesitate to use for almost any task. It just works the way a Windows tablet should. In fact, I used my Surface Pro 3 for all the examples in this book.

The following sections introduce a number of the newest features of Windows 10, including some familiar behavior that has been missing for some time, and some new things that make using Windows 10 across several platforms seamless. I have included this information for those that have yet to experience Windows 10 or those that have delayed upgrading. In order to use this book, you need a machine running Windows 10. If you have not upgraded yet, the following sections will be helpful.

However, if you are already using Windows 10 or have been for some time, you may want to skim this section so that you are familiar with the newest features. I have found that it is always helpful to read the impressions of others because I often discover features that I wasn't aware of or had yet to encounter. Plus, it gives me a greater depth of knowledge on the subject.

Overview of Windows 10 Features

This section explores the major advances and new features of Windows 10. If you are thinking about upgrading to Windows 10, this section should convince you to do so, because it covers what the latest Windows operating system has become. I cover the most important features related to developing IoT applications. Thus, this is not a complete list of the many features of the new version. For a complete list, see the Microsoft Windows 10 site (`www.microsoft.com/en-us/windows/features`). Let's begin with the most amazing feature of all: Windows is now free to download and install.

Free Windows? That's Insane!

Is it really such an insane concept to make an operating system free to users? If you consider that Linux distributions have been free (technically open source) for decades, and Apple's macOS, formerly known as Mac OS X, (and iOS) have been free for several years, it was only a matter of time before Microsoft had to follow suit and make Windows 10 free. Fortunately, all that you need (software-wise) is free to download and use. While there are some items that have fee-based alternatives or versions with greater features, you can experience Windows 10 in the IoT without buying any software. Indeed, I demonstrate how to do so in later chapters.

Getting free versions is not limited to the operating system, which is a huge part of the solution. The other half of that equation is the development suite of tools—all of which are free. Not only can you download a free version of Visual Studio 2015 called the Community Edition[10] (`visualstudio.com/products/visual-studio-community-vs`) that looks and works like its fee-based alternatives, you can also download and install all the libraries, examples, and sample tools needed to develop a complete IoT solution that runs on Windows 10.

The Return of the Desktop

One of the evolutions I found to be most unappealing was the shift from a desktop with a Start menu to that of a panel of small application interfaces. While I understand the reason for the evolution (the rise of the touch screen and tablets), I found the dual interface of Windows 8 confusing. It was as if the operating system had two heads: one for "legacy" users, complete with an abbreviated Start menu, and another for "new" touch-enabled applications (which seemed to only include the latest Office applications). Switching from one to another—particularly on a typical desktop without a touch screen—was awkward and often frustrating.

---

[10]Free only for personal and non-commercial use. You can use the paid version of Visual Studio if you already have it installed.

Fortunately, Windows 10 brings back the desktop concept with an all-new design that incorporates the best of the Windows 8 Start screen with a much improved Start menu. That is, we once again have the familiar menu, floating windows, tray, and more. Figure 1-8 shows a snapshot of the new desktop. Does this look familiar?

Figure 1-8. *Windows 10 desktop*

If you use a tablet or a machine that can switch from laptop to tablet (also called a 2-in-1), Windows allows you to use the new desktop when in desktop mode (a keyboard is attached) and the more tablet-friendly Start screen when in tablet mode (the keyboard is removed). Of course, you can configure this behavior to your liking.

■ **Tip** You can access the power user menu by pressing Windows key + X.

After having used the new desktop for some time, I must say it just works the way a PC should. That may seem like false praise, but it isn't. There are many things about Windows 10 that work like it should—as it should have in previous versions.

One feature of the desktop I welcome more than any other is the use of virtual desktops. The other platforms I use have had this feature for some time. Having it in Windows 10 allows me to use my PCs in a very familiar manner: by placing my most frequently used (and running) applications maximized in their own desktops.

To create a new virtual desktop, click the task view icon on the taskbar (see the red arrow in Figure 1-9). You will then see a pop-up pane that shows a thumbnail of all the virtual desktops that are active. To add a new virtual desktop, click the plus sign to the right. You can close a virtual desktop by clicking the X icon on the thumbnail. Figure 1-9 shows how to access the virtual desktop feature. You may find using virtual desktops to be very helpful when developing applications or working on productivity applications alongside your mail and other communication applications.

Figure 1-9. *Virtual desktops in Windows 10*

The task view is also used in tablet mode. In fact, the virtual desktop is very similar to the tablet mode of the desktop. You can switch from one to the other by using the task view icon.

Compatibility

With so many changes and new features, it is reasonable to expect Windows 10 to have problems running older applications. However, Microsoft has worked very hard to make the new version run the older applications. In fact, I have several rather old (Windows XP era) applications that I've installed on Windows 10 and they all work well. There are a few things that you can do to adjust compatibility, but most applications should run unaltered and without jumping through menus to customize. Thus, if you are concerned about being able to run your older applications, you should not have to worry.

Notifications and Action Center

The Action Center is an interesting feature that allows application developers to display notices. The Action Center is accessed by a right edge swipe or by clicking the Action Center icon on the system tray.

When applications trigger a notification, the Action Center displays the notification briefly as a small fly-in dialog. I like this new feature, especially since the other platforms I've used have their own implementation. Not only is it convenient to know what's going on—such as getting new e-mails or receiving a bid on your auction—it is also a great way to take a look at the events of the day. Just swipe and view all of your notifications in a list (see Figure 1-10).

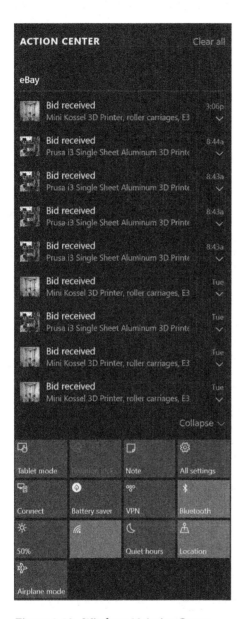

Figure 1-10. *Windows 10 Action Center*

At the bottom of the Action Center are a number of shortcut buttons for many common tasks, which puts the action in Action Center. Here we find buttons for switching to tablet mode, brightness, Bluetooth, network connections, and more. The Action Center, coupled with the new desktop and Start menu, complete the user experience.

Fortunately, you can develop your IoT solutions to include notification to users. I encourage you to do so in your own Windows applications.

Edge: The New Web Browser

There is a new web browser in town. Yes, Microsoft has abandoned Internet Explorer with the release of Edge in Windows 10. Edge is a complete rewrite rather than an evolution of IE. Many of the latest features that Internet users have grown to expect are being incorporated into Edge.

Edge works well for most normal uses. It is faster and has much better security than IE, but for some of your IoT experiments and IoT solutions, you may to use another browser. I installed Firefox because I found Edge to be—well, not complete, and it acted a little strangely when connecting to some of my projects. You may not experience this, but if you do, try another browser. I do not consider this a compatibility problem, but rather more of a limited feature set issue. I am certain later updates to Windows 10 will improve Edge and refine the rough spots.

Windows 10 and the IoT

There are several versions of Windows 10—ranging from those that run on phones to those that run on desktop computers and laptop replacement tablets. Indeed, Microsoft Windows 10 runs on more devices than any previous version of Windows,[11] but that isn't the end of the Windows 10 proliferation. Windows 10 also comes in a version designed to run on low-cost computing hardware, such as single-board computers designed for integration with hardware and embedding in other solutions, which makes it an excellent choice for use in IoT solutions.

The Windows 10 version designed to run on low-cost computing boards is called the Windows 10 IoT Core. We use Windows 10 IoT Core in this book. The Windows 10 IoT Core is designed to run on smaller devices, such as the Raspberry Pi, MinnowBoard Max, and other small computing boards. In essence, it is optimized to run in smaller memory without the need for advanced processors or even a graphical user interface. Thus, it supports only console or background applications.

Windows 10 IoT Core supports the Universal Windows Platform (UWP), allowing you to create your applications and deploy them. As mentioned previously, we will use Visual Studio 2015 to build these applications. As you will see in Chapter 4, Visual Studio 2015 includes all the tools you need to build UWP applications for deployment to the Raspberry Pi. This includes APIs and drivers for accessing the general-purpose input/output (GPIO) pins, as well as interfaces such as I2C and SPI. Best of all, there are a host of example code that you can use for your own projects.

Windows 10 IoT also includes the Arduino Wiring API that permits the use of Arduino-like direct hardware access. You learn more about the Arduino in Chapter 15.

This really is an exciting element to Windows 10. Indeed, except for some rather limited exploration of Windows Embedded Compact (Windows CE), Windows 10 represents the first time that you can use the Windows 10 IoT Core to leverage the power of Windows on your smaller devices. You explore the Windows 10 IoT Core in greater detail in the next chapter.

Summary

The Internet of Things is an exciting new world for us all. Those of us young at heart but old enough to remember *The Jetsons* TV series recall seeing a taste of what is possible in the land of make believe. Talking toasters, flying cars that spring from briefcases, and robots with attitude notwithstanding, television fantasy of decades ago is now coming true. We have wristwatches that double as phones and video players. We can unlock our cars from around the world, find out if our dog has gone outside, and even answer the door from across the city. All of this is possible and working today with the advent of the IoT.

---

[11]http://www.winbeta.org/news/as-2015-comes-to-a-close-windows-10-surpasses-200-million-installs

In this chapter, we discovered what the IoT is and saw some examples of well-known IoT solutions. We also discovered how Microsoft is opening doors for Windows users by expanding its Windows 10 operating system to the IoT via the Raspberry Pi and similar low-cost computing hardware. This is a very exciting opportunity for people who do not want to learn the nuances of a Linux-based operating system to explore the world of hardware and IoT from a familiar and well understood platform.

In the next chapter, we will discover more about the Windows 10 IoT Core including what hardware it runs on and how to get started running Windows 10 on a Raspberry Pi. As you will see, it is not difficult. We will then explore the Raspberry Pi in more detail in Chapter 3 to complete our tour of getting started with Windows 10 IoT Core.

CHAPTER 2

■ ■ ■

Introducing the Windows 10 IoT Core

Windows 10 represents an exciting new entry in the IoT arena. Finally, Windows users have a native toolset to experiment with building IoT solutions. While some platforms such as the Arduino are very Windows-friendly, other platforms have forced users to learn about new, sometimes very different, operating systems or tools that are, by contrast to Visual Studio, very challenging to learn. In fact, I've heard of some people giving up altogether or not even trying because the operating system and tools seemed too intimidating.[1] All of these became roadblocks for those wanting a familiar and easy to use platform to develop IoT solutions.

Microsoft has risen to the occasion creating a unique way to develop applications for and deploy solutions to hardware that traditionally has been off limits for many Windows users who did not want to learn a new operating system, such as Linux, which is the most popular choice for embedded hardware development. I firmly believe you should understand the basics of these other platforms and I present a short primer on the Raspberry Pi in the next chapter, including a look at the base or preferred operating system. But don't worry; you need not become a Linux expert to use Windows 10 with the Raspberry Pi.

In this chapter, you discover a new version of Windows 10 called the Windows 10 IoT Core that runs on low-cost computers, such as the Raspberry Pi. You will discover the basic features of Windows 10, including how to prepare your PC and get started with the Windows 10 on your device. You will even see how to boot up the Raspberry Pi with Windows 10! Let's get started.

Windows 10 IoT Core Features

The latest sensation in the world of Microsoft Windows and in particular the IoT is the ability for Windows users to leverage their experience and knowledge of developing applications for Windows on smaller devices. This new offering is called Windows 10 IoT Core. While Microsoft has offered several products designated as "embedded" or "compact" or "embedded compact", which was a scaled-down version of the operating system, there were many differences and a few bridges that had to be crossed to use them. While highly touted, the offerings never really lived up to the "write the code once and deploy everywhere" mantra. That is, until now.

---

[1]Learning Linux isn't really so terrible, as you shall see, but if you've never used such an operating system, it can be frustratingly difficult to learn how to do even simple tasks.

© Charles Bell 2016
C. Bell, *Windows 10 for the Internet of Things*, DOI 10.1007/978-1-4842-2108-2_2

WHAT ABOUT WINDOWS EMBEDDED?

You may have heard about a version of Windows called Windows Embedded. This was one of the early attempts to make the Windows operating system lightweight enough to run on low-power, low capability hardware. Unfortunately, there wasn't much direction or clear path for developers wanting to leverage it in their solutions. Fortunately, Windows 10 IoT Core is the answer (replacement) for embedded Windows applications.

Unlike the previous products meant for smaller platforms, Windows 10 IoT Core shares many of the same components as the flagship operating system for PCs. That is, it has the same core components, kernel, and even some of the middleware is based on the same core code. In fact, the code generated can be binary-compatible with the other platforms, which means you can write code that can run on either the IoT device or your PC. It should be noted that this capability is highly dependent on what the code does. For example, if your code access the general-purpose input/output (GPIO) hardware pins on the low-cost computing board, you cannot run the application on the PC (there are no GPIO pins on the PC).[2]

Interacting with Hardware

The ability to access hardware directly—such as the GPIO pins—is what makes Windows 10 IoT Core so attractive to hobbyists and IoT enthusiasts who want to build custom hardware solutions using small, inexpensive hardware.

For example, if you wanted to build a simple device to signal you when someone opens your screen door, you would likely not use a PC costing several hundreds of dollars. Not only would that be expensive and bulky, there isn't an easy way to connect a simple switch (sensor) to your PC, much less to a PC located elsewhere. It would be much more cost effective to use a simple switch connected to a small, inexpensive set of electronic components using a simple application to turn on an LED or ring a buzzer. What makes the Windows 10 IoT Core even more appealing is you don't have to relearn how to write software—you can write a Windows application to run on the small device.

Video Support

Since most IoT devices do not include a monitor (some may), Windows 10 IoT Core is designed to run headless (without a monitor) or headed (with a monitor). Headless solutions require less memory since they do not load any video libraries or subsystems. Headed solutions are possible if the hardware chosen supports video (all current hardware options have HDMI video capabilities).

Thus, you can create IoT solutions with visual components or interactive applications, such as those for kiosks, or even interactive help systems. You choose whether the application is headless or headed by the configuration of the device. In fact, the configuration is accessible from the device or remotely through a set of tools running on your PC. You'll see more about these features later in this chapter.

One Platform, Many Devices

For developers of Windows 10 applications, including IoT solutions, Microsoft has adopted a "one Windows" philosophy where developers can develop their code once and run it on any installation of Windows. This is accomplished with a technology called the Universal Windows Applications API (sometimes called UWP or universal applications or UWP apps).

---

[2]Well, most PCs. Some low-cost computing boards are simply a fancy case away from a fully functional PC.

Thus, developers can create an application that runs on phones, tablets, desktop, and even servers without having to change their code or exchange different libraries. A you will see once you start with the projects in this book, you are developing your applications (apps, scripts, etc.) on your Windows 10 desktop (tablet) and deploying them to the Raspberry Pi—all without having to move the code to the Raspberry Pi, alter it, compile it, and so on. This is a huge improvement for IoT developers over other development choices.

For example, if you chose to use a Raspberry Pi with its default operating system, you would have to learn how to develop Linux applications—complete with learning new development tools (if not a new code editor). With Windows 10, you use an old favorite—Visual Studio—to build and deploy the application. How cool is that?

The real power of the UWP API is discussed in Chapter 4 as you explore how a single application (code) can be compiled on your PC and deployed to the Raspberry Pi. Indeed, the UWP API allows you to write one solution (source code) and deploy it to any Windows 10 device from a phone, low-cost computing board, PC, tablet, and so forth. This opens the possibility of using any of the Windows 10 supported devices in your projects.

Supported Hardware

Windows 10 IoT Core is designed and optimized to run on smaller devices, such as low-cost computing boards. Furthermore, Windows 10 IoT Core can run headless[3] (without a display) thereby removing the need for sophisticated graphics (but still supports graphic applications with special libraries). All this is possible with the extensible Universal Windows Platform (UWP) API, as described earlier.

The hardware requirements for running Windows 10 IoT Core include the following.

- *Memory (headless)*: 256MB RAM (at least 128MB free for the operating system)

- *Memory (with display)*: 512MB RAM (at least 256MB free for the operating system)

- *Storage*: 2GB (can be SD card, non-volatile memory, or disk)

- *Processor*: 400MHz or faster ARM or Intel x86

Currently, the Windows 10 IoT Core runs on the Raspberry Pi, MinnowBoard Max–compatible boards, and the Arrow DragonBoard 410c. All of these boards are considered low-cost computing platforms. I describe each of these briefly in the upcoming sections.

■ **Note** Some early documentation, including web sites from Microsoft and Intel report and demonstrate using early releases of the Windows 10 IoT Core on the Intel Einstein and Galileo boards.[4] However, the latest releases of Windows 10 IoT Core have dropped support for these boards.

LOW-POWER COMPUTING PLATFORMS

Low-powered computing platforms, sometimes called low-cost computer boards or mini-computers, are built from inexpensive components designed to run a low-resource-intensive operating system. Most boards have all the normal features you would expect from a low-cost computer, including video, USB, and networking features. However, not all boards have all of these features.

---

[3]Oh, no, a harbinger for headless hardware!
[4]See https://software.intel.com/en-us/iot/home.

The reason they are sometimes called low power isn't because of their smaller CPUs or memory capabilities; rather, it is because of their power requirements, which are typically between 5V and 24V. Since they do not require a massive, PC-like power supply, these boards can be used in projects that need the capabilities of a computer with a real operating system but do not have space for a full-sized computer, cannot devote the cost of a computer, or must run on a lower voltage.

There are many varieties of low-cost computing boards. Some support the full features of a typical computer (and can be used as a pretty decent laptop alternative), while others have the bare essentials to make them usable as embedded computers. For example, some boards permit you to connect a network cable, keyboard, mouse, and monitor for use as a normal laptop or desktop computer while others have only networking and USB interfaces, requiring you to remotely access them in order to use them. Fortunately, all the low-cost computing boards available for Windows 10 have support for networking, video, and USB peripherals.

Raspberry Pi

The Raspberry Pi 3 Model B is the latest iteration of the Raspberry Pi (www.raspberrypi.org/products/raspberry-pi-3-model-b/). It has all the features of the original Raspberry Pi 2 but with a faster 64-bit quad core processor and onboard Wi-Fi (a first for the Raspberry Pi). However, the Raspberry Pi 2 is more than capable for running Windows 10 IoT Core solutions.

■ **Note** I use the term Raspberry Pi henceforth to refer to either the Raspberry Pi 2 or 3.

The Raspberry Pi is a popular board with IoT developers mainly because of its low cost and ease of use. Given the popularity of the Raspberry Pi, I cover it in greater detail in Chapter 3, including a short tutorial on how to get started using it with its native operating system. Thus, I briefly cover the highlights here and reserve a more detailed discussion on using the board for Chapter 3.

■ **Note** I describe the Raspberry Pi 2 here, but you can use either the Raspberry Pi 2 or Raspberry Pi 3 for this book.

The Raspberry Pi 2 hardware includes a 900MHz A7 ARM CPU, 1GB RAM, video graphics with HDMI output, four USB ports (up from just two on older boards), Ethernet, a camera interface (CSI), a display interface (DSI), a micro-SD card, and 40 GPIO pins. Figure 2-1 shows the Raspberry Pi 2 board.

Figure 2-1. *Raspberry Pi model 2*

The camera interface is really interesting. You can buy a camera module like the ones at Adafruit (http://adafruit.com/categories/177) and connect it to the board for use as a remote video-monitoring component. I've used this feature extensively by turning a couple of my Raspberry Pi boards into 3D-printing hubs where I can send print jobs over the network to print and check the progress of the prints remotely or as low-cost video surveillance devices.

The LCD interface is also interesting because there is now a 7-inch LCD touch panel that connects to the DSI port (http://element14.com/community/docs/DOC-78156/l/raspberry-pi-7-touchscreen-display). I have also seen a number of interesting Raspberry Pi tablets built using the new LCD touch panel. You can learn about one promising example (made by Adafruit, so I expect it to be excellent) at http://thingiverse.com/thing:1082431.

To date, the Raspberry Pi has been my go-to board for all manner of small projects due to its low cost and availability. There are also many examples from the community on how to employ the Raspberry Pi in your projects. For more information about the Raspberry Pi, see Chapter 3.

■ **Tip** There is a list of frequently asked questions (FAQ) on using Windows 10 IoT Core that includes a section on the Raspberry Pi (http://ms-iot.github.io/content/en-US/Faqs.htm). You may want to check it for answers if you encounter a problem using your Raspberry Pi.

WHAT ABOUT THE RASPBERRY PI ZERO?

Sorry, Windows 10 IoT Core does not work on the new and widely popular Raspberry Pi Zero board. The processor on the Zero is the older processor from one of the original Raspberry Pi boards. Perhaps in the future a "Zero 2" board is compatible, but for now, you can use only the Raspberry Pi 2 or 3 with Windows 10 IoT Core.

One of the things I like about the Raspberry Pi is you can run a number of operating systems on it by installing the operating system on a micro-SD card. This allows me to use a single Raspberry Pi for a host of projects; each with its own micro-SD card. In fact, the basic setup at `www.raspberrypi.org` includes a special boot loaded that permits you to install the operating system of your choice. Sadly, Windows 10 is not on that list (yet) but Windows 10 permits you to load it directly from your PC.

MinnowBoard Max–Compatible Boards

The MinnowBoard Max and compatible boards are a very interesting lot. They use an Intel processor with a wider array of features than the other boards, including an Intel Atom E3826 dual core 1.46GHz CPU, integrated HDMI output Intel HD Graphics with hardware-accelerated drivers (for Linux), 2GB of fast DDR3L 1067MT/s DRAM, 8MB SPI Flash memory, Ethernet, USB, SATA (e.g., a hard drive), micro-SD card drive, GPIO, and more. In many respects, this is the most powerful board of the lot with more features. About the only thing lacking is onboard Wi-Fi but that can be quickly remedied with any number of Wi-Fi USB dongles.[5]

One very interesting aspect to the MinnowBoard Max is that the developer has retained compatibility over several iterations of the board. The Microsoft web site lists the MinnowBoard Max as officially supporting Windows 10 IoT Core but actually any MinnowBoard Max derivative will work. In fact, the MinnowBoard Wiki page (`www.minnowboard.org`) refers to the line as simply "MinnowBoard Max–compatible" boards.

When I purchased my board, I bought the latest, most powerful MinnowBoard Max–compatible board available: the MinnowBoard Max Turbot. The Turbot offers a number of minor improvements over the older boards, including improved performance and many smaller improvements in the GPIO subsystem as well as a defect repair or two. Figure 2-2 shows my MinnowBoard Max Turbot. Now, that's a handsome board, isn't it?

Figure 2-2. *MinnowBoard Max Turbot*

---

[5]Just make sure that you get one that is compatible with Windows 10 IoT Core.

The MinnowBoard Max-compatible boards have one thing that may be very important for some IoT solutions that the Raspberry Pi lacks: an onboard Real Time Clock (RTC)—battery not included. You can see the battery holder in the photo between the power and Ethernet connectors on the left side of the board. Older boards did not come with this header installed but you can add it yourself if you know how to solder (or know someone who does). You can set the current date and time using the Unified Extensible Firmware Interface (UEFI) shell date command (see http://wiki.minnowboard.org/Shell_Commands).

WHAT IS A REAL TIME COCK AND WHY SHOULD I CARE?

A real-time clock allows you to keep accurate time of day for recording date and time of events (or simply reporting such to the user). Without an RTC, the Raspberry Pi needs either connection to the Internet to synchronize time or a seed value to keep time itself. However, without a RTC circuit, time keeping can become inaccurate over long periods.[6] I discuss time keeping in the later chapters with examples of recording events.

▒ **Note**   The other board supported by Windows 10 IoT Core, the DragonBoard 410c, has a RTC too but it does not require an external battery, which suggests it must be reset on boot.

Another interesting thing about the MinnowBoard Max is that it is open hardware so if you wanted to build one yourself or a derivative or perhaps an accessory board (called a *Lure*), you can find all the information to do so online. In fact, there are a number of vendors offering Lures (add-on boards) for the MinnowBoard Max-compatible boards. The wiki at http://wiki.minnowboard.org/Lures lists a number of available Lures and accessories (and a few retired ones).

I much prefer an open hardware (or open source) solution to proprietary offerings because I find it is often the case that there is more information available about the products if they have a strong (and growing) community to support it. This is the case with the MinnowBoard Max. You can find almost anything you want to know about this board on the http://wiki.minnowboard.org wiki.

▒ **Tip**   You can find additional information about the MinnowBoard Max–compatible boards at http://wiki. minnowboard.org.

As you can imagine, given the added performance, these boards do cost quite a bit more than the Raspberry Pi (about $150 vs. about $35) but the jump in performance may warrant the extra cost. If you search around, you may be able to save a bit by buying an older board rather than the newest board described here. Most sites I visited were out of stock but it is just a matter of time before they become more plentiful.

Thus, finding a MinnowBoard Max can be a bit of a challenge. Fortunately, the newer MinnowBoard Max Turbot is 100 percent-compatible (and a bit better) than the original MinnowBoard Max described on the Microsoft Windows 10 IoT web site. In the United States, you can buy a MinnowBoard Max Turbot at NetGate (http://store.netgate.com/Turbot.aspx). You can also find the MinnowBoard Turbot at Maker Shed at www.makershed.com/products/minnow-turbot. In the EU, you can find them at RS Components Ltd. (http://uk.rs-online.com/web/cpo/8842199/?searchTerm=minnowboard+max).

---

[6]There is much more to this than what I list, but suffice to say a typical clock on a computer cannot keep accurate time. That's the whole point of the RTC.

Arrow DragonBoard 410c

The Arrow DragonBoard 410c is a low-cost computing board that incorporates the Qualcomm quad core Snapdragon 410 processor. This processor is an ARM Cortex-based single-chip system supporting a wide variety of hardware from USB to networking. The processor runs up to 1.2GHz per core in either 32- or 64-bit mode, which is a bit more powerful than the Raspberry Pi.

The board is a fully featured low-cost computing platform complete with 1GB of RAM, 8GB onboard storage (eMMC), an HDMI 1080p display (with audio over HDMI), Wi-Fi, Bluetooth, GPS (yes, GPS!), USB ports, and even a micro-SD card. Figure 2-3 shows the DragonBoard 410c.

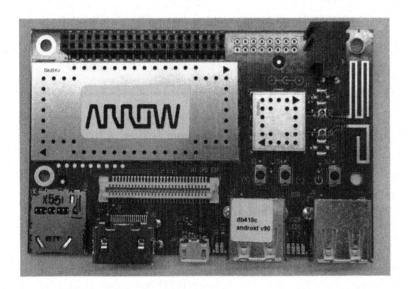

Figure 2-3. *Arrow DragonBoard 410c*

Interestingly, the DragonBoard 410c can be booted from the onboard memory using the Android 5.1 operating system; provided you haven't loaded Windows 10 IoT Core because you will overwrite the base operating system. However, you can recover the factoring settings by following the procedure at https:// github.com/96boards/documentation/wiki/Dragonboard-410c-Installation-Guide-for-Linux-and-Android. Figure 2-4 shows the default operating system (Android-based) of the DragonBoard 410c. Thus, you could use the DragonBoard 410c as an ultra-compact desktop or laptop computer.

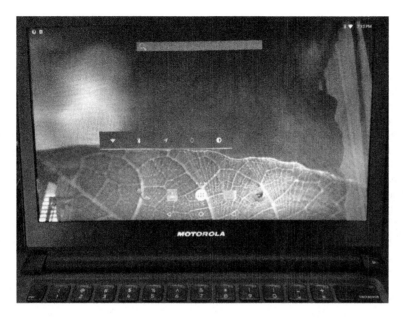

Figure 2-4. *Onboard Android OS on the DragonBoard 410c*

■ **Note**   The DragonBoard 410c does not use an SD card to boot Windows 10 IoT Core. I discuss these differences in a later section.

Given its small size, onboard Wi-Fi, USB, GPIO header, and more, the DragonBoard 410c is a good alternative to the Raspberry Pi. Yes, it does cost more (about $75 versus about $35 for the Raspberry Pi), but if you need the more powerful processor and convenience of onboard Wi-Fi, you may want to consider it for solutions that need a bit more processing power.

■ **Tip**   For more details on the DragonBoard 410c, visit the Arrow data sheet at www.arrow.com/en/ products/dragonboard410c/arrow-development-tools#partsDetailsDatasheet.

The best source for purchasing an Arrow DragonBoard 410c is from the manufacturer directly; go to www.arrow.com for details on ordering a board to complete your low-cost computer board arsenal. Note that the manufacturer stocks a host of additional electronic components making them another source for gathering components for your IoT project. You can also find it on Maker Shed at www.makershed.com/ products/dragonboard.

So Which One Should I Choose?

The three boards are those that are currently supported for use with Windows 10 IoT Core. Which you choose is largely up to you as each has their merits. Perhaps the most compelling reasons to choose one over the others for most hobbyists and enthusiasts are cost and availability.

At the time of this writing, the Raspberry Pi costs less than the other boards and is much easier to find. The Raspberry Pi costs about $35, the DragonBoard 410c about $75, and the MinnowBoard Max Turbot about $150 making the Raspberry Pi the most economic for initial cost.

Since most readers want to limit their investment (hardware can get expensive quickly once you start buying sensors and other bits and bobs you need), I focus on the Raspberry Pi in this book. Even if you plan to use one of the other boards, following along with the Raspberry Pi helps you learn the skills that you need without spending considerably more for the base component. However, if you need to use one of the other boards, the examples in the rest of this book can be adapted without much fuss.

Consider another possibility. If you are new to electronics, or you make a mistake with powering your board or components, you could damage the board.[7] Yikes! I have a small, sad box of components I've managed to destroy over the years. Fortunately, it is a small box with only a few inexpensive (but quite dead) components. I keep it around to remind me what a simple mistake reading a wiring diagram can do to your wallet. Wouldn't a $35 board that you can get from a host of vendors be a bit easier to accept if you kill it?

Although this book favors the Raspberry Pi for its economy and availability, the other boards are strong alternatives that you should consider especially for IoT solutions that need more powerful hardware. My own experience with these boards shows each to be a great solution. If cost were not an option, I would likely use the other boards more often. I particularly like the features of the MinnowBoard Max Turbot but the DragonBoard 410c's onboard Wi-Fi is a nice addition.

Things You'll Need

If you are just getting started with these boards, there are a number of things that you need, including some additional hardware (e.g., cables) to connect to and use the boards. You also need some software to write and deploy your software to the board.

Additional Hardware

To use these boards, you need, at a minimum, a power supply, network connection, and a micro-SD card. There is some optional hardware you may want to have on hand as well. I explain some of these in more detail.

Power Supply

The Raspberry Pi can be powered by a USB port on your computer via a USB type A male to a micro-A male cable (a commonly used cable for small electronic devices). Be careful with this cable, as the smaller end is rather fragile and easy to damage.

The DragonBoard 410c and the MinnowBoard Max–compatible boards must be powered by a dedicated power supply. The DragonBoard needs a power supply capable of delivering 6.5-18V whereas the MinnowBoard Max–compatible boards require a 5V 2.5A power supply. You can buy the correct power supply for each of these boards from the supplier, but you can use any power supply rated for the correct voltage and amperage.

I like to use universal power supplies with a variety of connectors that can be switched to different voltages. Figure 2-5 shows a typical universal power supply with several tips. If you want to minimize the gear knocking around on your workbench, get a universal power supply like this one. However, be sure to test the adapter at the proper setting. Some inexpensive universal power supplies are quite inaccurate and may produce more or less voltage or amperage than what is advertised. Thus, you should buy one that has been reviewed by others and has good reviews from buyers.

---

[7]Hey, it happens to everyone.

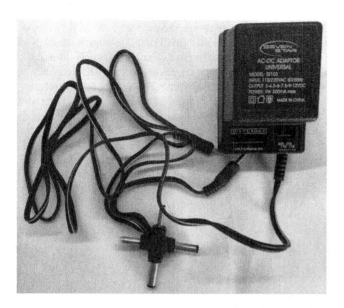

Figure 2-5. *Universal power supply*

Be sure to get one that can generate at least 3.3V, has a variety of tips (sizes), and the polarity of the plug can be selected. That is, some devices require the center pin to be positive. Having the ability to switch the polarity of the center pin makes the power supply usable on a wider variety of boards. I should note that most of the power supplies I've seen that have replaceable tips permit selecting polarity by plugging in the tip in one of two ways. However, this isn't always easy to tell which way the tip is oriented so check it twice before using it.

▪ **Caution**   Some universal power supplies may not generate the required amperage for your board. Be sure to check the power rating of the power supply to be sure it matches your board before you buy or use it to power the board.

Networking

The Raspberry Pi and the MinnowBoard Max–compatible boards have an Ethernet port. While you can use wireless connections with both boards (via a USB dongle for example), it is easier to simply use an Ethernet cable and plug it into your network. The DragonBoard 410c, on the other hand, has wireless networking onboard and can be used to connect to your wireless network. If you choose to use a Raspberry Pi 3, you can use the onboard WiFi.

Optional Hardware

There are a number of optional hardware components you may want to have on hand. In fact, they can be quite convenient for getting started with the board. Fortunately, all the boards have built-in video and USB host capabilities making it easy to set them up with interactive peripherals.

You need an HDMI-compatible monitor. The monitor doesn't have to be an expensive, 30-inch 4K display, since none of these boards has that sort of video capability. I recommend a small HDMI monitor of about 7 inches or larger. Naturally, you also need the appropriate HDMI cable. The Raspberry Pi and DragonBoard 410c use a standard HDMI cable, but the MinnowBoard Max–compatible boards use a micro-HDMI cable. When buying HDMI cables, be sure to purchase high-quality cables because not all cables are fully wired and they may not work. I've found the best are those designed to support audio and Ethernet but these do cost a bit more. The cables at www.mycablemart.com are of sufficient high quality and you can get whatever configuration of connectors you need including an assortment of adapters if you already have some HDMI cables.

If you want to interact with the device for setup or configuration, you should consider a small USB keyboard and mouse. Only the Raspberry Pi has a surplus of USB ports so you may want to consider a keyboard that has a USB hub or the mouse built in. Figure 2-6 shows an example of a compact wireless keyboard that I use for my low-cost computing boards (see https://www.adafruit.com/products/922). I like it for its small size, built-in mouse and even a small speaker for audio.[8]

Figure 2-6. *Mini wireless keyboard*

The keyboard is only about 6 inches long, making it easy to pack away in your kit bag but typing on one won't earn you any speed typing merit badges and can be a bit tedious. The keyboard comes with its own USB dongle that is compatible with Windows 10 IoT Core. You can find these under various vendor names on Amazon and other popular online computer vendors.

However, since these small keyboards are sold in many slight variations, you may want to buy one from a vendor that is willing to accept it as a return if it doesn't work. That said, a wired USB keyboard and mouse are the safer alternatives.

There is one other possibility you may want to consider. Motorola has been making a thin, compact clamshell keyboard and monitor, called a Lapdock, which permits certain mobile phones to connect to the Lapdock—turning the phone into a small laptop. The Lapdock has a small USB hub and battery. In fact, it has a micro-HDMI and micro-A USB port that you can use to connect to your low-cost computer board. Yes! This means you can turn your device into a laptop.

There is a catch, though. Depending on the power capabilities of your board, you have to find or build cables to connect the USB and micro-HDMI to your board. I have a detailed example of how to do this in my book *Beginning Sensor Networks* (Apress, 2013). In short, you must find a micro-A extension cable, cut the cable in the center and solder in a type-A male USB cable segment. This allows you to provide power to the Raspberry Pi and connect the keyboard to the USB host port. Of course, you can use the Lapdock and power your board with an external power supply, but powering it from the Lapdock makes it a bit more portable. Video connections are best done using a female-to-female micro-HDMI adapter and a micro-HDMI male to HDMI male cable. Finding these cables and connectors isn't difficult but can be a little frustrating since most of the vendors are based in Asia. If you live in the United States, shipping can take a few weeks.

---

[8]Not supported on all platforms but works great with Android OS.

However, the wait is worth it as the Lapdock is really thin and with the battery means you can use your low-cost computing board as a laptop just about anywhere. Figure 2-7 shows a Lapdock connected to a Raspberry Pi running the Raspbian operating system.

Figure 2-7. *Raspberry Pi laptop*

Fortunately, the Lapdock is still available and many can be found on online auction sites in good condition and even some new, unopened examples can be found. Expect to pay about $50 for a used one and up to $99 for a new one. If you want one, I recommend shopping around for the best prices. Be sure to get one with the power adapter because it is an odd voltage (19V) and can be hard to find (and expensive).

Software Development Tools

The software development tool of choice for Windows 10 IoT Core is Visual Studio 2015 (`www.visualstudio.com`). You can use any version of Visual Studio 2015, including the free community version. Yes, this means developing applications for the Windows 10 IoT Core uses a very familiar tool with a robust feature set. As you will see, you can leverage nearly all the features of Visual Studio when developing and deploying your IoT solutions.

Better still, there is a growing list of add-ons, sample applications, and more resources available for Visual Studio 2015 and Windows 10 IoT Core. You can develop your UWP applications in a variety of languages, including C++, Python, Arduino Wiring, and more. I show you the Python and Arduino extensions for Visual Studio as you explore example IoT projects. However, most examples are written in C++, which is the more popular choice among the examples from Microsoft and the community.

If you have never used Visual Studio before, do not despair—I include a step-by-step description of how to use the tools in each of the proceeding project chapters with a quick-start walk-through in Chapter 4. The following section presents an overview of how to get started with the Windows 10 IoT Core and your low-cost computing board. That is, the section demonstrates how easy it is to set up your PC and your board to begin developing an IoT.

Getting Started with Windows 10 IoT Core

Now that you know more about the Windows 10 IoT Core and the hardware it runs on and the accessories you need to hook things up, let's get your hands into the hardware and boot up Windows 10 IoT Core for the first time. As you may imagine, there are a few things that you need to do to get things going.

In this section, you see how to get all the prerequisites settled in order for you to start using the Windows 10 IoT Core. As you will see, this requires configuring your computer as well as setting up your board. I walk you through all of these steps for each of the boards available. Although this book focuses on the Raspberry Pi, I include all three boards so that when you want to work with one of the other boards, you will have everything that you need to get started.

Let's begin with setting up your computer.

Setting up Your Computer

While most would expect this, the first thing you must know about using the Windows 10 IoT Core is that you need to have a Windows 10 PC. Moreover, you must be running Windows 10 version 10.0.1.10240 or greater. To check your Windows version, go to the search box next to the Start button and enter **System Information**. Click the menu item shown. You see the System Information dialog, as shown in Figure 2-8. I have Windows 10 version 10.0.1.10240, which is the minimal required version.

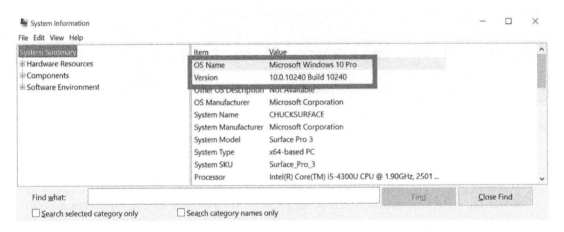

Figure 2-8. *System Information*

I've read some criticism about requiring a Windows 10 PC to use Windows 10 IoT Core, but again for Windows 10 users wanting to explore the IoT, it's a non-issue. However, if you normally use another platform for your PC (or Mac), that you need a Windows 10 machine going forward.

But there is more. That you need to configure your PC for use with the Windows 10 IoT Core tools. Briefly, this includes the following. I explain each of these steps in more detail. Once all of these steps are complete, your PC is ready to set up and use Windows 10 IoT Core on your low-cost computing board.

- Enable developer mode

- Install Visual Studio 2015 and the sample templates for the IoT

- Install the Windows 10 IoT Core Dashboard

Enable Developer Mode

This step is one that is often overlooked and easily forgotten, especially if you have more than one Windows 10 PC. Windows 10 has initiated a new licensing mechanism for developing applications. Rather than require a special developer license to develop and deploy your applications, you simply enable your Windows 10 PC to turn on developer mode, which allows you to compile, deploy, and test applications for Windows 10 IoT Core.

To enable developer mode, enter **use developer features** in the search box next to the Start button. Choose the Settings menu item by the same name. Alternatively, you can open the settings application, click **Update & Security**, and then click **For developers**. Once you have the dialog open, select the **Developer mode** radio button. Once you click the button, you are asked for confirmation to turn on developer mode. The message explains that using developer mode may increase your security risk. Be sure to take appropriate actions to ensure that you are protected while online. Click **Yes** in that dialog. Figure 2-9 shows the developer mode dialog with the correct settings selected.

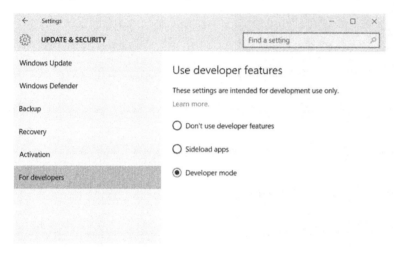

Figure 2-9. *Enabling Developer Mode*

Next, you need to install the software development tools. In this case, you want to install Visual Studio 2015. You also need to install the IoT templates and add-ons. You won't use Visual Studio in this chapter but since it is required, you'll discover how to install it so that you can complete the process to prepare your PC for developing IoT solutions.

Install Visual Studio Community 2015

If you do not have Visual Studio 2015, you can go to `https://www.visualstudio.com/en-us/visual-studio-homepage-vs.aspx` and click **Download Community 2015**. This version has a license making it free to use for individuals, open source projects, academic institutions, students, and small project teams. Visual Studio 2015 Community has everything that you need to develop Windows 10 IoT solutions.

The download consists of a small executable named `vs_community_ENU.exe` or similar, which you can execute once the download is complete. This is the Internet installation version, which downloads the components needed during installation. You use this version in this chapter since it is the easiest to do and requires less download time.[9]

---

[9]This could be a big deal if your data plan is limited to a fixed amount of data per month. The full download with all options is approximately 5.8GB or more.

■ **Tip**   If you would prefer to install from a local file, you can download an `.iso` file with all the components from `https://www.microsoft.com/en-us/download/details.aspx?id=49988`.

Once Visual Studio Community Edition is downloaded, you can double-click the executable to begin the installation. Once the installer launches, you see the splash screen followed shortly by the installation type page. You also may need to authorize changes to the system via a pop-up dialog.

Next, you see a page that asks you what type of installation you want to choose: default or custom. You can also change the location of the installation but I recommend accepting the default location. If your Internet download speed is very slow, you may want to choose a custom installation and uncheck everything to install the basic components and install components, as you need them. Figure 2-10 shows the installation type dialog page.

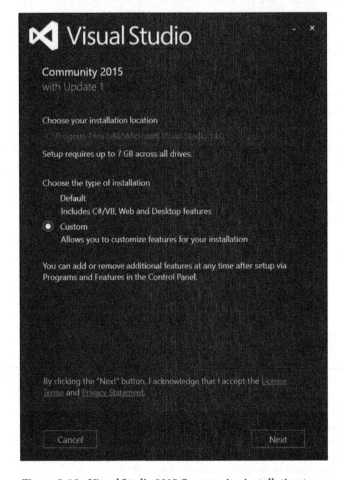

Figure 2-10. *Visual Studio 2015 Community: installation type*

■ **Tip**   If you already have Visual Studio 2015 (either the Professional edition or the Enterprise edition), you can use that, but you have to install the Universal Windows App Development tools. You can do so by choosing a custom installation or by running the installer again and modifying the installation.

The basic components that you need for this book include the UWP and Python features. You can select these by clicking **Custom** and then clicking **Next**. Expanding the tree and check the entries you want in the list, as shown in Figure 2-11. Here you see I have selected the UWP feature along with the Python tools and Web development tools. The default installation choice is to install Visual C++, which you use in this book. The components that you need are listed next.

- Visual C++ (selected by default)

- Python Tools for Visual Studio

- Universal Windows Platform (UWP)

- Web Developer Tools (optional)

Be advised, if you check **Select All**, the installation could exceed 15GB and require over 6GB of download data.

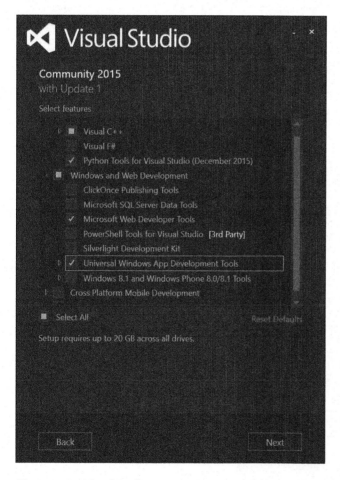

Figure 2-11. Visual Studio 2015 Community: custom installation

ADDING NEW FEATURES TO VISUAL STUDIO

You can add optional components and features by running the installer again. You can simply launch the executable or use the programs and features application. To launch the application, use the search tool to search for **Program and Features** and then open the application. Once open, select the Microsoft Visual Studio Community entry and click **Change**. The installer will run again and you can select the components you want.

For example, if you did not select the UWP module, you can do so by launching the installer again. Just choose **Modify** when the installer dialog opens, select the UWP feature, as shown, and then select **Next** and then **Update** on the confirmation screen. Wait for the installation to complete.

Once you've selected the features and components you want, click the **Next** button. You are presented with a summary of the installation, as shown in Figure 2-12. Click **Install** to accept the license and install the components.

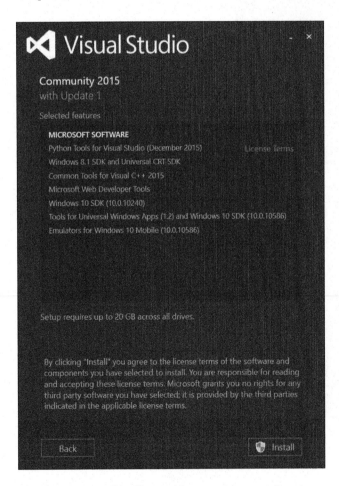

Figure 2-12. *Visual Studio 2015 Community: accept license*

Now comes the stage where the installer begins downloading the components you selected from the Internet and installs them. Depending on what options you chose, this could be a long list. Also, depending on your Internet download speed, downloads can take some time to complete.

Further, the installation of the components can also take a long time. It is not unusual to take several hours to complete the installation. Again, if this is a concern, you can use a custom installation and choose one component at a time. Once underway, you see a progress page like the excerpt shown in Figure 2-13.

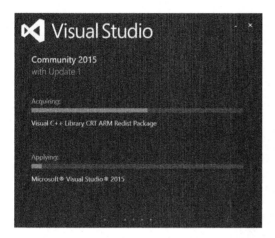

Figure 2-13. *Visual Studio 2015 Community: progress*

■ **Tip**   Do not cancel the installation. It may look like it is hung, but it may take several hours to complete the installation during which time it may appear as if nothing is happening (but it is). If you have a slow Internet connection, you may want to start the installation before you go to bed and let it run overnight.

When the installation completes, you see a dialog page that permits you to launch Visual Studio 2015 Community for the first time, as shown in Figure 2-14.

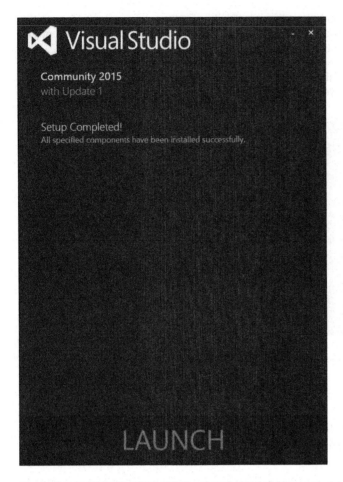

Figure 2-14. *Visual Studio 2015 Community: installation complete*

If you have been following along and installing Visual Studio 2015 Community, go ahead and launch it. When you first start Visual Studio 2015 Community, you may a delay as the application configures your system and the options you chose during installation. Figure 2-15 shows a typical layout of Visual Studio. Don't worry about what all the panels, menus, and hundreds of options are at this time. You learn the essentials of what you need to know in Chapters 4 through 6.

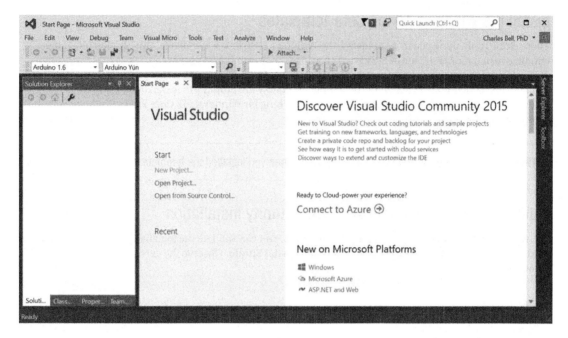

Figure 2-15. *Visual Studio 2015 Community user interface*

Update Visual Studio 2015 Community

Next, you may need to update Visual Studio. If you just downloaded Visual Studio, this may not be necessary, but I will explain the steps in case you need to do this in the future. To update Visual Studio, open the **Extensions and Updates** dialog in Visual Studio via the **Tools** menu. Click **Updates** to connect to the Internet and check for updates. If an update is available (for example, Update 1), go ahead and select it for installation. Figure 2-16 shows the Extensions and Updates dialog.

Figure 2-16. *Extensions and Updates dialog: Updates*

Install the Windows 10 IoT Core Templates

There is one more thing you need to install: the Windows 10 IoT Core templates. You can download the template installation file at `https://visualstudiogallery.msdn.microsoft.com/55b357e1-a533-43ad-82a5-a88ac4b01dec`. Once downloaded, double-click the file named `WindowsIoTCoreTemplates.vsix` or similar. This launches the installer for the templates. Follow the prompts to install the templates.

Alternatively, the templates can be found by searching for Windows IoT Core Project Templates in the Visual Studio Gallery.

■ **Tip**   You must restart Visual Studio if it was open when you installed the templates.

Validate Your Visual Studio 2015 Community Installation

OK, now you have Visual Studio installed and updated. You can validate the installation by opening Visual Studio and then selecting **Help ➤ About Microsoft Visual Studio**. Observe the version information, as shown in Figure 2-17.

Figure 2-17. *Validating Visual Studio: About dialog*

This lists a number of entries, but what you're checking is the version of Visual Studio which should be 14.0.24720.00 Update 1 or later. Figure 2-17 shows an example dialog from my machine. I have the correct versions installed.

Troubleshooting Visual Studio Installation Problems

While the installation of Visual Studio can take a very long time to complete, there are sometimes cases where the installation fails. This is most often caused by loss of connectivity to the Internet or simply failed download of one or more components. You can recover from this form of failure by simply restarting the installation. Indeed, most installation failures can be fixed in this manner.

However, if something goes really wonky, you may need to repair the installation. Fortunately, you can restart the installation or open the **Program and Features** application. Select the **Visual Studio 2015** entry and then click **Change**. You are presented with a dialog that allows you to modify or repair the installation. Click **repair** to recover from the installation failure.

On very rare occasions, the installation could fail in such a way that some components will not install, resulting in a failed installation. In these rare occasions, you can view the log of the failed installation (there is a link on the final dialog page for the installation) and try to determine the cause. Since there are so many things that could go wrong, it isn't possible to list them. Thus, you must examine the log and fix each error as described. I recommend doing a search for the error on the Internet and read the suggested solutions. Be sure to read several solutions thoroughly before you attempt them. Also, make a restore point before continuing.

■ **Tip**   Always make a restore point when attempting to fix installation problems. If the solution fails or makes it worse, you can restore the system to the last checkpoint and try another solution.

When you find yourself spending a lot of time trying to fix really odd errors for which you can find no solutions on the Internet, you should first attempt to uninstall using the command line with the /uninstall /force options with vs_community_ENU.exe /uninstall /force and then restart the installation. If this does not work or gives the same errors, you can try deleting the contents of the ProgramData/Program Cache folder. Use this is a last resort because not only will it force installation to download all the packages again, but it also removes cached packages from other applications. However, I've found that this trick works very well.

■ **Caution**   Deleting cached packages from ProgramData/Package Cache may affect your other applications forcing them to download packages again. Use with care and only as a last resort.

Now that you have Visual Studio installed, you need only one more thing on your PC: the Windows 10 IoT tools.

Install the Windows 10 IoT Tools

The last step is to install the Windows 10 IoT Core development tools. You need these tools to complete the installation of Windows 10 IoT Core on your low-cost computing board. To download the installation, go to http://go.microsoft.com/fwlink/?LinkID=708576. This link downloads an installation named setup.exe. Once downloaded, launch the executable and follow the prompts. The installation begins a small download of the tools, as shown in Figure 2-18.

Figure 2-18. *Windows 10 IoT Core Dashboard installation*

Once the installation is complete, the dashboard launches. Figure 2-19 shows the Windows 10 IoT Core Dashboard, which launches the **Set up a new device** page by default. This page is used to configure a bootable SD card for your device. You see how to use this in the next section. On subsequent launches, the application checks for updates and automatically downloads and installs them.

Figure 2-19. *Windows 10 IoT Core Dashboard*

There are three tags on the left side of the window. The **My devices** page lists the devices running Windows 10 IoT Core on your network. There is also a **Try some samples** link to a set of sample applications that you can use to get started. You see one of these in action in Chapter 4. Finally, the **Settings** link displays the version and builds information for the application.[10] You explore this tool in more detail in the following sections as you set up your hardware for use with the Windows 10 IoT Core.

Now that your PC is ready to go, let's see how each of the low-cost computing boards is set up to install Windows 10 IoT Core and boot for the first time.

---

[10]Oddly, there are no settings that you can change.

Getting Started with Your Board

Windows 10 IoT Core is very easy to set up and get your board running. The general process is as follows. I describe each in more detail through examples of each board. I recommend reading through the setup of all the boards, especially if you haven't decided which one you want to use. But first, I present some tips that may be helpful in setting up your board.

1. Download the Windows 10 board-specific installation package (via the .iso download) and install it. For the .iso images for the development boards, see http://ms-iot.github.io/content/en-US/Downloads.htm.

2. Download and write the Windows 10 IoT Core boot image to the SD card (or memory for the DragonBoard 410c).

3. Connect your hardware to power, monitor, keyboard, and mouse.

4. Power on the device.

5. Configure basic settings and connect to the network.

6. Connect to your device with the IoT Core Dashboard.

Tip: Be Patient and Thorough

I should encourage you to exercise patience and perseverance should you encounter problems. Although I will explain the steps you need to perform in detail, there are some things that could go wrong. I have included as many pitfalls as I can, but my experience has shown troubleshooting problems with the hardware may still arise.

For example, I spent quite some time trying to boot one of my boards only to discover one of the cables was defective. I neglected to consider the cable as the culprit because it was new. Thus, you should approach problems with an iterative mind-set where you check each component (SD card, cable, power supply, etc.) one at a time for correct working order and, more importantly, change only one thing at a time. If it doesn't solve the problem, return it to the original setting. That is, if you swap the HDMI cable and it doesn't fix the problem, go back to the original one.

Tip: Downloading the .iso Image

Downloading a bootable image can be accomplished in two ways: you can download the boot image (as an .iso file) or you can use the Windows 10 IoT Dashboard to download the image. There are also two ways to write the image to the SD card (for the Raspberry Pi and MinnowBoard Max-compatible boards). You can either use the Windows 10 IoT Dashboard or a helper application that comes with the .iso boot image. You'll see how to do both in the following sections.

Tip: Use Class 10 SD Cards

Microsoft recommends using class 10 SD cards or higher.[11] Currently, only two SD cards have been tested—Ultra cards from SanDisk and the EVO cards from Samsung. I've found other class 10 SD cards work well and some of the slower classes may work but results are mixed. If you observe your device booting slowly or the startup sequence seems jumpy (the screen flickers), you may need to use a faster SD card. Incompatible SD cards can result in unacceptable performance or failure to boot. If you are having problems with your device after initial setup, try a faster SD card. I prefer the SanDisk Ultra SD cards since they are more plentiful and thus can be a bit cheaper.

---

[11]https://en.wikipedia.org/wiki/Secure_Digital#Speed_class_rating

Tip: Double-Check Your Power Supply

Be sure to double-check your power supply to ensure that it has the proper rating for your board. The Raspberry Pi needs a 5V 1.5A power supply with a micro-USB connector, the MinnowBoard Max–compatible boards need a 5V 2A power supply with a 5mm connector, and the DragonBoard 410c works best with a 12V 2A power supply with a 4mm tip. If your board does not boot or powers off while running, it is possible the power supply is faulty or insufficient for the board and its peripherals.

■ **Caution**   Using the wrong tip can damage the power connector on the board.[12]

Now that you understand the basic process and have foreknowledge of some of the pitfalls, let's see how to set up each of the boards in turn starting with the Raspberry Pi. You see how to connect to the board after you learn how to configure all the boards.

Raspberry Pi Configuration

This section demonstrates how to install and boot Windows 10 IoT Core on the Raspberry Pi. You see the specific steps needed to get your board ready for Windows 10 IoT Core, including the hardware that you need, how to connect to the board once Windows 10 IoT Core boots, and how to configure the board for your network.

Let's begin with the prerequisite hardware that you need.

Prerequisites

The following are the miscellaneous hardware that you need to use Windows 10 IoT Core with your board. You will see where these cables are plugged into your board later in this section. For now, just gather the items you need.

- Raspberry Pi

- 5V micro-USB power adapter, like the one at www.adafruit.com/products/1995

- USB wired or wireless (not Bluetooth) keyboard and mouse

- HDMI monitor with HDMI cable or a suitable adapter for use with a DVI monitor

- Ethernet cable

- micro-SD Card 8GB or larger: class 10 or better

- SD card reader[13] (if your computer doesn't have one)

Download the .iso and Install the Board-specific Windows IoT Core Tools

The first step is to download the .iso file for the Raspberry Pi, mount the .iso file, and then install the board-specific setup using the .msi file. The installation installs a number of tools that you need, including the Windows IoT Core Watcher that monitors your devices, the binary boot image for your board, and a tool named the Windows IoT Image Helper that is an alternative tool for creating the SD image. Indeed, some older Windows web sites show procedures for using this application. You will use the newer Windows IoT Core Dashboard instead.

---

[12]Can you guess how I know this? It is far too easy to bend the tabs inside the connector. This is more likely when using universal power supplies.
[13]A standard micro-SD to SD adapter is included with the micro-SD card.

■ **Note** The Windows IoT Core Watcher application loads on startup but you can change that behavior in the Startup tab on the task manager.

You can download the file from `http://go.microsoft.com/fwlink/?LinkId=691711`. Once the `.iso` file is downloaded, simply double-click it. This mounts as a virtual drive that opens automatically. You see a file named `Windows_10_IoT_Core_RPi2.msi` or similar. Double-click that file to start the installation. You'll need to do the usual steps for most installations, including accepting the license, permitting the change on your computer, and so on.

The binary image for the Raspberry Pi is named `flash.ffu` and installed in the `c:\Program Files (x86)\Microsoft IoT\FFU\RaspberryPi2`. Once the installation is complete, you can unmount the virtual drive. If you changed the installation folder, be sure to note the correct path—you need it in the next step.

Creating the SD Card Image

Now you are ready to write the Windows 10 IoT Core bootable files to the micro-SD card. You can use the Windows 10 IoT Core dashboard to automatically download the `.iso` and build the image using the drop-down box selections, but since you downloaded and installed the tools manually, you can use the custom option in the dashboard. Let's see how to do that.

■ **Caution** The SD card image overwrites all data on the SD card. Be sure to copy any data on the card before you start the load.

Open the Windows 10 IoT Core Dashboard and insert a suitable SD card into your SD card reader. Make sure that you've backed up and data on the card before you proceed because the next step overwrites the contents.

Begin by selecting the **Set up a New Device** tab. Figure 2-20 shows the initial setup screen.

Figure 2-20. *Windows 10 IoT Core Dashboard: Set up new device*

Next, click the **Set up a new device** button to start the build. On the next screen, you see a drop-down box that allows you to choose the board for automatic download. However, since we downloaded the image, you will use the custom installation option. Choose the **Custom** selection from the drop-down box. Once selected, you see a text box that contains the path to the image (the flash.ffu file).

Use the **Browse** button to locate the file and select it. If you have more than one SD card or USB device connected to your computer, use the **Drive** drop-down box to select the correct drive. Finally, check the license agreement checkbox at the bottom. Figure 2-21 shows the Windows 10 Core Dashboard with the correct settings.

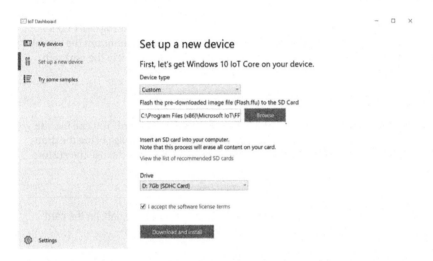

Figure 2-21. *Windows 10 IoT Core Dashboard: custom image build*

When you are ready, click the **Download and install** button. Since you choose the custom installation, it copies the file from your computer and does not download anything. Once the process begins, you see a new console window open running the Deployment Image Servicing and Management tool (named dism.exe), which provides feedback during the copy. Figure 2-22 shows an example of the console.

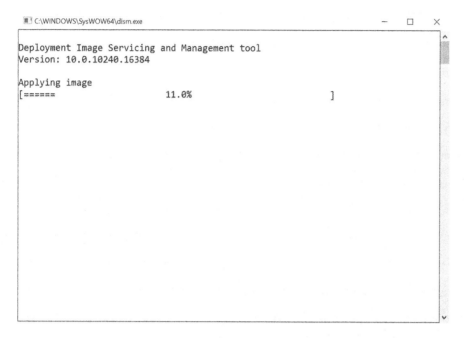

Figure 2-22. *Windows 10 IoT Core Dashboard: custom image build*

Once the image build is complete, the window closes and the dashboard displays the process complete screen, as shown in Figure 2-23.

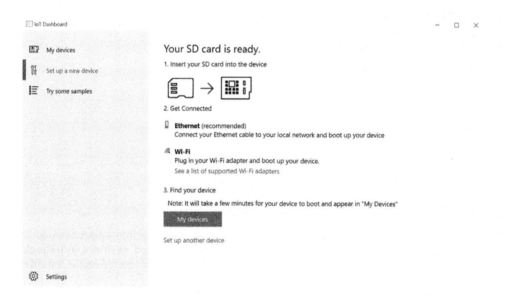

Figure 2-23. *Windows 10 IoT Core Dashboard: image build complete*

You can now take the SD card out of your computer. Be sure to eject it properly like you would any other USB or removable drive by using the Safely Remove Hardware and Eject Media icon on the system tray. Wait until you see a message that it is safe to remove the media before removing it. You're now ready to connect the hardware and boot the image for the first time. Before you do that, click the **My devices** tab (or button) in the Windows 10 IoT Core Dashboard. You'll be using this screen to connect to the Raspberry Pi in a later step.

Connecting the Hardware

If this is your first time using a Raspberry Pi, orient the board on the table with the Raspberry Pi logo facing you. Figure 2-24 helps you locate the connection points. Insert the micro-SD card into the Raspberry Pi SD card reader located on the bottom of the left side. Connect the HDMI monitor to the HDMI port located on the bottom. You can connect your Ethernet cable and USB mouse and keyboard to the ports on the right side of the board.

Figure 2-24. Connections for the Raspberry Pi

OK, you're now ready to power the board and boot up Windows 10 IoT Core!

Booting Windows 10 IoT Core for the First Time

Now you can power on the board and boot from the SD card. Plug your AC adapter into a power source and then insert the micro-USB power located in the bottom-left corner of the board, as shown. You should see the power LED illuminate and the SD card activity LEDs blink. These LEDs are located on the left side of the board.

You see the Windows logo and an activity cursor appear on the monitor connected to the Raspberry Pi. The first boot may take some time but eventually you be asked to choose the default language. Use the mouse or keyboard to select your language.

Next, you see the boot up screen, as shown in Figure 2-25. You can configure your board using this screen, as well as shut down or reboot the board. The **Device Settings** button is the small gear located in the upper right portion of the screen. If you click it, you see the **Device Settings** screen that allows you to change the default language or connect to a Wi-Fi network if you have a wireless network adapter plugged into the Raspberry Pi. Finally, you can shut down or restart the board by clicking the power button in the upper right of the screen.

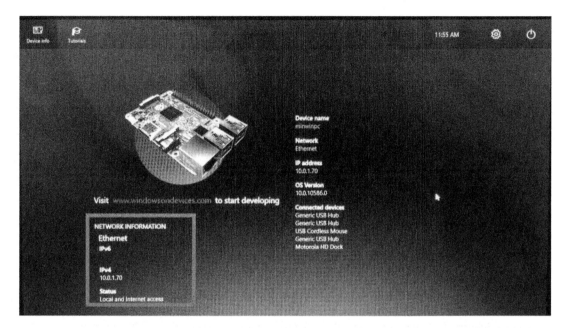

Figure 2-25. *Windows 10 IoT Core boot screen: Raspberry Pi*

I have highlighted the network information located on the left side of the screen. Take note of the IP address as you may need this if you want to connect to the Raspberry Pi from your computer. Since the methods for connecting are the same for all the boards, you will see how to connect to the board after you learn how to set up the other two boards.

MinnowBoard Max Turbot Configuration

This section demonstrates how to install and boot Windows 10 IoT Core on MinnowBoard Max–compatible boards as demonstrated with the MinnowBoard Max Turbot. You will see the specific steps needed to get your board ready for Windows 10 IoT Core, including the hardware you need and how to connect to the board once Windows 10 IoT Core boot as well as how to configure the board for your network.

Since many of the steps are similar for this board as the Raspberry Pi, some of the following is familiar if you read the section on the Raspberry Pi. However, I demonstrate an alternative method for building the SD card image. You will also see two additional steps that you need to perform on the MinnowBoard Max; you must ensure that the board has the 32-bit version of the firmware loaded. You do not need to do this step for the other boards and you need only do it once when setting up the board for the first time.

Let's begin with the prerequisite hardware that you need.

Prerequisites

The following are the miscellaneous hardware that you need to use Windows 10 IoT Core with your board. You see where these cables are plugged into your board later in this section. For now, just gather the items you need.

- MinnowBoard Max–compatible board

- 5V 2.5A power adapter with a 5.5mm × 2.1mm barrel male plug that is configured with the center pole positive and the outer casing negative. If you use a universal power adapter, you should double-check the power ratings.

- USB wired or wireless (not Bluetooth) keyboard and mouse

- HDMI monitor with HDMI cable or a suitable adapter for use with a DVI monitor. The board uses a micro-HDMI connector so make sure that you have the correct cable or a HDMI to micro-HDMI adapter.

- Ethernet cable

- micro-SD Card 8GB or larger: class 10 or better

- SD card reader (if your computer doesn't have one)

- USB thumb drive, 2GB or larger (for loading the firmware)

Download the .iso and Install the Board-specific Windows IoT Core Tools

The first step is to download the .iso file for the MinnowBoard Max, mount the .iso file, and then install the board-specific setup using the .msi file. The installation installs a number of tools that you need, including the Windows IoT Core Watcher that monitors your devices, the binary boot image for your board, and a tool named the Windows IoT Image Helper that is an alternative tool for creating the SD image. Indeed, some older Windows web sites show procedures for using this application.

You can download the file from http://go.microsoft.com/fwlink/?LinkId=691712. Once the .iso file is downloaded, simply double-click it. This mounts as a virtual drive that opens automatically. You see a file named Windows_10_IoT_Core_Mbm.msi or similar. Double-click that file to start the installation. You'll need to do the usual steps for most installations, including accepting the license, permitting the change on your computer, and so on.

The binary image for the MinnowBoard Max is named flash.ffu and installed in the c:\Program Files (x86)\Microsoft IoT\FFU\MinnowBoardMax. Once the installation is complete, you can unmount the virtual drive. If you changed the installation folder, be sure to note the correct path—you need it in the next step.

Creating the SD Card Image

Now you are ready to write the Windows 10 IoT Core bootable files to the micro-SD card. Rather than use the Windows 10 IoT Core Dashboard to create the SD card image, you will use the older Windows IoT Image Helper application. Let's see how to do that.

Open the Windows IoT Image Helper (search for the app named WindowsIOTImageHelper) and insert a suitable SD card into your SD card reader. Make sure you've backed up and data on the card before you proceed because the next step overwrites the contents. Simply select the drive associated with the SD card and then click the Browse button to locate the flash.ffu file. This is located in the c:\Program Files (x86)\Microsoft IoT\FFU\MinnowBoardMax folder. Figure 2-26 shows the Windows IoT Image Helper application with the drive and flash file selected.

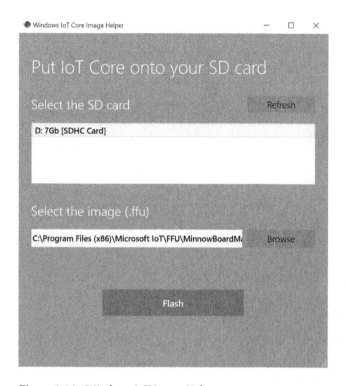

Figure 2-26. *Windows IoT Image Helper*

■ **Caution** The SD card image overwrites all data on the SD card. Be sure to copy any data on the card before you start the load.

When you are ready, click the Flash button to begin writing the image to the SD card. You see a new console open that runs the Deployment Image Servicing and Management tool (dism.exe) to write the image. Figure 2-27 shows an example of the output.

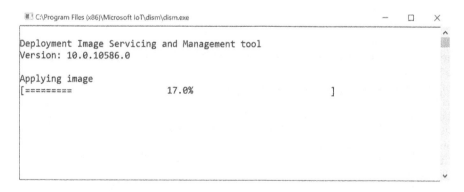

Figure 2-27. *Deployment Image Servicing and Management tool*

Once that is done, Windows mounts the card. Eject the SD card and set it aside. You will use it in a future step but first let's see how to connect the hardware and prepare the firmware on the board for use with Windows 10 IoT Core.

Connecting the Hardware

If this is your first time using a MinnowBoard Max–compatible board, orient the board on the table with the Ethernet port on the left. Figure 2-28 helps you locate the connection points. The micro-SD card drive is located on the right side of the board. Don't insert the card at this time. You must do one additional step before you can boot Windows 10 IoT Core.

Figure 2-28. *Connections for the MinnowBoard Turbot*

Connect the HDMI monitor to the HDMI port located on the left side of the board. This is a micro-HDMI connector so you need to make sure that you have the correct cable. You can connect your Ethernet cable on the left side as well. Connect the USB mouse and keyboard to the USB ports on the right side of the board. Now that all the cables are connected, you can connect the power and let the board boot up.

Load the 32-bit Firmware

The MinnowBoard Max–compatible boards are preloaded with a 64-bit version of its firmware. Unfortunately, you need to use the 32-bit version of the firmware before you can boot Windows 10 IoT Core because Windows 10 IoT Core does not work with the 64-bit version. Loading (also called flashing) the new firmware requires downloading the firmware files, copying them to a USB thumb drive, booting the board and using commands entered in a special administration tool called the UEFI shell.

However, you should first check the current firmware version displayed on the boot configuration screen (also called BIOS like on your PC). If your board has the 32-bit firmware loaded, you may be able to skip this section. However, most new boards have the 64-bit version loaded. You may also want to load the latest version of the firmware since the newer firmware typically is more stable. Moreover, you need to know whether the current firmware is 32-bit or 64-bit.

Go ahead and connect your monitor, keyboard, network, and power as described in the previous section. Be sure to remove any SD card before you turn the power on. Once powered on, you see the boot up sequence. You can enter the BIOS by pressing **F2** when the board boots up. If you miss it, you can simply wait until the UEFI shell is launched and then enter the **EXIT** command. Figure 2-29 shows the location of the firmware version on the screen (at the top). This statement is located in the upper-left corner of the BIOS screen.

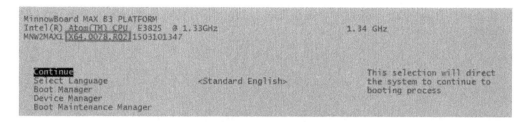

Figure 2-29. *MinnowBoard Max firmware version (courtesy of minnowboard.org)*

In Figure 2-29 the BIOS is running a 64-bit version as noted by the X64 in the name. You can now power off the board and prepare the USB thumb drive for the next step.

You need to download several files. Visit the MinnowBoard Max firmware download page at http://firmware.intel.com/projects/minnowboard-max. Locate the latest version of the firmware and download both the 32-bit and 64-bit versions. The files are .zip archives, which you expand before placing on the USB thumb drive. For example, I downloaded version 0.90. You want the release versions of the software (there are links for both debug and release, choose release). Thus, I downloaded the MinnowBoard MAX 0.90 64-Bit and MinnowBoard MAX 0.90 32-Bit files.

Next, unzip the files and copy the contents to the USB thumb drive. The thumb drive does not need to be empty (so long as there is room for the files) but it must be formatted as a FAT drive (usable from a Windows PC). If you are using a larger SD card, you may need to use extFat. For example, my USB thumb drive has the following files loaded.

```
MinnowBoard.MAX.0.90.BIN-ReleaseNotes.txt
MinnowBoard.MAX.FirmwareUpdateIA32.efi
MinnowBoard.MAX.FirmwareUpdateX64.efi
MinnowBoard.MAX.I32.90.R01.bin
MinnowBoard.MAX.X64.90.R01.bin
```

Note that the release notes file is the same in both archives. You only need the one file, and only for reference. Once you have the files copied, eject the drive, insert it into the MinnowBoard Max USB port, and power it on. Allow the board to boot into the UEFI shell. Once the shell is open, you can flash the firmware.

The command needed depends on whether your board has the 32-bit or 64-bit firmware loaded. More specifically, if it has the 32-bit firmware, you must use the 32-bit flashing tool to load the firmware. Similarly, if it has the 64-bit firmware, you must use the 64-bit flashing tool even though you are loading the newer 32-bit version.

The first step is to change to the USB thumb drive. This is either fs0: or fs1:. Enter the **map -u** command to see the drives attached. The following shows an example where my board associated the fs0: drive with the USB thumb drive.

```
EFI Shell version 2.40 [1.0]
 Current running mode 1.1.2
 Device mapping table
```

```
fs0       :Removable HardDisk - Alias hd15a0b blk0
          PciRoot(0x0)/Pci(0x14,0x0)/USB(0x0,0x0)/HD(1,MBR,0x3CAE3447,0x800,0x1DD0000)
blk0      :Removable HardDisk - Alias hd15a0b fs0
          PciRoot(0x0)/Pci(0x14,0x0)/USB(0x0,0x0)/HD(1,MBR,0x3CAE3447,0x800,0x1DD0000)
blk1      :Removable BlockDevice - Alias (null)
          PciRoot(0x0)/Pci(0x14,0x0)/USB(0x0,0x0)
```

You can change to the drive by entering its name, as shown next. You can also list the contents of the drive, as shown.

```
Shell> fs0:
fs0:\> ls
  Directory of: fs0:\

  01/27/16  06:20p              8,388,608  MinnowBoard.MAX.I32.090.R01.bin
  03/09/15  06:19p              8,388,608  MinnowBoard.MAX.X64.090.R01.bin
  08/14/14  07:54p                 23,040  MinnowBoard.MAX.FirmwareUpdateX64.efi
  08/14/14  08:11p                 14,144  MinnowBoard.MAX.FirmwareUpdateIA32.efi
  02/05/16  01:49p                  9,806  MinnowBoard.MAX.0.90.BIN-ReleaseNotes.txt
              5 File(s)     16,824,448 bytes
              0 Dir(s)
```

Next, you flash the firmware. If you have the 32-bit version of the firmware, execute the following command substituting the version number of the firmware you downloaded. I highlight the executable in bold. Listing 2-1 shows the complete execution of the firmware update for the 32-bit version. You see the progress of the flashing and when it is done, the board reboots. Once rebooted, check the firmware version to make sure that the flashing was successful.

Listing 2-1. Flashing the MinnowBoard Max with 32-bit Firmware Loaded

```
fs0:\> MinnowBoard.MAX.FirmwareUpdateA32.efi MinnowBoard.MAX.A32.090.R01.bin
 Intel(R) UDK2014 Firmware Update Utility for the Intel(R) Server Board S1200V3RPS
 Version 0.97
 Copyright(c) Intel Corporation 2006 - 2014

 Reading file MinnowBoard.MAX.A32.090.R01.bin

 Updating Firmware.  This may take a few minutes.
 ..............................................................................................
 .................
 ..............................................................................................
 .................
 ..........................................................................................
 Update successful
 Shutdown system in 5 seconds ...
```

If you have the 64-bit version of the firmware, execute the following command substituting the version number of the firmware you downloaded. I highlight the executable in bold. Listing 2-2 shows the complete execution of the firmware update for the 64-bit version. You will the progress of the flashing and when it is done, the board reboots. Once rebooted, check the firmware version to make sure that the flashing was successful.

Listing 2-2. Flashing the MinnowBoard Max with 64-bit Firmware Loaded

```
fs0:\> MinnowBoard.MAX.FirmwareUpdateX64.efi MinnowBoard.MAX.A32.090.R01.bin
Intel(R) UDK2014 Firmware Update Utility for the Intel(R) Server Board S1200V3RPS
Version 0.97
Copyright(c) Intel Corporation 2006 - 2014

Reading file MinnowBoard.MAX.A32.090.R01.bin

Updating Firmware.  This may take a few minutes.
.............................................................................................
..................
.............................................................................................
..................
.....................................................................................
Update successful
Shutdown system in 5 seconds ...
```

Now, there is just one more step to accomplish and you need to do this in the BIOS so while you're there, let's complete the BIOS setup.

Configure the BIOS

If you haven't booted into the BIOS, do that now. Press F2 as the board boots or you can enter the **EXIT** command on the UEFI shell to enter the BIOS setup. In this step, that you need to change the boot order so that the SD card gets checked first and turn off a couple of BIOS settings.

First, enter the BIOS and use the down arrow to select **Boot Maintenance Manager** and press **Enter**. Choose the **Boot Options** option and then the **Change Boot Order** item. Use the arrow keys to select the **EFI MISC Device** option. Use the + key to move it to the top and then press **Enter**. This moves the SD card to the top. To save your changes, press F10 and reply Y to confirm the changes.

Next, you need to turn two settings off (disabled). Navigate to the main menu and select **Device Manager ➤ System Setup ➤ South Cluster Configuration ➤ LPSS & SCC Configuration**. Use the arrow keys to locate the following pulse-wave modulation extensions and set each to **Disable**. You do not need these for Windows 10 IoT Core.

```
LPSS PWM #1 Support
LPSS PWM #2 Support
```

When complete, press **Esc** until you are asked to save the changes. Save them and allow the board to reset. OK, you're now ready to power the board and boot up Windows 10 IoT Core!

Booting Windows 10 IoT Core for the First Time

Now you can power on the board and boot from the SD card. Plug your AC adapter into a power source and then power on the board. You see the Windows logo and an activity cursor appear on the monitor. The first boot may take some time but eventually you be asked to choose the default language. Use the mouse or keyboard to select your language.

Once selected, you see the boot up screen, as shown in Figure 2-30. Recall from the "Raspberry Pi Configuration" section that you can configure your board using this screen, as well as shut down or reboot the board. The **Device Settings** button is the small gear located in the upper right portion of the screen. If you click it, you see the **Device Settings** screen that allows you to change the default language or connect to a Wi-Fi network if you have a wireless network adapter plugged into the USB port. Finally, you can shut down or restart the board by clicking the power button in the upper right of the screen.

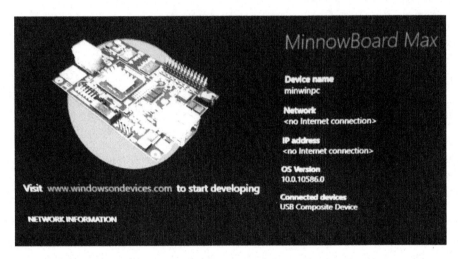

Figure 2-30. *Windows 10 IoT Core boot screen excerpt: MinnowBoard Turbot*

Now let's see how you can configure the DragonBoard 410c. After that section, you learn how you can connect to your board from your PC.

DragonBoard 410c Configuration

This section demonstrates how to install and boot Windows 10 IoT Core on the DragonBoard 410c. You see the specific steps needed to get your board ready for Windows 10 IoT Core, including the hardware that you need, how to connect to the board once Windows 10 IoT Core boot, and how to configure the board for your network.

Many of the steps are similar for this board as the Raspberry Pi. However, you do not use an SD card to boot the DragonBoard 410c. Instead, you will use a special application to download the boot image into a special memory drive (re-writable, non-volatile memory) on the DragonBoard 410c. Thus, that step is completely different that either of the other boards, but as you will see, you still download and install the firmware tools.

Let's begin with the prerequisite hardware that you need.

■ **Note** The DragonBoard 410c does not use an SD card to boot Windows 10 IoT Core.

Prerequisites

The following are the miscellaneous hardware that you need to use Windows 10 IoT Core with your board. You see where these cables are plugged into your board later in this section. For now, just gather the items you need.

- DragonBoard 410c

- +6.5V to +18Vpower adapter with a 4.75 x 1.75mm barrel male plug, like the one at www.arrow.com/en/products/wm24p-12-a-ql/autec-power-systems#page-1

- USB wired or wireless (not Bluetooth) keyboard and mouse

- HDMI monitor with HDMI cable or a suitable adapter for use with a DVI monitor

- micro-SD Card 8GB or larger: class 10 or better

- SD card reader (if your computer doesn't have one)

Download the .iso and Install the Board-specific Windows IoT Core Tools

The first step is to download the .iso file for the DragonBoard 410c, mount the .iso file, and then install the board-specific setup using the .msi file. The installation installs a number of tools that you need, including the Windows IoT Core Watcher that monitors your devices, the binary boot image for your board, and a tool named the DragonBoard Update Tool, which permits you to download the binary image to the board.

You can download the file from http://go.microsoft.com/fwlink/?LinkId=691713. Once the .iso file is downloaded, simply double-click it. This mounts as a virtual drive that opens automatically. You see a file named Windows_10_IoT_Core_QCDB410C.msi or similar. Double-click that file to start the installation. You'll need to do the usual steps for most installations, including accepting the license, permitting the change on your computer, and so on.

The binary image for the Raspberry Pi is named flash.ffu and installed in the c:\Program Files (x86)\Microsoft IoT\FFU\QCDB410C. Once the installation is complete, you can unmount the virtual drive. If you changed the installation folder, be sure to note the correct path—you need it in the next step.

Connecting the Hardware

If this is your first time using a DragonBoard 410c, orient the board on the table with the Arrow logo facing you. Figure 2-31 helps you locate the connection points.

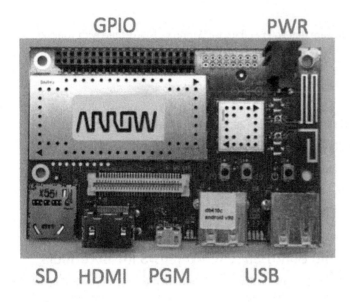

Figure 2-31. Connections for the DragonBoard 410c

Connect the HDMI monitor to the HDMI port located on the bottom of the board. Connect the USB mouse and keyboard to the USB ports on bottom of the board to the right of the HDMI connector. Now that all the cables are connected, you can connect the power and let the board boot up.

Downloading the Image to the DragonBoard 410c

The process to download the image to the DragonBoard 410c requires you to power it off, change one switch on the bottom of the board, and then use a USB to micro-USB cable to connect the board to your PC. Let's begin with setting the micro switch.

If you turn the board over, you see a set of four small switches located in the lower-right corner (reorient the board so you can view it this way). If you look closely, the switches are labeled from the bottom up as the following and referred to in this order starting from switch 1. The default setting for all switches is off (tab oriented to the right): (1) USB BOOT, (2) SD BOOT, (3) USB HOST, and (4) HDMI SEL.

When the USB BOOT switch is turned on, the board boots from a USB host connected to the programming port (labeled PGM in Figure 2-31). Similarly, when the SD BOOT switch is turned on, the board boots from the SD drive. When both of these switches are off, the board boots from memory.

To download the Windows 10 IoT Core image to the board, power off the board and set the USB BOOT switch to the on position, as shown in Figure 2-32.

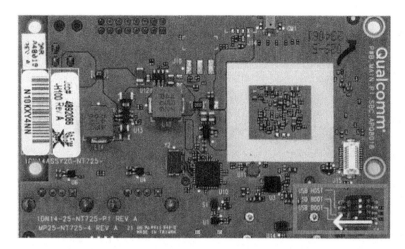

Figure 2-32. *DragonBoard 410c USB boot mode*

Next, you can connect the micro-USB to USB cable to the DragonBoard 410c and your PC. Launch the DragonBoard Update Tool (you can search for this string in the search box) on your PC and power on the board. Once your PC recognizes the board, the DragonBoard Update Tool connects and then prompts you for the image file. Click the **Browse** button and locate the flash.ffu file. This is located in the c:\Program Files (x86)\Microsoft IoT\FFU\QCDB410C folder and select it. You should now see the **Program** button and connection status light turn green, as shown in Figure 2-33.

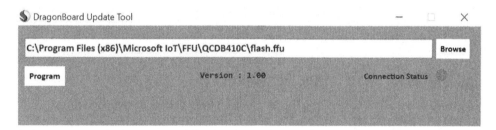

Figure 2-33. *DragonBoard Update Tool: connection established*

■ **Tip** If the DragonBoard Update Tool does not connect, try using a different USB cable and that the board is powered on. If that doesn't work, reinstall the DragonBoard 410c tools and reboot your computer. If you are still having problems, check the Device Manager for issues with the USB driver.

When you are ready, press the **Program** button. This begins the transfer of the image to the board. This overwrites the base Android image. Figure 2-34 shows the DragonBoard Update Tool in progress of downloading the image to the board.

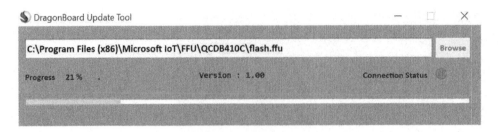

Figure 2-34. DragonBoard Update Tool downloading

When the process is complete, you are prompted to power off the board and reset the USB BOOT switch, as shown in Figure 2-35.

Figure 2-35. DragonBoard Update Tool: complete

Click **OK** and then power off the board and disconnect the USB cable from your computer. Then turn the USB BOOT switch to the off position. You will not need the USB cable for the next step.

Booting Windows 10 IoT Core for the First Time

Now it's time to boot the board with Windows 10 IoT Core. Connect the HDMI monitor, USB keyboard and mouse, and then the power adapter. Power on the device. It takes a couple of minutes for Windows to boot the first time but subsequent boots are a bit faster. You see the Windows logo and an activity cursor appear on the monitor. The first boot may take some time but eventually you be asked to choose the default language. Use the mouse or keyboard to select your language.

Once selected, you see the bootup screen, as shown in Figure 2-36. Recall from the Raspberry Pi section, you can configure your board using this screen, as well as shut down or reboot the board. The **Device Settings** button is the small gear located in the upper right portion of the screen. If you click it, you see the **Device Settings** screen that allows you to change the default language and connect to a Wi-Fi network. Finally, you can shut down or restart the board by clicking the power button in the upper right of the screen.

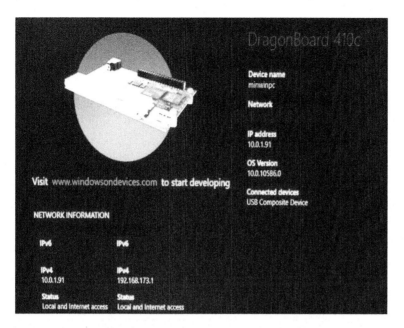

Figure 2-36. *Windows 10 IoT Core boot screen excerpt: DragonBoard 410c*

The DragonBoard 410c has Wi-Fi, so you can connect your board to your Wi-Fi network by clicking the Device Settings icon (looks like a gear) in the upper-right portion of the screen.

On the Device Settings page, select **Network & Wi-Fi** via the menu on the left. The board begins to search for Wi-Fi networks. When your Wi-Fi network name (SSID) appears in the list, select it and click the **Connect** button to connect. You can then close the Device Settings and view the network name and IP address on the home screen. Be sure to note this IP address as that you need it to connect to the board remotely.

Connecting to Your Board

Now that your board is booted and connected to your network, you can log in remotely from your computer. Indeed, once you've set up the board, you normally would not connect it to a monitor and keyboard. That is, it is more common that would deploy your board in your solution and run it headless.

Connecting to your board can be done in a number of ways, including the Windows 10 IoT Core Dashboard, a secure shell (SSH) connection via the command line, or using Windows PowerShell. There are other ways, but these are the most common and most useful. I'll discuss each of these in the following sections. These methods are the same for all the boards—there are no special steps for the MinnowBoard Max–compatible or DragonBoard 410c boards.

Connect with the Windows 10 IoT Core Dashboard

The Windows 10 IoT Core Dashboard offers two ways to connect to your board, both of which are initiated from the My Devices panel. Once the board is booted, click the **My Devices** tab to see the list of all of your Windows 10 IoT Core devices. It could take a few minutes for your board to appear in the list. If it does not, make sure that the board is connected to the same network as your computer. If you have multiple boards running, you see all of them in the list. Figure 2-37 shows an example of the **My Devices** list.

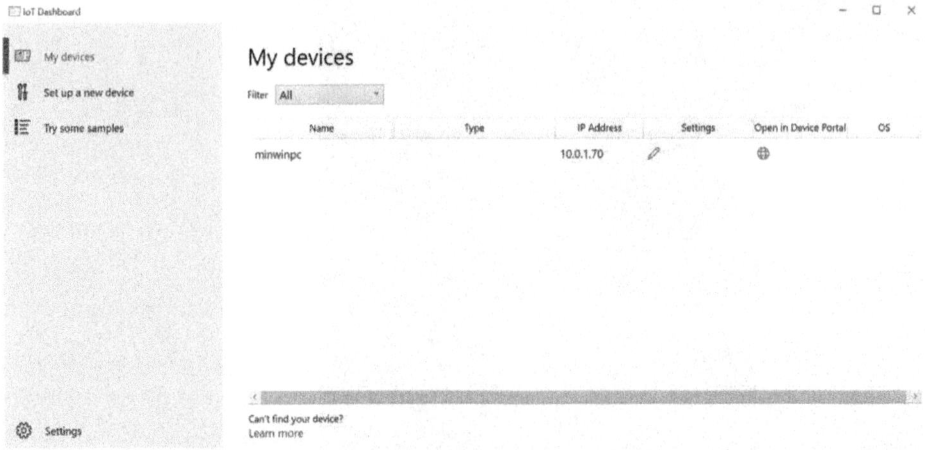

Figure 2-37. *Windows 10 IoT Core Dashboard: My Devices list*

■ **Note**   Once the board is booted, it could take a few minutes for the board to show up in the My Devices list.

There is one device in the list named miniwpc (the default name for all boards—but you can change that). You also see the IP address of the board and two small icons.

The first icon (shown as a pencil icon) permits you to connect to the board and change its name and password. I recommend you do this as your first action to secure your board. Simply click the pencil icon on the row that shows your board and enter the default password (**p@ssw0rd**). Optionally, you change the name of the board, choose a new password, and save the settings. Figure 2-38 shows the settings dialog.

Figure 2-38. *Windows 10 IoT Core Dashboard: Settings dialog*

■ **Tip** The default password for the Windows 10 IoT Core is p@ssw0rd (the 0 is a zero). You should change the administrator password when your device boots for the first time.

The second method is where all the fun begins. You see the Windows Device Portal link (shown as a world icon). This option connects to a special web page running on the board that allows you to configure all manner of things on the board, including the following features, organized as a series of tabs or panels.

- *Apps*: Allows you to install or uninstall packages on your device.

- *Processes*: Allows you to see which processes are currently running as well as the memory used.

- *Performance*: Displays real-time diagnostics of the CPU, I/O usage, and memory.

- *Debugging*: Provides tools to help you diagnose problems with your application.

- *ETW*: An advanced tool for Windows event tracing.

- *Perf Tracing*: Allows you to create a log of system changes and settings.

- *Devices*: Lists the devices connected to the board.

- *Bluetooth*: Allows you to configure Bluetooth connections.

- *Audio*: Allows you to configure the audio options (if audio capable).

- *Networking*: Displays the IP address and description of the network.

- *Windows Update*: Allows you to update the system files on the board (just like your computer).

Don't worry about learning what each and every one of these do; you'll discover the most frequently used throughout this book. As you can see, some of these are advanced features for performance testing, debugging, and event tracing. Most hobbyists and enthusiasts do not use these advanced features, but they're there if you need them. Sadly, there isn't a lot of documentation for how to use some of them.

To connect to the Windows Device Portal, click the world icon. You are asked to enter the user account and password. You want to use the administrator account and the new password you set previously. Figure 2-39 shows an example of the web page that you see once connected.

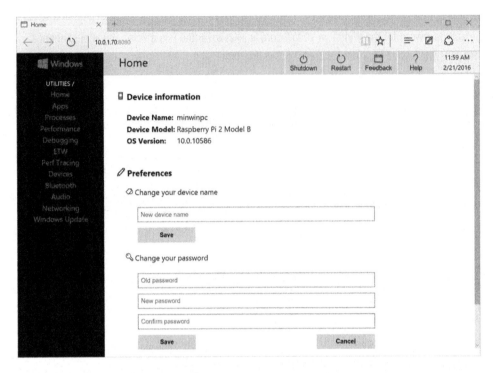

Figure 2-39. *Windows Device Portal*

You can also shut down and restart the board from this web page. Interestingly, there is help available for the options but the content is still rather terse. Again, you will see how to use some of these features in later chapters, starting with how to deploy applications in Chapter 4.

Connecting with SSH

Another, popular method for connecting to your board and indeed a method I use often is to use secure shell (SSH). Sadly, Windows, unlike other platforms, does not come with a SSH client.[14] Fortunately, there are several varieties available as third-party applications. One of the ones I like is called PuTTY, which is actually a general terminal session application. The SSH feature in PuTTY is very easy to use. What I like most about PuTTY is it is open source software and therefore free to download and use.

You can download PuTTY from `www.chiark.greenend.org.uk/~sgtatham/putty/download.html`. The download is not an installation package (`.msi`), rather, it is simply the PuTTY executable (`putty.exe`). Simply download the file and place it in a folder that is in your path environment variable. You can also put it in your documents folder and simply execute it from there or by referencing the folder.

There are other applications available from the download site, including an installation package that installs most of the tools. You only need the PuTTY executable for connecting to your boards but I encourage you to check out the other tools.

■ **Tip** For more information about PuTTY, visit `www.putty.org`.

---

[14]I've never understood why that is.

To connect to your board, you need the IP address from the Windows 10 IoT Core boot screen on the board. You captured this when you booted the board for the first time. When you launch the putty.exe executable, you see the initial dialog, as shown in Figure 2-40. Select the **SSH** radio button and enter the IP address of your board in the **Host Name (or IP address)** box. Leave the port as 22.

Figure 2-40. *PuTTY dialog*

When you are ready, click the **Open** button. Once the connection completes, you see a new command window open and be asked to log in. Enter **Administrator** as the user and press **Enter**. Then enter the password that you set in a previous step (or the default **p@ssw0rd**, if you have not changed it) and press **Enter**. Figure 2-41 shows an example of the SSH session.

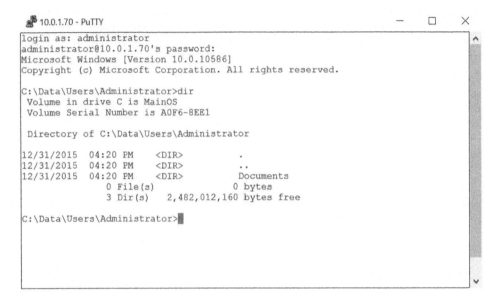

Figure 2-41. *SSH connection example*

There are a variety of commands you can use. You can perform nearly all the operations on the home screen via command-line utilities, such as boot configuration, startup applications, shutdown, restart, and more.

Thus, you can do everything you need to do remotely via the command line rather than use the Windows Device Portal (but I've found the portal more convenient). For more details about the command utilities available, see https://ms-iot.github.io/content/en-US/win10/tools/CommandLineUtils.htm.

Connecting with the Windows PowerShell

You can also connect to your board with Windows PowerShell. However, it isn't quite as obvious or intuitive unless you've used PowerShell previously. You can launch PowerShell, but you should do so as the administrator since some commands require elevated privileges. Simply search for **powershell**. When the application appears in the list, right-click it and select the **Run as administrator** option.

The steps you need to perform require starting the Windows remote management service, creating a trust relationship between your PC and the board, and then opening a session. See, not so intuitive, is it? Let's see how to do this.

■ **Note**   Connecting to your board may take several seconds from the PowerShell.

Once the PowerShell launches, start the Windows remote management service with the following command. Since this is a network operation, you use the net command. You see messages stating that the service has started.

```
> net start WinRM
```

Next, you create the trust relationship with the set-item command. Type the following command substituting the IP address from the board as you recorded earlier when you booted Windows 10 IoT Core for the first time. You are asked to confirm the change. Enter **Y** to continue.

```
> Set-Item WSMan:\localhost\Client\TrustedHosts -Value 10.0.1.70
```

■ **Tip**   If you want to connect to more than one board, you can place the list of IP addresses inside double quotes separated by commas, such as "10.0.1.70,10.0.1.71".

Finally, you can start a session with your Windows IoT Core device using the Enter-PSSession. Yes, another unintuitive command.[15] Here you provide the computer name (IP address is fine) and user name (Administrator). You are prompted for the password in a pop-up dialog. The following shows the command I used to open a session to my board at 10.0.1.70. Figure 2-42 shows the PowerShell session once login succeeds.

```
> Enter-PSSession -ComputerName 10.0.1.70 -Credential 10.0.1.70\Administrator
```

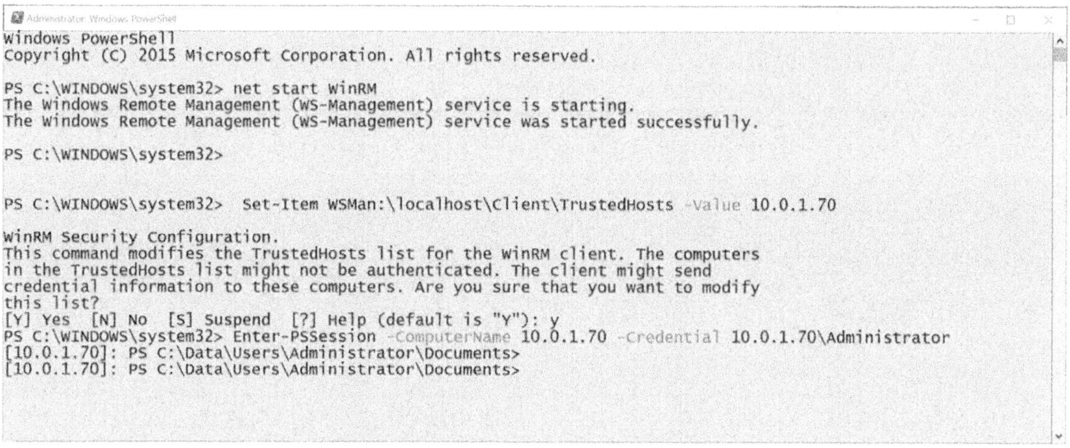

Figure 2-42. *Connecting with the Windows PowerShell: session established*

The IP address of the board is in square brackets. This lets you know which board you are currently connected to, should you connect to more than one at the same time. From here you can use the command-line utilities described in the previous section.

---

[15]For muggles at least.

Summary

The Windows 10 IoT Core offers an option to explore electronics and the IoT for the rest of us. While there have been and continue to be options for exploring electronics and IoT for users of other platforms, Windows users have had to learn alternative operating systems and (seemingly) arcane commands in order to explore even the most basic of hobbyist electronics. With Windows 10, those days are gone!

This chapter explored the Windows 10 IoT Core and the hardware it runs on. You discovered some of the key features of the platform along with the details of the three low-cost computing boards that support Windows 10 IoT Core. You also discovered how to get your PC and your hardware configured to run Windows 10 IoT Core along with a walk-through for each of the available boards; from creating the boot image to booting and configuring the board to connecting to the board from your PC.

In the next chapter, you take a short break to examine the Raspberry Pi in more detail. You will learn more about the features and specifications of the Raspberry Pi, including a quick overview of how to boot it in its native operating system as well as how to write a simple program to illuminate an LED from code written in Python. While the chapter does not use Windows 10, you learn a great deal about the Raspberry Pi and a little bit about Python, which you will use in later chapters to write applications for Windows 10 IoT Core. How cool is that?

CHAPTER 3

Introducing the Raspberry Pi

The Raspberry Pi is one of the latest disruptive devices that have changed the way that we think about and design embedded solutions and the IoT. In fact, the Raspberry Pi has had tremendous success among hobbyists and enthusiasts. This is partly due to its low cost but also because it is a full-fledged computer running an open source operating system that has a wide audience: Linux.

Fortunately for us, Windows 10 runs on the Raspberry Pi. However, given the popularity of the Raspberry Pi, it is likely that you will encounter examples and resources that are written for or only work with Linux. Thus, learning more about the Raspberry Pi and its native environment allows you to leverage the plethora of data for the Raspberry Pi and adapt, or helps you develop similar solutions for Windows 10. Plus, it gives you a brief insight into the non-Windows world of Raspberry Pi.[1]

This chapter explains how to set up and configure the Raspberry Pi using the Linux operating system. You'll also discover a few key concepts of how to work with Linux and even a brief look at writing Python scripts, which you will use to write Windows 10 IoT applications in later chapters. Let us begin with an in-depth look at the Raspberry Pi.

Note   While there are several versions of the Raspberry Pi, including the compute module, Zero, Model 1 A+, 2 Model B, and 3 Model B, I discuss only those versions that run Windows 10 IoT Core: the Raspberry Pi 2 Model B and Raspberry Pi 3 Model B. For more details about all of the Raspberry Pi boards, see www.raspberrypi.org.

Getting Started with the Raspberry Pi

The Raspberry Pi is a small, inexpensive personal computer, also called a *low-cost computing board*. Although it lacks the capacity for memory expansion and can't accommodate on-board devices such as CD, DVD, and hard drives, it has everything a simple personal computer requires. That is, it has USB ports, an Ethernet port, HDMI, and even an audio connector for sound. The latest version, the Raspberry Pi 3 Model B, even has Bluetooth and Wi-Fi!

The Raspberry Pi has a micro-SD drive that you can use to boot the computer into any of several Linux operating systems (and Windows 10 IoT Core). All you need is an HDMI monitor (or DVI with an HDMI-to-DVI adapter), a USB keyboard and mouse, and a 5V power supply—and you're off and running.

---

[1]This should reinforce how well Windows 10 IoT Core works!

© Charles Bell 2016
C. Bell, *Windows 10 for the Internet of Things*, DOI 10.1007/978-1-4842-2108-2_3

■ **Tip** You can also power your Raspberry Pi using a USB port on your computer. In this case, you need a USB type A male to micro-USB type B male cable. Plug the type A side into a USB port on your computer and the micro-USB type B side into the Raspberry Pi power port.

The Raspberry Pi costs as little as $35.[2] It can be purchased online from electronics vendors such as SparkFun and Adafruit. Most vendors have a host of accessories that have been tested and verified to work with the Raspberry Pi. These include small monitors, miniature keyboards, and cases for protecting the board.

This section explores the origins of the Raspberry Pi 3, tours the hardware connections, and covers the accessories needed to get starting using the Raspberry Pi.

Raspberry Pi Origins

The Raspberry Pi was designed to be a platform to explore topics in computer science. The designers saw the need to provide inexpensive, accessible computers that could be programmed to interact with hardware such as servomotors, display devices, and sensors. They also wanted to break the mold of having to spend hundreds of dollars on a personal computer, and thus make computers available to a much wider audience.

The designers observed a decline in the experience of students entering computer science curriculums. Instead of having some experience in programming or hardware, students are entering their academic years having little or no experience working with computer systems, hardware, or programming. Rather, students are well versed in Internet technologies and applications. One of the contributing factors cited is the higher cost and greater sophistication of the personal computer, which means parents are reluctant to let their children experiment on the family PC.

This poses a challenge to academic institutions, which have to adjust their curriculums to make computer science palatable to students. They have had to abandon lower-level hardware and software topics due to students' lack of interest or ability. Students no longer wish to study the fundamentals of computer science such as assembly language, operating systems, theory of computation, and concurrent programming. Rather, they want to learn higher-level languages to develop applications and web services. Thus, some academic institutions are no longer offering courses in fundamental computer science.[3] This could lead to a loss of knowledge and skillsets in future generations of computer professionals.

To combat this trend, the designers of the Raspberry Pi felt that, equipped with the right platform, today's youth could return to experimenting with personal computers as in the days when PCs required a much greater commitment to learning the system and programming it in order to meet your needs. For example, the venerable Commodore 64, Amiga, and early Apple and IBM PC computers had very limited software offerings. Having owned a number of these machines, I was exposed to the wonder and discovery of hardware and programming at an early age. Perhaps that is why I find low-cost computing boards so fascinating.

WHY IS IT CALLED RASPBERRY PI?

The name was partly derived from design committee contributions and partly chosen to continue a tradition of naming new computing platforms after fruit (think about it). The Pi portion comes from Python, because the designers intended Python to be the language of choice for programming the computer. However, other programming language choices are available.

---

[2]Although history shows new releases of the board can command higher prices until supply catches up with demand.
[3]Sadly, my alma mater is a fine example of this decline.

The Raspberry Pi is an attempt to provide an inexpensive platform that encourages experimentation. The following sections further explore the Raspberry Pi, discussing topics such as the required accessories and where to buy the boards.

Versions that Work with Windows 10 IoT Core

There are currently several versions of Raspberry Pi boards. The latest boards that support Windows 10 IoT Core are the Raspberry Pi 2 Model B and Raspberry Pi 3 Model B. The model nomenclature has to do with the layout and ports available. Essentially, there is a model A variant that does not include Ethernet, whereas the model B variant does.

WINDOWS 10 IOT CORE FOR THE RASPBERRY PI 3

Microsoft's latest stable release of Windows 10 IoT Core may not support the latest boards. However, for those of us who want to use the latest boards or the latest features (such as the Raspberry Pi 3), Microsoft provides a preview release called the Windows 10 IoT Core Insider Preview. Sometimes these preview releases are built for specific boards, so choose the one that matches the board or other feature that you want.

To download this release, you must have a valid Microsoft account. You can initiate the download by visiting `https://ms-iot.github.io/content/en-US/Downloads.htm` and clicking the link for the Insider Preview. Once you log in, the download begins. You can create an account from the page if you do not have one.

This downloads a new `.iso` file that you mount. Install the `.msi` and create the SD image in the same manner as the base images, as described in Chapter 2. However, the preview may overwrite existing files, so be sure to backup any files you want to keep.

The Raspberry Pi 2 and 3 are very similar. In fact, they are very hard to tell apart without reading the label on the top of the board. This is because they share the same layout (model B) with the same connectors. The hardware differences are very minor and difficult to spot. Figure 3-1 shows the two boards together, with the Raspberry Pi 2 on the left.

Figure 3-1. *Raspberry Pi 2 and 3 top side*

Can you tell them apart? Hint: look at the lower-left corner of each board. The Raspberry Pi 3 has a small antenna, whereas the Raspberry Pi 2 has LEDs in the same spot.

While the boards appear nearly identical, the Raspberry Pi 3 has a much faster 64-bit quad core processor (Windows 10 IoT Core runs only in 32-bit mode currently) that has shown to be as much as ten times faster than the Raspberry Pi 2. Furthermore, the Raspberry Pi 3 has both Bluetooth and Wi-Fi onboard, whereas the Raspberry Pi 2 has neither. There are a number of smaller changes, but these are by far the most significant features you should use to decide which board to buy.

The underside is easier to distinguish the boards. Figure 3-2 shows the underside of both boards, again, with the Raspberry Pi 2 on the left. Here you can see the difference at the top-left corner of the boards, where the Raspberry Pi 3 has a small rectangular chip that contains the Wi-Fi and Bluetooth components.

Figure 3-2. *Raspberry Pi 2 and 3 bottom side*

A Tour of the Board

Not much larger than a deck of playing cards, the Raspberry Pi board contains a number of ports for connecting devices. This section presents a tour of the board. If you want to follow along with your board, hold it with the Raspberry Pi logo face up. I work around the board clockwise. Figure 3-3 depicts a drawing of the board with all the major connectors labeled.

Figure 3-3. *Raspberry Pi 3 Model B*

Let's begin by looking at the bottom edge of the board (looking from above). In the center of the bottom side, you see an HDMI connector. To the left of the HDMI connector is the micro-USB power connector. The power connector is known to be a bit fragile on some boards, so take care plugging and unplugging it. Be sure to avoid putting extra strain on this cable while using your Raspberry Pi. To the right of the HDMI connector is the camera ribbon cable connector and next to that is the audio connector.

On the left side of the board is the LCD ribbon cable. You can use this connector with the Raspberry Pi 7-inch Touch LCD and similar devices. On the underside of the board is the micro-SD card drive. When installed, the SD card protrudes a few millimeters out of the board. If you plan to use a case for your Raspberry Pi, be sure the case provides access to the SD card drive (some do not).

■ **Caution**　Because the board is small, it is tempting to use it in precarious places, like in a moving vehicle or on a messy desk. Ensure that your Raspberry Pi is in a secure location. The micro-USB power, HDMI, and SD card slots seem to be the most vulnerable connectors.

On the top edge of the board is the general-purpose input/output (GPIO) header (a double row of 20 pins each), which can be used to attach to sensors and other electronic components and devices. You will work with this connector later in this chapter.

On the right side of the board are two USB connectors with two USB ports each and the Ethernet connector. An external powered USB hub connected to the USB ports on the Raspberry Pi can power some boards, but it is recommended that you use a dedicated power supply connected to the micro-USB connector.

Take a moment to examine the top and bottom faces of the board. As you can see, components are mounted on both sides. This is a departure from most printed circuit boards (PCB) that have components on only one side. The primary reason the Raspberry Pi has components on both sides is that it uses multiple layers for trace runs (the connecting wires on the board). Stacking the trace runs on multiple levels means that you don't have to worry about crossing paths. It also permits the board to be much smaller and enables the use of both surfaces. This is probably the most compelling reason to consider using a case—to protect the components on the bottom of the board and thus avoid shorts (accidental connection of contacts or pins) and can lead board failure.

Required Accessories

The Raspberry Pi is sold as a bare system board with no case, power supply, or peripherals. Depending on how you plan to use the Raspberry Pi, you need a few commonly available accessories. If you have been accumulating computer and electronic spares like me, a quick rummage through your stores may locate most of what you need.

If you want to use the Raspberry Pi in console mode (no graphical user interface), you need a USB power supply, a keyboard, and an HDMI monitor. The power supply should have a minimal rating of 700mA or greater. If you want to use the Raspberry Pi with a graphical user interface, you also need a pointing device (such as a mouse).

If you have to purchase these items, stick to the commonly available brands and models without extra features. For example, avoid the latest multifunction keyboard and mouse because they may require drivers that are not available for the various operating system choices for the Raspberry Pi.

You also must have a micro-SD card. I recommend an 8GB or higher version. Recall that the micro-SD is the only on-board storage medium available. You need to put the operating system on the card, and any files you create are stored on the card.

If you want to use sound in your applications, you also need a set of powered speakers that accept a standard 3.5mm audio jack. Finally, if you want to connect your Raspberry Pi to the Internet, you need an Ethernet cable, or if you are using a Raspberry Pi 3, you need a Wi-Fi network.

Recommended Accessories

I highly recommend, at a minimum, adding small 5mm to 10mm rubber or silicone self-adhesive bumpers to the bottom side of the board over the mounting holes to keep the board off your desk. On the bottom of the board are many sharp prongs that can come into contact with conductive materials, which can lead to shorts, or worse, a blown Raspberry Pi. They can also damage your skin and clothing. Small self-adhesive bumpers are available at most home-improvement and hardware stores.

If you plan to move the board from room to room or you want to ensure that your Raspberry Pi is well protected against accidental damage, you should consider purchasing a case to house the board. Many cases are available, ranging from simple snap-together models to models made from laser-cut acrylic or even milled aluminum.

■ **Tip** If you plan to experiment with the GPIO pins, or require access to the power test pins or the other ports located on the interior of the board, you may want to consider either using the self-adhesive bumper option or ordering a case that has an open top to make access easier. Some cases are prone to breakage if opened and closed frequently.

Aside from a case, you should also consider purchasing (or pulling from your spares) a powered USB hub. The USB hub power module should be 700mA to 1000mA or more. Even though the Raspberry Pi 2 and 3 have four USB ports, a powered hub is required if you plan to use USB devices that draw a lot of power, such as a USB hard drive or a USB toy missile launcher.

Where to Buy

The Raspberry Pi 2 has been available in Europe for some time, so it is relatively easy to find. However, at the time of this writing, the Raspberry Pi 3 was in very short supply. It is getting easier to find, but very few brick-and-mortar stores stock the Raspberry Pi 3. Fortunately, a number of online retailers stock it, as well as a host of accessories that are known to work with the Raspberry Pi. The following are some of the more popular online retailers with links to their Raspberry Pi catalog entry.[4]

- Adafruit: www.adafruit.com/category/105

- SparkFun: www.sparkfun.com/categories/233

- Maker Shed: www.makershed.com/collections/raspberry-pi

The next section presents a short tutorial on getting started using the Raspberry Pi. If you have already learned how to use the Raspberry Pi, you can skip to the following section to begin learning how to use your board.

Setting up the Raspberry Pi

The Raspberry Pi is a personal computer with a surprising amount of power and versatility. You may be tempted to consider it a toy or a severely limited platform, but that is far from the truth. With the addition of onboard peripherals like USB, Ethernet, and HDMI video, (as well as Bluetooth and Wi-Fi for the Raspberry Pi 3), the Raspberry Pi has everything you need for a lightweight desktop computer. If you consider the addition of the GPIO header, the Raspberry Pi becomes more than a simple desktop computer and fulfills its role as a computing system designed to promote hardware experimentation.

---

[4]Although some vendors do not currently have the Raspberry Pi 3 in stock, they should have them in stock by the time this book is published.

The following sections present a short tutorial on getting started with your new Raspberry Pi, from a bare board to a fully operational platform. A number of excellent works cover this topic in much greater detail. If you find yourself stuck or wanting to know more about beginning to use the Raspberry Pi and more about the Raspbian operating system, read *Learn Raspberry Pi with Linux* by Peter Membrey and David Hows (Apress, 2012). If you want to know more about using the Raspberry Pi in hardware projects, an excellent resource is *Practical Raspberry Pi* by Brendan Horan (Apress, 2013).

As mentioned in the "Required Accessories" section, you need a micro-SD card, a USB power supply rated at 700mA or better with a male micro-USB connector, a keyboard, a mouse (optional), and an HDMI monitor. However, before you can boot your Raspberry Pi and bask in its brilliance, you need to create a boot image for your micro-SD card.

Choosing a Boot Image (Operating System)

The first thing you need to do is decide which operating system variant you want to use. There are several excellent choices, including the standard Raspbian "Jessie" variant. Each is available as a compressed file called an image or card image. You can find a list of recommended images along with links to download each on the Raspberry Pi foundation download page: `www.raspberrypi.org/downloads`. The following images are available at the site.

- *Raspbian (Jessie).* A Debian-based official operating system and contains a graphical user interface (Lightweight X11 Desktop Environment [LXDE]), development tools and rudimentary multimedia features.

- *Ubuntu Mate.* Features the Ubuntu desktop and a scaled-down version of the Ubuntu operating system. If you are familiar with Ubuntu, you will feel at home with this version.

- *Snappy Ubuntu Core.* The developer's edition of core Ubuntu system. It is the same as Mate with addition of the developer core utilities.

- *Windows 10 IoT Core.* Windows 10 for the IoT. Microsoft's premier IoT operating system.

- *OSMC* (Open Source Media Center). Build yourself a media center.

- *OpenELEC* (Open Embedded Linux Entertainment Center). Another media center option.

- *PiNet.* A classroom management system. A special edition for educators using the Raspberry Pi in the curriculum.

- *RISC OS.* A non-Linux, Unix-like operating system. If you know what IBM AIX is, or you've used other Unix operating systems, you'll recognize this beastie.

■ **Tip**   If you are just starting with the Raspberry Pi and haven't used a Linux operating system, you should use the Raspbian image as it is the most popular choice and more widely documented in examples. Plus, it is the base or default image for Raspberry Pi.

There are a few other image choices, including a special variant of the Raspbian image from Adafruit. Adafruit calls their image "occidentals" and includes a number of applications and utilities preinstalled, including Wi-Fi support and several utilities. Some Raspberry Pi examples—especially those from Adafruit—require the occidentals image. You can find out more about the image and download it at `http://learn.adafruit.com/adafruit-raspberry-pi-educational-linux-distro/overview`.

Wow! That's a lot of choices, isn't it? As you can see, the popularity of the Raspberry Pi is very wide and diverse. This makes Windows 10 a huge deal for the platform and Windows users alike. While you may not use these operating systems, it is good to know what choices are available should you need to explore them.

Now let's see how to install the base operating system. As you will see, it is very easy.

Creating the Boot Image

There are two methods for installing the boot image. You can use the automated, graphical user interface platform called New Out Of Box Software (NOOBS[5]) or you can install your image from scratch onto a micro-SD drive. Both require downloading and formatting the micro-SD drive.

Since you are curious about using the Raspberry Pi native operating system as an experiment or for research, let's stick to the easier method and use NOOBS. Aside from formatting the micro-SD card, everything is automated; nothing requires any complicated commands.

With NOOBS, you download a base installer image that contains Raspbian Jessie. You can choose to install it or configure NOOBS to download another image and install it. But first, you have to get the NOOBS boot image and copy it to your micro-SD drive.

Begin by downloading the NOOBS installer from www.raspberrypi.org/downloads/noobs/. You see two options: an offline and network installer that includes the Raspbian image or a network installer that does not contain any operating systems (called NOOBS Lite). The first option (the one with the base image) is what you should use if you are following along with this chapter.

Once you've downloaded the installer (to date about 1.4GB), you need to format a micro-SD card of at least 8GB. You'll use the SD Formatter 4.0 utility for Windows (www.sdcard.org/downloads/formatter_4/). Simply download the application and install it. Then insert your micro-SD card in your card reader and launch the application. Once you verify that you've selected the correct media, enter a name for the card (I used RASPI) and click the **Format** button. Figure 3-4 shows the SD Formatter application.

Figure 3-4. *SDFormatter 4.0*

---

[5]An unfortunate resemblance to the slightly derogatory slang *noob* or *newbie*.

Once you've formatted the micro-SD card, you now must copy the contents of the NOOBS image to the card. Right-click the file that you downloaded and choose the option to unzip or unarchive the file. This creates a folder containing the NOOBS image. Copy all of those files (not the outside folder) to the SD card and eject it. You are now ready to boot into NOOBS and install your operating system. When this process has finished, safely remove the SD card and insert it into your Raspberry Pi.

Booting the Board

You are now ready to hook up all of your peripherals. I like to keep things simple and only connect a monitor, keyboard, and (for NOOBS) a mouse. If you want to download an operating system other than Raspbian, you also need to connect your Raspberry Pi to your network. If you are planning to use the Wi-Fi option on the Raspberry Pi 3, you'll need to set up your Wi-Fi configuration after you boot up. I'll show you how to do that in a moment.

Once your Raspberry Pi powers on, you see a scrolling display of various messages. This is normal and may scroll for some time before NOOBS starts. When NOOBS is loaded, you see a screen similar to Figure 3-5.

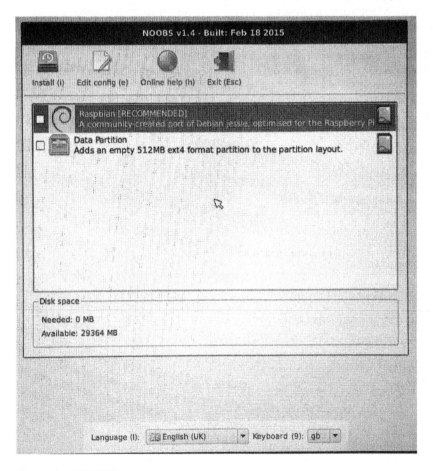

Figure 3-5. *NOOBS startup screen*

Notice the Raspbian image in the list of operating systems. To install it, just tick the check box beside the thumbnail and then click the **Install** button. Note the two boxes at the bottom. They set the language and keyboard for use in NOOBS, which does not affect the Raspbian setup.

Once you initiate the install, you see a series of dialogs as Raspbian begins its installation to the SD card. This could take a while. The good news is that the dialogs provide a lot of useful information to help you get started. You learn about how to log in to Raspbian, tips for configuring and customizing, and suggestions on how to get the most out of your experience.

When installation finishes, click **OK** on the completed dialog and then wait for the Raspberry Pi to reboot into Raspbian. The system boots and automatically logs on as the **pi** user. You can change this in the Raspberry Pi Configuration dialog. The configuration dialog (click the **Edit config** button) is used to set the time and date for your region, enable hardware such as a camera board, create users, change the password, and more. Figure 3-6 shows the configuration dialog. If you want to change the auto login feature, you can open the dialog from the **Menu ➤ Preferences ➤ Raspberry Pi Configuration** menu option.

■ **Tip**   The default password for the pi user is `raspberry`.

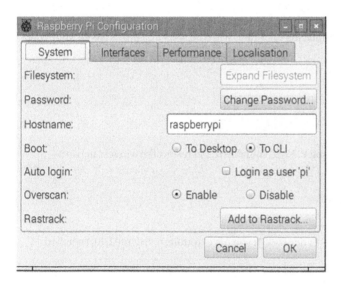

Figure 3-6. *Raspbian configuration dialog*

There are four tabs that you can use to change settings for the system. The following list briefly explains each and includes the recommended settings for each. Once you have made your changes, click **OK** to close the dialog. Depending on which settings you choose, you may be asked to reboot.

- *System*. The board controls for the system. Use this panel to change the root password (highly recommended), hostname (optional), type of boot (use command-line interface [CLI] if you want to set up the Raspberry Pi to boot headless, and automatic login (not recommended).

- *Interfaces*. Used to enable system and hardware services such as the camera, SSH (recommended), and hardware interfaces for the GPIO header.

- *Performance*. Used to make changes to how the processor performs. You can choose to overclock (run the CPU faster) but I do not recommend this setting for a Raspberry Pi that hosts IoT solutions.

- *Localization*: Used to set the default language, keyboard, and date and time. If you change nothing else, be sure to set these to your local settings.

To shut down or reboot Raspbian, click **Menu** and then choose **Shutdown**. You see a prompt for rebooting, shutting down, or returning to the command line. If you are at the command line, use the command shutdown -h now to shut down the system.

Setting up Wi-Fi on the Raspberry Pi 3

You can use the Wi-Fi feature of the Raspberry Pi 3 with the latest version of Raspbian. In fact, there are three ways to configure the Wi-Fi (www.raspberrypi.org/documentation/configuration/wireless/). I prefer the command-line option, which requires editing a file to add your Wi-Fi information. Let's begin by editing the wpa_supplicant.conf configuration file used to read network configurations. You can do this from the terminal using the following command.

```
sudo nano /etc/wpa_supplicant/wpa_supplicant.conf
```

The sudo portion of the command tells the operating system to run as a super user, which is roughly equivalent as run as administrator for Windows.

Add the following to the end of the file, replacing the values with the SSID and password for your Wi-Fi network.

```
network={
    ssid="YOUR SSID GOES HERE"
    psk="YOUR PASSWORD GOES HERE"
}
```

Save the file by pressing **Ctrl-X** and replying **Y**. Next, shut down and restart the wireless network connector with the following commands.

```
sudo ifdown wlan0
sudo ifup wlan0
```

You should now be connected to your Wi-Fi network. If you have trouble, revisit the URL specified earlier for updated instructions.

A Brief Linux Primer

OK, now you have a Raspberry Pi booting Linux (Raspbian) into the desktop environment. And although it looks cool, it can be a bit confusing and intimidating. The best way to learn the GUI is to simply spend some time clicking your way through the menus. You'll find the most basic of features, including productivity tools.

However, working with hardware typically requires knowledge of basic commands used in a terminal (also called the *command line*). This section describes a number of the more basic commands you need to use. This is by no means meant to be a complete or thorough coverage of all of the commands. Rather, it gives you the basics that you need to get started.

Thus, this primer is more like a 15-minute guided tour of an automobile engine. You cannot possibly learn all of the maintenance requirements and internal components in 15 minutes. You would need to have an automotive technician's training or years of experience before you could begin to understand everything. What you get in a 15-minute lightning tour is more of a bird's eye view with enough information to permit you to know where the basic maintenance items are located, not necessarily how they work.

The latest versions of Raspbian boot into the desktop, but you can boot directly into the command line. Just open the Raspberry Pi Configuration dialog and select **To Cli** (command-line interface). In fact, I normally do this since most of my Raspberry Pi projects run headless and thus there is no need to start the desktop. You can always restart the desktop with the startx command from the command line.

I recommend you read through the following sections to familiarize yourself with the commands that you may need. You can refer back to these sections should you need to recall the command name. Often times it is simply a matter of learning a different name for the same commands (conceptually) that you're familiar with from Windows. As you will see, many of these commands are familiar in concept, as they also exist on Windows albeit with a different name and parameters.

■ **Tip** If you want to master the Linux command-line commands, tools, and utilities, read the book *Beginning the Linux Command Line* by Sander van Vugt (Apress, 2015).

Let's begin with how to get help about commands.

Getting Help

Linux provides help for all commands by default. While it can be a bit terse, you can always get more information about a command by using the manual command as shown in Listing 3-1. Here you want more help with the list directory command (ls).

Listing 3-1. Getting Help for a man (manual) Command

```
pi@raspberrypi:~ $ man ls

LS(1)                          User Commands                          LS(1)

NAME
       ls - list directory contents

SYNOPSIS
       ls [OPTION]... [FILE]...

DESCRIPTION
       List  information  about  the FILEs (the current directory by default).
       Sort entries alphabetically if none of -cftuvSUX nor --sort is speci-
       fied.

       Mandatory  arguments  to  long  options are mandatory for short options
       too.

       -a, --all
             do not ignore entries starting with .

       -A, --almost-all
             do not list implied . and ..

...
```

File and Directory Commands

Like any operating system, some of the most basic commands are those that allow you to manipulate files and directories. These include operations such as copying, moving, and creating files and directories. I list a few of the most common commands in the following sections and provide an example of each. If you want to know more about each, try using the manual command (man) to learn about each. Just use the name of the command you want to know more about as the option. For example, to learn more about ls, enter man ls.

List Directories and Files

The first command you will likely need is the ability to list files and directories. In Linux, we use the ls (list files and directories) command. Without any options, the command lists all of the files and directories in the current location. There are many options available, but the ones I find most helpful are show long listing format (-l), sort the output (-s), and show all files (-a). You can combine these options in a single string, such as -lsa.

The command uses color and highlighting to help distinguish directories from files, executable files, and more. The long listing format also shows you the permissions for the file (the series of rwx values). The first character in the directory list refers to the file type (d means a directory and - is a regular file), the next three characters refer to file owner permissions, the next three are group permissions, and the final three are for other users' permissions. Figure 3-7 shows an example of the ls -lsa command output.

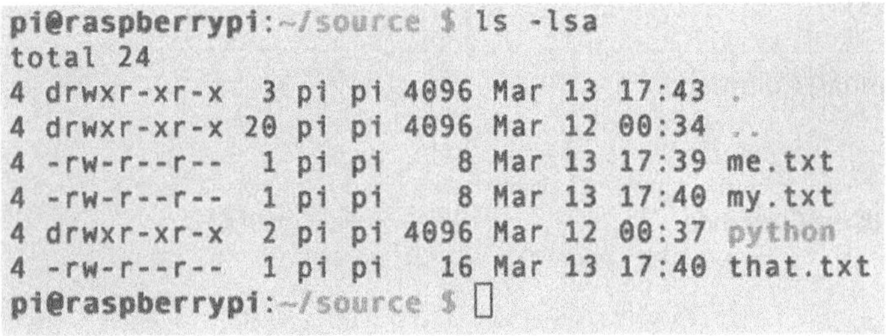

```
pi@raspberrypi:~/source $ ls -lsa
total 24
4 drwxr-xr-x  3 pi pi 4096 Mar 13 17:43 .
4 drwxr-xr-x 20 pi pi 4096 Mar 12 00:34 ..
4 -rw-r--r--  1 pi pi    8 Mar 13 17:39 me.txt
4 -rw-r--r--  1 pi pi    8 Mar 13 17:40 my.txt
4 drwxr-xr-x  2 pi pi 4096 Mar 12 00:37 python
4 -rw-r--r--  1 pi pi   16 Mar 13 17:40 that.txt
pi@raspberrypi:~/source $ 
```

Figure 3-7. Output of list directories (ls) command

Change Directory

You can change from one directory to another by using the cd command, which is quite familiar.

```
pi@raspberrypi:~ $ cd source
pi@raspberrypi:~/source $
```

■ **Tip** The Linux path separator is a /, which can take some getting used to.

Copy

You can copy files with the cp command with the usual expected parameters of <from_file> <to_file>, as shown next. You can also use full paths to copy files from one directory to another.

```
pi@raspberrypi:~/source $ ls
```

```
me.txt   python
pi@raspberrypi:~/source $ cp me.txt my.txt
pi@raspberrypi:~/source $ ls
me.txt   my.txt   python
```

▓ **Tip**   Use the * symbol as a wildcard to specify all files (synonymous with *.* in Windows). For example, to copy all of the files from one folder to another, use the `cp ./old/* ./new` command.

Move

If you want to move files from one folder to another, you can use the `mv` command with the usual expected parameters of `<from> <to>`, as shown next. You can also use full paths to move files from one directory to another.

```
pi@raspberrypi:~/source $ ls
me.txt   my.txt   python   this.txt
pi@raspberrypi:~/source $ mv this.txt that.txt
pi@raspberrypi:~/source $ ls
me.txt   my.txt   python   that.txt
```

Create Directories

Creating directories can be accomplished with the `mkdir` command. If you do not specify a path, the command executes in the current directory.

```
pi@raspberrypi:~/source $ ls
me.txt   my.txt   python   that.txt
pi@raspberrypi:~/source $ mkdir test
pi@raspberrypi:~/source $ ls
me.txt   my.txt   python   test   that.txt
```

Delete Directories

If you want to delete a directory, use the `rmdir` command. This command requires that the directory be empty. You will get an error if the directory contains any files or other directories.

```
pi@raspberrypi:~/source $ ls
me.txt   my.txt   python   test   that.txt
pi@raspberrypi:~/source $ rmdir test
pi@raspberrypi:~/source $ ls
me.txt   my.txt   python   that.txt
```

Create (Empty) Files

Sometimes you may want to create an empty file for use in logging output or just to create a placeholder for editing later. The `touch` command allows you to create an empty file.

```
pi@raspberrypi:~/source $ ls ./test
pi@raspberrypi:~/source $ touch ./test/new_file.txt
pi@raspberrypi:~/source $ ls ./test
new_file.txt
pi@raspberrypi:~/source $ rmdir test
rmdir: failed to remove 'test': Directory not empty
```

Delete Files

If you want to delete a file, use the rm command. There are a number of options for this command, including recursively deleted files in subfolders (-r) and options for more powerful (thorough) cleaning.

```
pi@raspberrypi:~/source $ rm ./test/new_file.txt
pi@raspberrypi:~/source $ ls ./test
```

■ **Caution**   You can use the rm command with the force option to remove directories, but you should use such options with extreme caution. Executing sudo rm * -rf in a directory will permanently delete all files!

System Commands

The Linux operating system provides a huge list of system commands to do all manner of operations on the system. Mastering all of the system commands can take quite a while. Fortunately, there are only a few that you may want to learn to use Linux with a minimal of effort.

Show (Print) Working Directory

The system command I use most frequently is the print working directory (pwd) command. This shows you the full path to the current working directory.

```
pi@raspberrypi:~/new_source $ pwd
/home/pi/new_source
```

Command History

The one system command that you may find most interesting and helpful is the history command. This command lists the commands that you have entered over time. So if you find that you need to issue some command you used a month ago, use the history command to show all of the commands executed until you find the one you need. This is especially helpful if you cannot remember the options and parameters! However, this list is only for the current user. The following is an excerpt of the history for my Raspberry Pi 3.

```
pi@raspberrypi:~/source $ history
    1  sudo apt-get update
    2  sudo apt-get upgrade
    3  sudo shutdown -r now
    4  rpi-update
    5  sudo
    6  sudo rpi-update
```

```
 7  sudo apt-get dist-upgrade
 8  sudo shutdown -r now
 9  startx
10  ls /lib/firmware/brcm
...
```

■ **Tip** Use the Up and Down keys on the keyboard to call back the last command issued, and scroll forward and backward through the history one command at a time.

Archive Files

Occasionally, you may need the ability to archive or unarchive files, which you can do with a system command (utility). The tape archive (tar) command shows the longevity of the Linux (and its cousin/predecessor, Unix) operating system from the days when offline storage included tape drives[6] (no disk drives existed at the time). The following shows how to create an archive and extract it. The first tar command creates the archive and the second extracts it.

```
pi@raspberrypi:~ $ tar -cvf archive.tar ./source/
./source/
./source/test/
./source/my.txt
./source/that.txt
./source/python/
./source/python/blink_me.py
./source/me.txt
pi@raspberrypi:~ $ mkdir new_source
pi@raspberrypi:~ $ cd new_source
pi@raspberrypi:~/new_source $ tar -xvf ../archive.tar
./source/
./source/test/
./source/my.txt
./source/that.txt
./source/python/
./source/python/blink_me.py
./source/me.txt
pi@raspberrypi:~/new_source $ ls
source
```

There are a host of options for the tape archive command. The most basic are the create (-cvf) and extract (-xvf) option strings, as shown in the preceding code. See the manual for the tape archive command if you want to perform more complicated operations.

Administrative Commands

Like the system commands, there is a long list of administrative operations that you may need to perform. I list those operations that you may need to perform for more advanced operations, starting with the run as administrator equivalent command.

---

[6]Anyone remember punch cards?

87

Run as Super User

To run a command with elevated privileges, use the sudo command. Some commands and utilities require sudo. For example, to ping another computer, install software, or change permissions, and so forth, you need elevated privileges.

```
pi@raspberrypi:~/new_source $ sudo ping localhost
PING localhost (127.0.0.1) 56(84) bytes of data.
64 bytes from localhost (127.0.0.1): icmp_seq=1 ttl=64 time=0.083 ms
64 bytes from localhost (127.0.0.1): icmp_seq=2 ttl=64 time=0.068 ms
64 bytes from localhost (127.0.0.1): icmp_seq=3 ttl=64 time=0.047 ms
^C
--- localhost ping statistics ---
3 packets transmitted, 3 received, 0% packet loss, time 1998ms
rtt min/avg/max/mdev = 0.047/0.066/0.083/0.014 ms
```

Change File/Directory Permissions

In Linux, files and directories have permissions, as described in the previous section. You can see the permissions with the list directory command. To change the permissions, use the chmod command as shown in the following code. Here we use a series of numbers to determine the bits of the permissions. That is, 7 means rwx, 6 means rw, and so forth. For a complete list of these numbers and an alternative form of notation, see the manual for chmod.[7]

```
pi@raspberrypi:~/source $ ls -lsa
total 12
4 drwxr-xr-x  3 pi pi 4096 Mar 13 18:07 .
4 drwxr-xr-x 21 pi pi 4096 Mar 13 18:02 ..
0 -rw-r--r--  1 pi pi    0 Mar 13 18:07 cmd
4 drwxr-xr-x  2 pi pi 4096 Mar 12 00:37 python
pi@raspberrypi:~/source $ chmod 0777 cmd
pi@raspberrypi:~/source $ ls -lsa
total 12
4 drwxr-xr-x  3 pi pi 4096 Mar 13 18:07 .
4 drwxr-xr-x 21 pi pi 4096 Mar 13 18:02 ..
0 -rwxrwxrwx  1 pi pi    0 Mar 13 18:07 cmd
4 drwxr-xr-x  2 pi pi 4096 Mar 12 00:37 python
```

Change Owner

Similarly, you can change ownership of a file with the chown command, if someone else created the file (or took ownership). You may not need to do this if you never create user accounts on your Raspberry Pi, but you should be aware of how to do this in order to install some software such as MySQL.

```
pi@raspberrypi:~/source $ ls -lsa
total 12
4 drwxr-xr-x  3 pi pi 4096 Mar 13 18:07 .
4 drwxr-xr-x 21 pi pi 4096 Mar 13 18:02 ..
```

---

[7]For more information, see the numerical permissions section at https://en.wikipedia.org/wiki/Chmod.

```
0 -rwxrwxrwx  1 pi pi     0 Mar 13 18:07 cmd
4 drwxr-xr-x  2 pi pi 4096 Mar 12 00:37 python
pi@raspberrypi:~/source $ sudo chown chuck cmd
pi@raspberrypi:~/source $ ls -lsa
total 12
4 drwxr-xr-x  3 pi     pi 4096 Mar 13 18:13 .
4 drwxr-xr-x 21 pi     pi 4096 Mar 13 18:02 ..
0 -rw-r--r--  1 chuck pi     0 Mar 13 18:13 cmd
4 drwxr-xr-x  2 pi     pi 4096 Mar 12 00:37 python
pi@raspberrypi:~/source $ sudo chgrp chuck cmd
pi@raspberrypi:~/source $ ls -lsa
total 12
4 drwxr-xr-x  3 pi     pi    4096 Mar 13 18:13 .
4 drwxr-xr-x 21 pi     pi    4096 Mar 13 18:02 ..
0 -rw-r--r--  1 chuck chuck    0 Mar 13 18:13 cmd
4 drwxr-xr-x  2 pi     pi    4096 Mar 12 00:37 python
```

■ **Tip**   You can change the group with the chgrp command.

Install/Remove Software

The second most used administrative operation is installing or removing software. To do this on Raspbian (and similar Linux distributions), you use the apt-get command, which requires elevated privileges.

Linux maintains a list of header files that contain the latest versions and locations of the source code repositories for all components installed on your system. Occasionally, you need to update these references and you can do so with the following options. Do this before you install any software. In fact, most documentation for software requires you to run this command. You must be connected to the Internet before running the command and it could take a few moments to run.

```
sudo apt-get update
```

To install software on Linux, you use the install option (conversely, you can remove software with the remove option). However, you must know the name of the software you want to install, which can be a challenge. Fortunately, most software providers tell you the name to use. Interestingly, this name can be the name of a group of software. For example, the following command initiates the installation of MySQL, which involves a number of packages (shown in bold).

```
pi@raspberrypi:~/source $ sudo apt-get install mysql-server
Reading package lists... Done
Building dependency tree
Reading state information... Done
The following extra packages will be installed:
  libaio1 libdbd-mysql-perl libdbi-perl libhtml-template-perl libmysqlclient18
  libterm-readkey-perl mysql-client-5.5 mysql-common mysql-server-5.5
  mysql-server-core-5.5
Suggested packages:
  libclone-perl libmldbm-perl libnet-daemon-perl libsql-statement-perl
  libipc-sharedcache-perl mailx tinyca
The following NEW packages will be installed:
```

```
libaio1 libdbd-mysql-perl libdbi-perl libhtml-template-perl libmysqlclient18
libterm-readkey-perl mysql-client-5.5 mysql-common mysql-server
mysql-server-5.5 mysql-server-core-5.5
0 upgraded, 11 newly installed, 0 to remove and 0 not upgraded.
Need to get 8,121 kB of archives.
After this operation, 88.8 MB of additional disk space will be used.
Do you want to continue? [Y/n]
```

Shutdown

Finally, you want to shut down your system when you are finished using it or perhaps reboot it for a variety of operations. For either operation, you need to run with elevated privileges (sudo) and use the shutdown command. This command takes several options: use -r for reboot and -h for halt (shutdown). You can also specify a time to perform the operation but I always use the now option to initiate the command immediately.

To reboot the system, use this command:

```
sudo shutdown -r now
```

To shut down the system, use this command:

```
sudo shutdown -h now
```

Useful Utilities

There are a number of useful utilities that you need at some point during your exploration of Linux. Those that I use most often are described in the following list, which includes editors. There are, of course, many more examples but these will get you started for more advanced work.

- *Text editor*: nano (A simple, easy to use text editor. It has a help menu at the bottom of the screen. Some operations may seem odd after using Windows text editors, but it is much easier to use than some other Linux text editors.)

- *File search*: find (Locates files by name in a directory or path.)

- *File/text search*: grep (Locates a text string in a set of files or directory.)

- *Archive tools*: gzip, gunzip (A zip file archive tool (an alternative to tar)).

- *Text display tools*: less, more (less shows the last portion of a file; more shows the file contents a page (console page) at a time.)

Now that you know more about how to get around in Linux and use the command line, let's look at how you can write a simple program to run on the Raspberry Pi.

Working with Python: Blink an LED

Now that you know a little bit about how to use Raspbian, let's take an interesting diversion into the world of programming the Raspberry Pi with Python and working with the GPIO pins. This may seem a bit premature, but I provide this example for those readers who want to experience programming the Raspberry Pi, especially those who want to jump into working with hardware. Thus, I don't explain every detail about the electronic components; however, I present a primer on electronics in Chapter 7. More specifically, I use some components from the Microsoft Internet of Things Pack for the Raspberry Pi from Adafruit (`www.adafruit.com/products/2702`), which is discussed in more detail in Chapter 9.

If you prefer to wait until you've learned more about electronics and the Microsoft Internet of Things Pack for the Raspberry Pi, you can always read through the example and revisit it later once you read through the later chapters. However, you will find a very similar example in Chapters 5 and 6. Reading through this example gives you some insights about what you will accomplish later.

The programming language that you will use is a very easy scripting language called Python.[8] As you will see, the commands are quite intuitive and very expressive. For the purposes of this demonstration, you do not need to be an expert with the language. I provide all of the code and commands you need, explaining each as we go along. Once again, this is a lightning tour rather than a comprehensive study. Let's begin with a description of the project.

PYTHON? ISN'T THAT A SNAKE?

The Python programming language is a high-level language designed to be as close to like reading English as possible while being simple, easy to learn, and very powerful. Pythonistas[9] will tell you the designers have indeed met these goals.

Python does not require a compilation step prior to being used. Rather, Python applications (whose file names end in `.py`) are interpreted on the fly. This is very powerful, but unless you use a Python development environment, some syntax errors (such as incorrect indentation) are not discovered until the application is executed. Fortunately, Python provides a robust exception-handling mechanism.

If you have never used Python or you would like to know more about it, the following are few good books that introduce the language. A host of resources are also available on the Internet, including the Python documentation pages at `www.python.org/doc/`.

- *Programming the Raspberry Pi* by Simon Monk (McGraw-Hill, 2013)

- *Beginning Python from Novice to Professional*, 2nd Edition, by Magnus Lie Hetland (Apress, 2008)

- *Python Cookbook* by David Beazley and Brian K. Jones (O'Reilly Media, 2013)

Interestingly, Python was named after the British comedy troupe Monty Python, not the reptile. As you learn Python, you may encounter campy references to *Monty Python* episodes. I find these references entertaining. Of course, your mileage may vary.

You're going to build a very simple circuit that turns on an LED briefly in a loop that makes the LED appear to blink (you turn it on then off again repeatedly). This may sound like mad scientist work or something that requires years of electronics training, but it really isn't. You will use only two electronics components—an LED and a resistor—as well as a two wires and a breadboard to complete this project.

---

[8]A plethora of information is available about Python at `https://www.python.org`.
[9]Python experts often refer to themselves using this term. It is reserved for the most avid and experienced Python programmers.

■ **Tip**   If you do not feel comfortable working with electronics, that's OK! Just read through this section and come back to it once you've read the later chapters on electronics (Chapters 6 and 7).

Hardware Connections

Let's begin by gathering the hardware that you need. The following lists the components that are needed. All of these are available in the Microsoft Internet of Things Pack for the Raspberry Pi from Adafruit. If you do not have that kit, you can find these components separately on the Adafruit web site (www.adafruit.com), or from SparkFun (www.sparkfun.com), or any electronics store that carries electronic components.

- 560 ohm 5% 1/4W resistor (green, blue, brown stripes[10])
- Diffused 10mm red LED (or similar)
- Breadboard (mini, half, or full sized)
- (2) male-to-female jumper wires

Take a look at the breadboard. You see a center divide (normally a groove or a thick line). This sections the breadboard into two sides. The holes running perpendicular to the center groove are connected together but are not connected to adjacent holes (rows). If you use a half or full sized breadboard, you may have power rails, which run horizontally to the center grove, which are connected. Thus the power rails run parallel to the center groove and the interior connections run perpendicular to the groove. Now that you know how the breadboard is wired, let's build our circuit.

The only component that is polarized is the LED. Take a look at the LED. You see that one leg (pin) of the LED is longer than the other. This longer side is the positive side.

Begin by placing the breadboard next to your Raspberry Pi and power the Raspberry Pi off orienting the Raspberry Pi with the label facing you (GPIO pins in the upper-left corner). Next, take one of the jumper wires and connect the female connector to pin 6 on the GPIO. The pins are numbered left-to-right starting with the lower left pin. Thus, the left two pins are 1 and 2 with pin 1 below pin 2. Connect the other wire to pin 7 on the GPIO.

Next, plug the resistor into the breadboard with each pin on one side of the center groove. You can choose whichever area you want on the breadboard. Then, connect the LED so that the long leg (sometimes shown as the leg with a bend in drawings) is plugged into the same row as the resistor and the other pin on another row. Finally, connect the wire from pin 6 to the same row as the negative side of the LED and the wire from pin 7 to the row with the resistor. Figure 3-8 shows how all of the components are wired together. Be sure to study this drawing and double-check your connections prior to powering on your Raspberry Pi. Once you're satisfied that everything is connected correctly, you're ready to power on the Raspberry Pi and write the code.

---

[10]See https://en.wikipedia.org/wiki/Electronic_color_code.

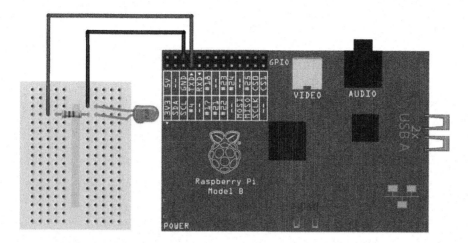

Figure 3-8. *Wiring the LED to a Raspberry Pi*

Writing the Code

The code (a Python script) that you need for this project manipulates one of the GPIO pins on the Raspberry Pi. Recall that you connected the negative side of the LED to pin 6, which is a ground pin. You connected the other side to pin 7. You will write a Python script to turn this pin on (applying power) and off (no power) through a simple command.

Now, create a new directory and open a text editor with the following commands. Use the name blink_me.py for the file.

```
pi@raspberrypi:~ $ mkdir source
pi@raspberrypi:~ $ cd source
pi@raspberrypi:~/source $ mkdir python
pi@raspberrypi:~/source $ cd python
pi@raspberrypi:~/source/python $ nano blink_me.py
```

As you can see, I like to place my source code in folders organized by language, but you can use whatever folder names you'd like. When the editor opens, enter the code as shown in Listing 3-2.

You can skip the comment statements (those that start with #) if you like, but I highly recommend that you get used to documenting your code. You can add your own name if that helps. Notice that I added comments to some of the lines to help understand what the code does. This is another excellent skill to hone.

Listing 3-2. Blink LED Script

```
#
# Windows 10 for the IoT
#
# Raspberry Pi Python GPIO Example
#
# This script blinks an LED placed with the negative lead on pin 6 (GND)
# and the pin 7 connected to a 220 resistor, which is connected to the
# positive lead on the LED.
#
# Created by Dr. Charles Bell
```

```
#
import RPi.GPIO as GPIO         # Raspberry Pi GPIO library
import sys                      # System library
import time                     # Used for timing (sleep)

ledPin = 7                      # Set LED positive pin to pin 7 on GPIO
GPIO.setmode(GPIO.BOARD)        # Setup the GPIO numbering mode
GPIO.setup(ledPin, GPIO.OUT)    # Set LED pin as output
GPIO.output(ledPin, GPIO.LOW)   # Turn off the LED pin

print("Let blinking commence!")

for i in range(1,20):
  GPIO.output(ledPin, GPIO.HIGH) # Turn on the LED pin
  time.sleep(0.25)
  sys.stdout.write(".")
  sys.stdout.flush()
  GPIO.output(ledPin, GPIO.LOW)  # Turn off the LED pin
  time.sleep(0.25)

GPIO.cleanup()                  # Shutdown GPIO

print("\nThanks for blinking!")
```

■ **Tip**   Indentation is important in Python. Indented statements form a code block. For example, to execute multiple statements for an `if` statement, indent all the lines that you want to execute when the conditions are evaluated as true.

What you see here is a series of comment lines (again, the ones that start with #) followed by some import statements that tell Python which modules you want to use. In this case, you use the GPIO, sys (system), and time modules. Following that, you see code to identify the GPIO pin (7) and set up the GPIO code module to initiate pin 7 as an output pin.

On the next line, you print a greeting message and place statements inside a loop that turn on the GPIO pin for the LED (set to high) for a period of time using a delay (in seconds), and then turn off the GPIO pin for the LED (set to low). You loop through 20 times, printing a dot to the screen using the system stdout class mechanism. You do this so that you can write the character directly to the screen without buffering (buffering can delay the display output). Finally, you display a message that the process is complete so that you know that it finished.

As you can see, the code is really easy to read. And even if you've never written Python before, you can understand what it is doing. Double-check your code and then save the file (**Ctrl+X**) and reply **Y**. You are now ready to run the code!

Running the Script

Once you've entered the script as written, you are ready to run it. To run the Python script, launch it as follows:

```
pi@raspberrypi:~/source/python $ python ./blink_me.py
```

You should see the following in the command-line terminal.

```
pi@raspberrypi:~/source/python $ python ./blink_me.py
Let blinking commence!
..................
Thanks for blinking!
```

Figure 3-9 shows a photo of the program running. Now, did the LED blink? If so, congratulations—you're a Raspberry Pi Python GPIO programmer!

Figure 3-9. *Running the* `blink_me.py` *Python script*

If something went wrong, it is likely it's just staring back at you with that one dark LED—almost mockingly. If the LED did not illuminate, shut down the Raspberry Pi (`sudo shutdown -h now`) and check your connections against Figure 3-7. Be sure the wires are connected to the correct pins and the LED is oriented correctly with the longer pin connected to the resistor and the resistor connected to GPIO pin 7. Also make sure that you are using the same rows on the breadboard (it is easy to get off by one row or pin). The other pin on the LED should be wired to GPIO pin 6. Once you've corrected any wiring issues, reboot your Raspberry Pi and try the project again.

Once it is working, try the project a few times by running the `python ./blink_me.py` command until the elation passes. If you're an old hand at the Raspberry Pi or electronics, that may be a very short period. If this is all new to you, go ahead and run it again and again, basking in the glory of having built your very first Python script and hardware project!

Summary

There can be little argument that the Raspberry Pi has contributed greatly to the world of embedded hardware and the IoT. With its low-cost, GPIO headers, and robust peripheral support, the Raspberry Pi is an excellent choice for building your IoT solutions. Due to its increasing popularity, there is tons of information available for those who want to learn how to work with hardware.

In this chapter, you explored the origins of the Raspberry Pi, including a tour of the hardware and a short primer on how to use its native operating system. You also explored how easy it is to write programs to control hardware on the Raspberry Pi using a Python script.

You learned these things about the Raspberry Pi to help with learning more about the origins of the Raspberry Pi and its native environment so that, once you learn how to write applications in Windows 10 and deploy them to the Raspberry Pi, you can leverage the host of examples written for Linux to implement them in Windows 10. I hope that you have found this chapter aligned toward this goal.

The next chapter returns to Windows 10. You learn how to write the example program in Visual Studio. As you will see, the hardware connections will be similar, but the code and the way you work with the Raspberry Pi will be much more familiar, and in my opinion, easier.

CHAPTER 4

■ ■ ■

Developing IoT Solutions with Windows 10

Microsoft has produced one of the most advanced integrated development environments (IDE) that easily rivals all competition. Indeed, IDEs on other platforms are often compared to Visual Studio for their depth of features, refinement of tools, and breadth of languages supported.

The feature set in Visual Studio is so vast in fact that it would require a book several times the size of the one you're holding to cover the basics of every feature. Moreover, each language supported (over 7 and counting) would require its own book of similar size. Clearly, mastering all of the features of Visual Studio would require a dedication that few would endure outside of a vocation or research requirement.

Fortunately, most IoT hobbyists and enthusiasts never need to learn every nuance of Visual Studio to develop IoT applications. As you will see, you need only to learn a few of the features, including writing the code, building (compiling), deploying, and debugging.[1] Do not let the sheer size of the features in Visual Studio intimidate you. You're likely to find mastery of the basics is all that you'll ever need. Should you ever need to use the advanced features, you can always learn them when you need them. I find it nice to know that there are advanced tools that can help me develop solutions more easily.

In this chapter, you'll see a demonstration of how to get started using Visual Studio 2015. You will also learn the layout of the GPIO headers for the three compatible boards and even see how to build, deploy, and test your first Windows 10 IoT Core application. Let's begin with a look at the GPIO headers from all three boards.

Working with GPIO Headers

The general-purpose input/output (GPIO) headers permit you to connect hardware, such as electronic circuits, devices, and more, to your board, which you then access via special libraries from your applications. However, you must know what pins are available and what features they support. As you will see, not all pins can be programmed in the same way.

Once you've identified the pins you want to use, you can use those pin numbers or nomenclature to write the code that you need to set up and access the pin and read (or write) to the pin. That is, you can turn pins on or off (applying power or no power), read analog values, write analog values, and more. This allows you to work with both analog and digital sensors. You will discover more about sensors in Chapter 7.

---

[1]Much like most people will never need or use more than 20% of the features of Microsoft Word.

© Charles Bell 2016

C. Bell, *Windows 10 for the Internet of Things*, DOI 10.1007/978-1-4842-2108-2_4

■ **Caution** Whenever you want to connect sensors or circuits to the GPIO header—either directly (not recommended) or via a breakout board (recommended)—you should first shut down your board. This may sound inconvenient when you're working through a project, but it is the best method for ensuring that you do not accidentally short some pins or make the wrong connections.

It is common to use the word "pins" when talking about the header, but you should refrain from use "GPIO pins" when referring to the header itself. As you will see, the pins in the header may be mapped to one of several interfaces (also called a bus) and the GPIO is just one type of interface available. Although you can use the interface pins in your code, the pins that you use to connect to devices (save those that use one of the supported interfaces) are named GPIO pins. However, the physical pin numbering and order may differ among the various boards.

Furthermore, when you use one of the pins in your Visual Studio application, you specify the GPIO number. For example, GPIO 13 on the Raspberry Pi is physically pin number 33. The Visual Studio libraries are designed to use the pin nomenclature rather than the header pin number. Thus, when you refer to GPIO 13, you use 13 as shown in the following code snippet.

```
const int LED_PIN = 13;   // GPIO13
auto gpio = GpioController::GetDefault();
...
pin_ = gpio->OpenPin(LED_PIN);
pin_->Write(pinValue_);
pin_->SetDriveMode(GpioPinDriveMode::Output);
```

I discuss each of the three boards supported by Windows 10 IoT Core in the following sections. As you will see, each board has a different GPIO header layout, including some very important power differences.

Raspberry Pi

The GPIO headers on the Raspberry Pi 2 and 3 have the same layout. The header is located in the upper left and consists of a double row of 20 pins, making a 40-pin header. They are numbered sequentially in pairs starting with the leftmost set of pins.

Figure 4-1 shows the Raspberry Pi next to the header layout. I have oriented the photo to make it easier to see the GPIO pins. Pin 1 is the leftmost pin at the left side of the header (top side shown in the photo). Notice that the GPIO named pins are arranged in a non-sequential pattern.

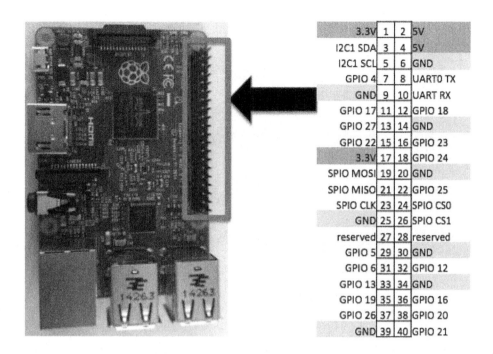

3.3V	1	2	5V
I2C1 SDA	3	4	5V
I2C1 SCL	5	6	GND
GPIO 4	7	8	UART0 TX
GND	9	10	UART RX
GPIO 17	11	12	GPIO 18
GPIO 27	13	14	GND
GPIO 22	15	16	GPIO 23
3.3V	17	18	GPIO 24
SPIO MOSI	19	20	GND
SPIO MISO	21	22	GPIO 25
SPIO CLK	23	24	SPIO CS0
GND	25	26	SPIO CS1
reserved	27	28	reserved
GPIO 5	29	30	GND
GPIO 6	31	32	GPIO 12
GPIO 13	33	34	GND
GPIO 19	35	36	GPIO 16
GPIO 26	37	38	GPIO 20
GND	39	40	GPIO 21

Figure 4-1. GPIO header (Raspberry Pi 2 and 3)

The Raspberry Pi GPIO header supports a number of interfaces including an I2C bus, SPIO bus, and Serial UART with pins devoted accordingly. You can see these in Figure 4-1. You also see two reserved pins (consider them unusable), eight ground pins, two 3.3V power and two 5V power pins. This leaves a total of 17 pins that you can use in your applications.

You must take care when using the GPIO pins for reading voltage. On the Raspberry Pi, all pins are limited to 3.3V. Attempting to send more than 3.3V will likely damage your Raspberry Pi. Always test your circuit for maximum voltage before connecting to your Raspberry Pi. You should also limit current to no more than 5mA.

MinnowBoard Turbot

The GPIO headers on the MinnowBoard Max–compatible boards (I use the Turbot in this section) is located on the bottom-left side of the board. The GPIO consists of a double row of 13 pins making a 26-pin header. They are physically numbered sequentially in pairs starting with the left-post pin.

Figure 4-2 shows the MinnowBoard Turbot next to the header layout. Pin 1 is the leftmost pin at the left side of the header (top side shown in the photo). I have oriented the photo to make it easier to see the GPIO pins. Notice that the GPIO named pins are arranged in a sequential pattern unlike the Raspberry Pi. That is, the GPIO pins are grouped together.

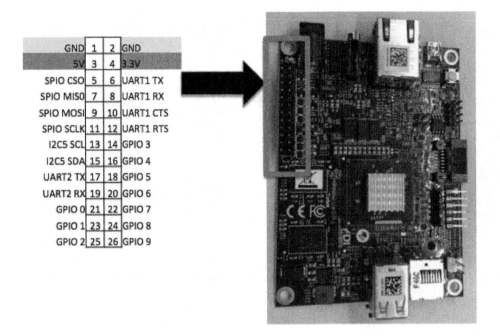

Figure 4-2. *GPIO header (MinnowBoard Turbot)*

The MinnowBoard Turbot GPIO header supports a number of interfaces including an I2C bus, SPIO bus, and Serial UART with pins devoted accordingly. You can see these in the Figure 4-2. You also see two ground pins, one 3.3V power, and one 5V power pin. This leaves a total of 10 pins that you can use in your applications.

DragonBoard 410C

The GPIO headers on the DragonBoard 410C are located on the top-left side of the board. The GPIO consists of a double row of 20 pins making a 40-pin header. They are physically numbered sequentially in pairs starting with the left-post pin. Unlike the Raspberry Pi and MinnowBoard Max–compatible boards, the DragonBoard 410C uses female header pins.

Figure 4-3 shows the DragonBoard 410C next to the header layout. Pin 1 is the leftmost pin at the left side of the header (top side shown in the photo). I have oriented the photo to make it easier to see the GPIO pins. The GPIO named pins are arranged in a non-sequential pattern like the Raspberry Pi.

The DragonBoard 410C GPIO header supports a number of interfaces including an I2C bus, SPIO bus, and two serial UART with pins devoted accordingly. You can see these in Figure 4-3. You also see four ground pins, one 1.8V power, and one 5V power pin. This leaves a total of 11 pins that you can use in your applications. GPIO 24 can be used for input only. Also, the pins marked SYS DC in can be used to power the board.

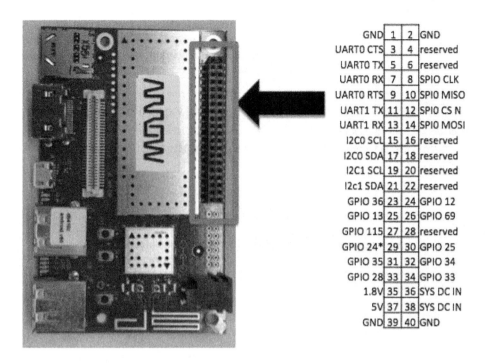

GND	1	2	GND
UART0 CTS	3	4	reserved
UART0 TX	5	6	reserved
UART0 RX	7	8	SPIO CLK
UART0 RTS	9	10	SPIO MISO
UART1 TX	11	12	SPIO CS N
UART1 RX	13	14	SPIO MOSI
I2C0 SCL	15	16	reserved
I2C0 SDA	17	18	reserved
I2C1 SCL	19	20	reserved
I2c1 SDA	21	22	reserved
GPIO 36	23	24	GPIO 12
GPIO 13	25	26	GPIO 69
GPIO 115	27	28	reserved
GPIO 24*	29	30	GPIO 25
GPIO 35	31	32	GPIO 34
GPIO 28	33	34	GPIO 33
1.8V	35	36	SYS DC IN
5V	37	38	SYS DC IN
GND	39	40	GND

Figure 4-3. *GPIO header (DragonBoard 410C)*

Now that you know more about the GPIO headers on your boards, you can learn how to get started using Visual Studio 2015.

Visual Studio 2015 Primer

The Visual Studio product line is very long lived. In fact, it has been around since the early days of Windows. As you can imagine, the product has undergone a great deal of changes—with new languages, frameworks, platforms, and more added every few years. As I mentioned earlier, Visual Studio 2015 offers a huge array of features for developing applications for Windows using a variety of languages. Visual Studio also supports several platforms.[2]

With all of these features and the many languages, it can be quite intimidating getting started with Visual Studio. In fact, books devoted to Visual Studio (just the features, not languages) can easily exceed hundreds of pages in length. However, you can accomplish quite a lot with a small amount of knowledge.

■ **Tip**   If you want to know more about Visual Studio 2015, check out *Professional Visual Studio 2015* by Bruce Johnson (Wrox, 2015). The book contains over 1,000 pages of detailed explanations about every feature of Visual Studio.

---

[2]There's even support for Android if you're into that kind of thing.

This section gives you a brief introduction to how to get around in Visual Studio for the purposes of developing applications for Windows 10 IoT Core. Don't feel like you have to learn everything about Visual Studio to enjoy working with Windows 10 IoT Core. This section and the next chapters provide you a guide to getting started writing Windows 10 IoT Core applications by example. But first, let's explore the major features, interface features, and project templates that you will use to write your Windows 10 IoT Core applications.

■ **Note**   The following provides a high-level overview of the features that you need to know to write Windows 10 IoT Core applications. As such, it is not a complete tutorial of Visual Studio. If you need more information about Visual Studio, refer to the help system inside the application and the Microsoft Developer Network (MSDN) library of documentation.

Major Features

Visual Studio 2015 is truly the do everything tool for Windows software development. In that respect, Microsoft has established the bar for which all others are measured. Visual Studio is an IDE that places all the tools that you need to develop applications in one interface, making Visual Studio the only tool that you need to develop applications for Windows.

Another aspect to its superiority has to do with the languages supported. For example, there are a variety of programming languages, including C, C#, C++, Visual Basic, Python, and more. Each of these languages can be used to build a host of different applications in one of several frameworks.

■ **Note**   Most of the examples for building Windows 10 IoT Core applications are written in either C#, C++, or Python. Once you master these languages and mechanisms for building projects, learning to do the same in other languages requires only learning the syntax and semantic nuances of the language.

For example, you can build desktop applications (GUI or command line) in several frameworks, including Windows 32 and .NET, create web applications, dynamic libraries for reuse, and more. Of course, you can also build applications for Windows 10 as a Universal Windows Platform or one of several special project templates (prebuilt collections of files and specific settings).[3] Project templates create a special file called a solution (with a filename of <project_name>.sln) that contains several types of files.

To create an application, you select a project template and name the solution, write your code, compile the application into an executable, test and debug the application (optional), and finally, deploy the application.

I guide you through selecting the right project template later in this section. Writing the source code (developing the functionality) requires modifying one or more of the files in the solution. I discuss the major types of Windows 10 IoT Core projects later in this section.

The compilation step is easy to initiate but can be an iterative process. This is because the compiler performs intensive syntax and semantic checks on the code flagging anything that isn't quite up to the language or framework rules as warnings or errors. Fortunately, you can click each warning or error to zoom to the line of code. But don't worry; compiler errors are normal and are not a sign of inexperience or lack of knowledge—they're just a part of the learning experience. That said, as you learn the language and frameworks, you should encounter fewer issues and errors. Also, do not disregard warnings. While they may not prohibit your code from compiling, it is always a good practice to remove the cause of the warning before completing your solution.

---

[3]Project templates are quite extensive and specific to a particular language, framework, and platforms. Indeed, you will find project templates are organized in that order.

I think it is the interactive debugger where Visual Studio shines the most. In fact, I find the interactive debugger the most useful of all features in Visual Studio. Not only can you run your applications stepping through code one line at a time, you can do so while inspecting variables, memory, the stack (the order of method calls), and more. This helps you test your application and improve quality while finding logic or data errors. You can even remotely debug your Windows 10 IoT Core applications.

Finally, when you are satisfied with the quality of your code, you can use Visual Studio to deploy your application to your Windows 10 IoT device.

The Interface

At first glance, the Visual Studio 2015 interface appears with several smaller windows arranged inside a larger window. The layout of the windows can change depending on settings you used when installing Visual Studio and can vary slightly from one project template to another. Figure 4-4 shows the layout of the IDE using C++ environment settings. I have placed numbers next to the major components of the IDE. I explain each in more detail.

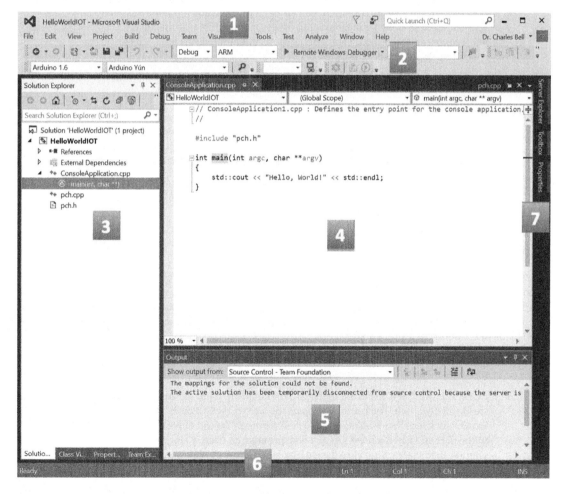

Figure 4-4. *The Visual Studio 2015 interface (C++ settings)*

The IDE components shown include the following.

1. *Menus*: This area contains the system of menus for all the standard features. The features are categorized into several areas, including operations on the project, building, debugging, and more.

2. *Toolbars*: In this area you see a variety of toolbars that have buttons and other controls for commonly used functions for particular features. In this image, you see toolbars for debugging and Arduino. You can add and remove (show and hide) using the Tools menu.

3. *Solution Explorer*: This window lists the various files included in the solution. You can double-click any of the files to open the appropriate window to work with that file. You can also use the tree control to drill down into the specifics for each file type. For example, I opened the `ConsoleApplication.cpp` node and double-clicked the `main()` method, which opened the code editor window and zoomed (placed the cursor) to the method.

4. *Code Editor*: This window is where you enter all the code for your source code. You can use this area to edit all manner of files and the IDE will change its behavior based on context. That is, it performs operations such as language-specific automatic code completion.

5. *Output Window*: This window is used to communicate messages to you from the compiler and other features. For example, look here when you compile your application for warnings and errors.

6. *Information Bar*: The bar down at the bottom of the IDE is also used to provide contextual information. Since the focus of the IDE is on the source code editor, data such as cursor location (position on line), code line number, and more, is shown.

7. *Dock*: To the right (in this layout) are additional windows that are docked or minimized. You can click any of the tabs to open the windows.

Windows

The windows in the IDE can be repositioned and resized. There are a number of states each window can become, including the following. When you expand a window by clicking the window title bar to expand the window, you see three small icons on each window title bar. The first one, a small down arrow, when clicked, opens a context menu with options for the disposition of the window. I describe how the other two icons are used in the descriptions of window states.

- *Float*: The window is free to be moved around and floats above (outside) the confines of the IDE. This is also called *unpinning*.

- *Dock*: The window is restricted to the confines of the IDE but can be repositioned inside the IDE itself. The window remains the size you set. Click the thumbtack (or pin) icon to dock the window. This is also called *pinning*. You can move the window around in the IDE, and when it hovers near a docking area, an overlay allows you to dock the window in that area. Much like the side-by-side feature of Windows.

- *Dock as Tabbed Document*: The window is reduced to a tab on the tab bar associated with its function. There are multiple tab areas throughout the IDE.

- *Auto Hide*: The window opens when you hover over its tab. The window stays open as long as the mouse pointer is within its borders. When you move the mouse away from the window, it reverts to a tab on the tab bar. To auto hide, click the small thumbtack (sometimes called the pin).

- *Hide*: The window closes and is removed from the tab bar. You have to use the menu options to reopen it. Click the X icon to hide the window.

Environment Settings

Recall that during the installation of Visual Studio in Chapter 2, you chose the C++ option for the environment settings. This is another great feature of Visual Studio. Environment settings configure the layout of the IDE for a specific language or framework. There are generic (general) environment settings for language-agnostic layout as well as environment settings for languages such as C# and C++. Figure 4-5 shows the environment settings that are available.

Figure 4-5. *Visual Studio 2015 environment settings*

You can change the environment settings any time you want. Use the **Tools ➤ Import and Export Settings...** menu to open the settings management dialog. You can use this dialog to save the settings (export) or restore settings you've saved (import) by selecting the appropriate radio button.

To change the default environment settings, choose the **Reset all settings** radio button and then click **Next**. The dialog gives you another chance to save your settings. If you do not want to save them, or once you have saved the settings, choose the **No** radio button and then click **Next**. On the next screen, choose the environment settings (see Figure 4-5) that you want and then click **Finish**.

Common Menu Items

While there are a number of menus in the interface, each with dozens of subitems and submenus, there are a few that you use quite often. The following is a brief overview of the commonly used menu items categorized by the entries on the main menu bar. I add the keyboard shortcut in parenthesis where available. Note that *<project>* changes to the name of the currently opened or selected project.

- *File*: File and project operations

  - *New ➤ Project* (Ctrl+Shift+N): Start a new project

  - *Save Selected Items* (Ctrl+S): Save the file(s) selected

  - *Save All* (Ctrl+Shift+S): Save all files

- *Project*: Operations on the current project

  - *<project> Properties* (Alt+F7): Open the properties dialog for the project

- *Build*: Compilation and deployment with two sections: one for the entire solution, and one for the current project (a solution can have multiple projects)

  - *Build Solution* (F7): Build (compile) all projects in the solution

  - *Rebuild Solution* (Ctrl+Alt+F7): Rebuild (compile) all files in the solution

  - *Deploy Solution*: Deploy the compiled solution to the destination specified in the debug settings

  - *Clean Solution*: Remove all compiled files and headers and generated files in the solution

  - *Build <project>*: Build (compile) the currently selected project

  - *Rebuild <project>*: Rebuild (compile) all files in the project

  - *Deploy Selection*: Deploy the compiled project to the destination specified in the debug settings

  - *Clean <project>*: Remove all compiled files and headers and generated files in the project

  - *Compile* (Ctrl+F7): Compile the currently selected file/project

- *Debug*: Interactive debugger

  - *Start Debugging* (F5): Start the interactive debugger

  - *Start Without Debugging* (Ctrl+F5): Run the project without the debugger (adds a pause at the end of execution for console applications)

  - *Step Into* (F11): While debugging, enter any method calls one line at a time

  - *Step Over* (F10): While debugging, execute current line (do not step into methods)

  - *Toggle Breakpoint* (F9): Turn a breakpoint on/off

Now that you know a bit more about the interface, let's look at the templates you will use for your Windows 10 IoT Core projects.

Windows 10 IoT Core Project Templates

A project template is a special set of files and settings that are configured for a specific language, framework, or application target. Project templates are arranged by programming language and by project type. The following are some of the more common project types available in the standard Visual Studio installation.

- *Windows*: A very broad category that covers any application type for Windows

- *Web*: Applications built using ASP.NET

- *Office/SharePoint*: Applications for add-ins for the Microsoft productivity tools

- *Android*: Builds C# applications for Android devices

- *iOS*: Builds C# applications for some iOS devices

- *Cloud*: Builds applications for Windows Azure

- *Windows IoT Core*: Builds applications for deployment to Windows 10 IoT Core devices

---

■ **Note**   Some project template categories are only available for certain languages. Similarly, some languages may list fewer project template categories.

See the Visual Studio online documentation for a complete list of project types available. Check the online help for more information about the project types from the following resources.

- Microsoft Developer Network (MSDN) (https://msdn.microsoft.com/library/dd831853%28v=vs.140%29.aspx)

- Visual Studio Getting Started (https://www.visualstudio.com/get-started/get-started-vs)

To start a new project, click the **File ➤ New ➤ Project...** menu item. The **New Project** dialog is displayed, as shown in Figure 4-6.

You can see the programming languages installed in the tree view on the left. You can use this view to drill down to the specific language and framework/platform you want. If you click the language itself, you see all the project templates for that language. You can see this in Figure 4-6.

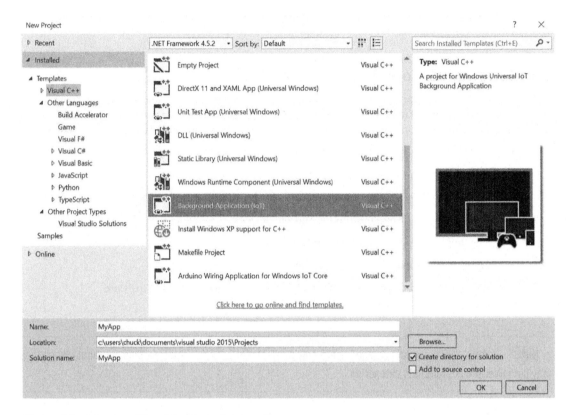

Figure 4-6. *Selecting a Visual Studio project template*

Once you select the project template you want, you can name the project in the area at the bottom, including placing the project and all of its files in a specific location. By default, the name of the application and the solution name are the same but you can change them. Once you are ready to create the project, click **OK**. The IDE opens the project (a project's settings are saved in a file called a *solution*) and creates the basic source files for you.

▪ **Tip**   You may see a Windows for IoT category once you have installed the Arduino feature. This is only for projects that work with the Arduino. It is not the same as the Windows IoT Core project template category.

There are three basic project templates that you use in this book to create applications for Windows 10 IoT Core. They aren't the only ones you can use, but they are the templates you should use to get started. If you want to build more complicated applications, you may want to consider some of the other project types. The three project templates include the following ordered by complexity. I explain each in the following sections.

- Blank Windows IoT Core Console Application

- Background Application (IoT)

- Blank App (Universal Windows)

Again, there are other project templates that you may want to use, but these three are the ones that you use in this book.

Blank Windows IoT Core Console Application

This project template is the most basic and simplistic of all project templates for Windows 10 IoT Core. It permits you to create a simple text-only console application. If you want to experiment with writing applications or want to create a solution that provides data or produces a report for a headed device, choose this project template. You will use this template in the walkthrough of building your first application.

The code for this project template is very simple. The template creates a source file named ConsoleApplication.cpp, which you can use to write your application. Use this source file to write your application. You can create additional source files if you want, need, or desire to do so. For example, if you are modeling concepts or creating abstractions for hardware, you may want to add additional source files to contain these models (e.g., classes).

Listing 4-1 shows this file. The project template fills in the main() function with a simple, Hello, World! print statement. You can find this project template under the **Windows ➤ Windows IoT Core** category.

Listing 4-1. Blank Windows IoT Core Console Application

```
// ConsoleApplication1.cpp : Defines the entry point for the console application.
//

#include "pch.h"

int main(int argc, char **argv)
{
    std::cout << "Hello, World!" << std::endl;
}
```

Background Application (IoT)

This project template is used to create an application that runs in the background (or headless) on your device. Typically, you'd use this project template to create an application that communicates to other devices, interacts with hardware (and then communicates with other devices), or drives hardware to display or report sensor data to the user.

The project files for this project template are quite different and there are a few extra files. The one source file that you work with most is named StartupTask.cpp, which is where you place the code you want to run when the application starts. There is also a corresponding header file named StartupTask.h where you can place function and class primitives for your design.

Listing 4-2 shows the StartupTask.cpp file as created by the new project dialog. You can find this project template under the **Windows ➤ Windows IoT Core** category.

Listing 4-2. Background Application (IoT)

```
#include "pch.h"
#include "StartupTask.h"

using namespace BackgroundApplication6;

using namespace Platform;
using namespace Windows::ApplicationModel::Background;
```

```
// The Background Application template is documented at http://go.microsoft.com/fwlink/?Link
ID=533884&clcid=0x409

void StartupTask::Run(IBackgroundTaskInstance^ taskInstance)
{
    //
    // TODO: Insert code to perform background work
    //
    // If you start any asynchronous methods here, prevent the task
    // from closing prematurely by using BackgroundTaskDeferral as
    // described in http://aka.ms/backgroundtaskdeferral
    //
}
```

There is a link in the code for more information about the project template. Also, there are hints in the code comments (designated by //) to help you get started.

Blank App (Universal Windows)

This project template is the most sophisticated of the three project templates that you will use. This project template allows you to create simple user interfaces using Microsoft's Extensible Application Markup Language (XAML). XAML is similar to XML except you use a special declarative language to create a user interface. It is a very expressive and powerful language that permits you to create sophisticated user interfaces without the need for high overhead graphical libraries.

■ **Tip** To learn more about XAML, see https://msdn.microsoft.com/en-us/windows/uwp/xaml-platform/xaml-overview?f=255&MSPPError=-2147217396.

There are several files for this project template. The main files that you will be working with are the XAML files (MainPage.xaml, MainPage.xaml.h, and MainPage.xaml.cpp), which include the XAML code. You also see files named App.xaml, App.xaml.h, and App.xaml.cpp, which are the entry or starting point for the application. There is no specific XAML code here; rather, it contains the code for the OnLaunched and OnSuspending events. You can use these events to initialize your application or cleanup when it is shut down or suspended. There are a number of other files, but these are explored in later chapters when you see examples of each project template.

Listing 4-3 shows the main page for the XAML source files. You can find this project template under the **Windows ➤ Universal** category.

Listing 4-3. Blank App (Universal Windows)

```
//
// MainPage.xaml.cpp
// Implementation of the MainPage class.
//

#include "pch.h"
#include "MainPage.xaml.h"

using namespace App8;

using namespace Platform;
```

```
using namespace Windows::Foundation;
using namespace Windows::Foundation::Collections;
using namespace Windows::UI::Xaml;
using namespace Windows::UI::Xaml::Controls;
using namespace Windows::UI::Xaml::Controls::Primitives;
using namespace Windows::UI::Xaml::Data;
using namespace Windows::UI::Xaml::Input;
using namespace Windows::UI::Xaml::Media;
using namespace Windows::UI::Xaml::Navigation;

MainPage::MainPage()
{
        InitializeComponent();
}
```

Now that you've had a brief overview of Visual Studio, let's see these features in action by writing your very first Windows 10 IoT Core application.

Example Project: Hello, World

Now let's get our hands into some code and see how to write a basic application for your Windows 10 IoT Core device. I know you've been itching to get started, so grab a stockpile of your favorite diet crushing snacks, kid-approved beverages of choice, recline your chair[4] and let's write some code!

The project is an adaptation of a classic computer science homework assignment: converting temperature from Celsius to Fahrenheit. But you're going to mix it up a bit and make the application a bit more interactive. That is, you ask the user which scale she wants to use as the base temperature, prompt for the temperature, and finally, convert and print the result.

In case you've forgotten or perhaps never gave it another thought after the mid-term and final exam, the formulas that you use include converting Celsius to Fahrenheit and Fahrenheit to Celsius as follows. You use these in the code that you write for the application.

```
Celsius = (5/9) X (fahrenheit - 32)
Fahrenheit = ((9/5) X Celsius) + 32.0
```

To keep things as simple as possible, you'll write a Windows IoT Core console application that you can deploy to your Windows 10 device and run from a remote login (SSH). This keeps the code simple and allows you to focus on the mechanics of building and deploying the application.

■ **Note**   In this example and in future projects I use the Raspberry Pi as my Windows 10 IoT Device (or simply device) and often use those terms interchangeably. You can perform all of these steps with the other boards. The only difference in the steps that follow concerns the architecture selection. For Raspberry Pi and DragonBoard 410C, choose **ARM**, for the MinnowBoard Max–compatible boards, choose **x86** in the Solutions Platform drop-down box. That is the only difference.

Now that you know what you want to build, let's get started!

---

[4]A common pose that is very bad for your posture but some insist it exudes the proper attitude of a serious coder. Recline at your own risk.

Create the Project

You need to create a new project in Visual Studio to contain your code. If you're following along, go ahead and launch Visual Studio and create a new project. You can use the **File ➤ New ➤ Project...** menu option or you can click the **New Project...** link in the welcome dialog or press **Ctrl+Shift+N**. Figure 4-7 shows the location of the link. I highlighted it so you can find it easier.

Figure 4-7. *Visual Studio: Create new project*

When the new project dialog appears, you need to select the language, platform, and project type. For this project, you want to choose Visual C++ as the language, Windows as the platform, and Windows IoT Core as the framework (category) in the tree view on the left. Then, choose the **Blank Windows IoT Core Console Application** entry from the list, as shown in Figure 4-8.

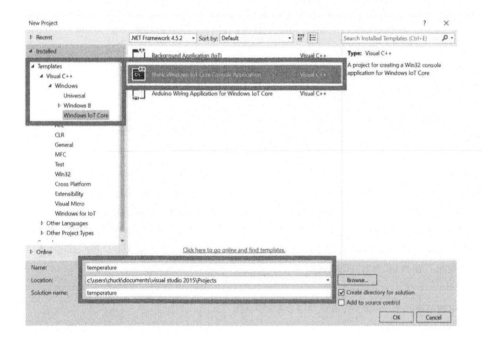

Figure 4-8. *New Project dialog*

At the bottom of the dialog you can choose the name of the application (temperature) and optionally a directory of where to store the solution (all the source code and related files). If you type **temperature** in the Name box, the dialog will fill in the same name for the solution name. Once you are satisfied with the name and location, click **OK** to create the project. You may also be asked for the target and minimum platform versions. You should choose the latest version of each. Once you click **OK**, the project creation process may take a moment to run. Once complete, you see the blank files for the solution, as shown in Figure 4-9.

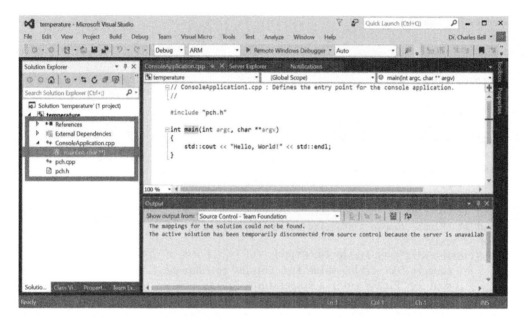

Figure 4-9. *Blank console application project*

Write the Code

Now it is time to write the code! In this project, you place all the source code in the main(int, char**) method inside the ConsoleApplication.cpp source file. To open this file (if it is not already open in the code editor), select the project in the Solution Explorer and navigate to the ConsoleApplication.cpp file. Then, open that branch and double-click the main(int, char**) method. Visual Studio opens and refocuses the code editor window to the method, as shown in Figure 4-9.

Did you notice something interesting in the source code? Yep, Microsoft has populated the main(int, char**) method with the ubiquitous "Hello, World" code. In this case, it is statement that sends the character string to the standard out (std::cout) or console. Thus, if you were to run this code without modification, the application will simply print that string and exit. You want to do something similar since you are only working with the console, but with a bit more sophistication than that simple output statement.

Don't worry too much about whether you've written C++ code before. I provide all the statements that you need in this chapter. I present a brief tutorial of the C++ language in the next chapter. For now, just enter the code as shown in Listing 4-4. I'll walk you through what the code does in a moment.

Listing 4-4. Temperature Example Code

```cpp
//
// Windows 10 for the IoT
//
// Example C++ console application to demonstrate how to build
// Windows 10 IoT Core applications.
//
// Created by Dr. Charles Bell
//
#include "pch.h"

using namespace std;

int main(int argc, char **argv)
{
        double fahrenheit = 0.0;
        double celsius = 0.0;
        double temperature = -.0;
        char scale{ 'c' };

        cout << "Welcome to the temperature conversion application.\n";
        cout << "Please choose a starting scale (F) or (C): ";
        cin >> scale;
        if ((scale == 'c') || (scale == 'C')) {
                cout << "Converting value from Celsius to Fahrenheit.\n";
                cout << "Please enter a temperature: ";
                cin >> celsius;
                fahrenheit = ((9.0 / 5.0) * celsius) + 32.0;
                cout << celsius << " degrees Celsius = " << fahrenheit <<
                        " degrees Fahrenheit.\n";
        }
        else if ((scale == 'f') || (scale == 'F')) {
                cout << "Converting value from Fahrenheit to Celsius.\n";
                cout << "Please enter a temperature: ";
                cin >> fahrenheit;
                celsius = (5.0 / 9.0) * (fahrenheit - 32.0);
                cout << fahrenheit << " degrees Fahrenheit = " << celsius <<
                        " degrees Celsius.\n";
        }
        else {
                cout << "\nERROR: I'm sorry, I don't understand '" << scale << "'.";
                return -1;
        }
}
```

The code begins with a number of comment lines. These are designated with the // symbol at the start of the line. In fact, you can place the // anywhere on the line making everything to the right a comment. The compiler ignores these statements so you can write whatever you like. The very first thing that you may notice is the following line of code. This is part of the preamble of the code. The first line is an include directive that tells the compiler to include the pch.h (precompiled header file), which Visual Studio supplies.

```
#include "pch.h"
```

Next is a shorthand notation I like to use. It allows you to avoid typing the namespace over and over again. In this case, you're using the namespace `std::` a lot. The following line of code permits you to omit that—provided that what you are referencing (`cout`, `cin`) can be found unambiguously in the namespace. That is, if there were multiple namespaces used, `cout` and `cin` would have to be unique to one of them otherwise, you'd still have to provide the namespace `std::`.

```
using namespace std;
```

Those two lines are the only code outside of the `main()` method. The `main()` method is the entry point for a C++ application. When the application is run, code inside this method is executed first. The code you want for this method is shown in Listing 4-1.

You begin by asking the user what scale she wants to use. You prompt the user to enter a **C** for Celsius or an **F** for Fahrenheit. Take a look at the code again. If the user specifies either a C or c, you calculate the conversion using Celsius as the base. If the user specifies either an F or f, you calculate the conversion using Fahrenheit as the base. If something else was entered, you exit with an error. You include additional prompts to ask the user for the base temperature. Following that, you use the formulas to perform the calculations.

You may be wondering what all of those `cout` and `cin` and `<<` and `>>` thingys do. Essentially, the `cout <<` string phrase tells the compiler you want to take a string and print it to the screen (standard out or stdout). The `cin >> variable` phrase tells the compiler to read information entered by the user and store it in the variable named. Go through the code again and read it until you understand how this works. Again, I explain C++ coding in more detail in the next chapter.

Once you've entered all the code, go ahead and save the project using either the save on the toolbar or the **File ➤ Save Solution** menu item. Next, let's build and test the code!

Build and Test Your Code

To build the code, you first must choose build type and the architecture. Figure 4-10 shows where these options are on the toolbar. Go ahead and choose Debug as the build type and x86 or x64 as the architecture to match your PC. You choose x86 or x64 here because you will first build and test the application on your PC before deploying it to the Raspberry Pi. This shows the true power of UWP—the ability to run the same code on different architectures.

Once selected, press **Ctrl+Shift+B** or choose **Build ➤ Build Solution** from the menu. This compiles your application and opens the Output window to show you the results of the build. As you can see in Figure 4-10, the solution built without errors. If you encounter errors, check the code against Listing 4-4 and correct any lines that do not compile.

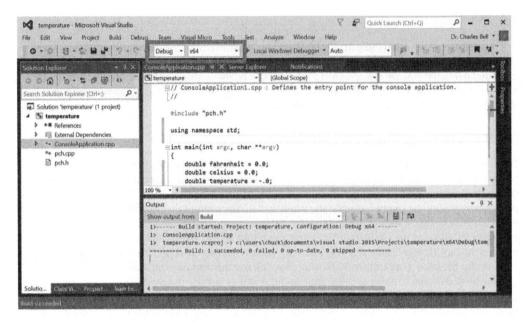

Figure 4-10. *Build and test your code*

To test the application, you can simply run it with the debugger (but without any breakpoints) by pressing **Ctrl+F5** or choosing **Debug ➤ Start without Debugging**. This opens a console and runs the application on your PC, and waits to close the console. Note that if you changed your code since the last build, you may be asked to rebuild the solution/project. If you choose the F5 option, the console would close when the application terminates. Figure 4-11 shows what the application should look like.

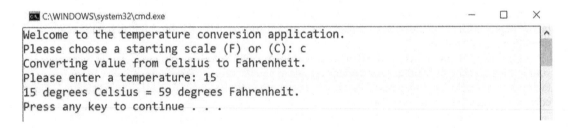

Figure 4-11. *Testing the temperature application*

You see what you expected: the application prompts you for some information and depending on your selection, converts the temperature entered. Go ahead and run the application several times until the joy has passed or you are convinced it is working correctly. When you're done, congratulate yourself on having written your first Windows 10 application! You are now ready to set up your Windows 10 IoT Device and deploy the application.

Set up your Windows 10 IoT Device

You're ready to power on your device and get it ready for deployment. All you need to do is set up the device as described in Chapter 2 and power it on.

COOL GADGET: MICRO-USB TO USB ADAPTER

If you're like me and have very few micro-USB cables and most of those devoted to specific devices, chances are that you have to borrow a cable to get your Raspberry Pi powered up. Rather than buying a whole new cable, you can buy a brilliant little adapter that lets you convert your normal USB type A cable to a micro-USB cable. The adapter (known by a variety of names) fits inside the male end of a USB cable converting it to a micro-USB plug. The following shows this adapter.

You can find it on most online auction sites. Just search for micro-USB adapter and you should find it among the more normal adapters. I like this adapter because it is small and removes the need for yet another cable to junk up my store of USB cables.

Once powered on, take note of the IP address. You need this to deploy your application in the next step.

Deploy and Test to your Windows 10 IoT Device

This is the step that most people have issues, especially when deploying a Windows IoT Core application for the first-time. This is mostly due to the fact that you must configure the project settings correctly or else the deployment will fail. Once you know what to change (and it isn't so obvious), deployment is really easy.

The first thing you should understand is the build configuration. You can choose to build an application for debugging (debug) or for releasing (release). Remember that you also need to choose the architecture. If you are working with the Raspberry Pi or DragonBoard 410C, you choose ARM; otherwise, you choose x86 for the MinnowBoard Max–compatible boards.

However, the project properties for these combinations of build and architecture are different! Thus, setting the values for debug won't change those for release and so on. This is perhaps the most difficult thing to remember. But do not despair, I will demonstrate the settings you need for both a debug and release build.

In fact, I will demonstrate one of the most powerful features of Visual Studio: the interactive debugger. But not just that, I'll also demonstrate how to debug your application running on your Windows 10 IoT device directly from your PC. How cool is that?

Using the Interactive Debugger

The interactive debugger allows you to step through your code one line at a time, view the state of variables, see the stack, and much more. You must set the build type to debug in order to use debugging.

The key concepts involved are the use of breakpoints or places in the code where you want the debugger to stop at when executing. You can set breakpoints on most lines of code. To do so, click in the leftmost portion of the code editor next to the line you want a breakpoint. You can also position the cursor on the line and press **F9** to set (turn on) the breakpoint. A small red circle appears, which indicates that a breakpoint has been set. To clear a breakpoint, just click the red circle again or position the cursor on the line and press **F9**. Figure 4-12 shows the code editor window with a breakpoint set on the first print statement.

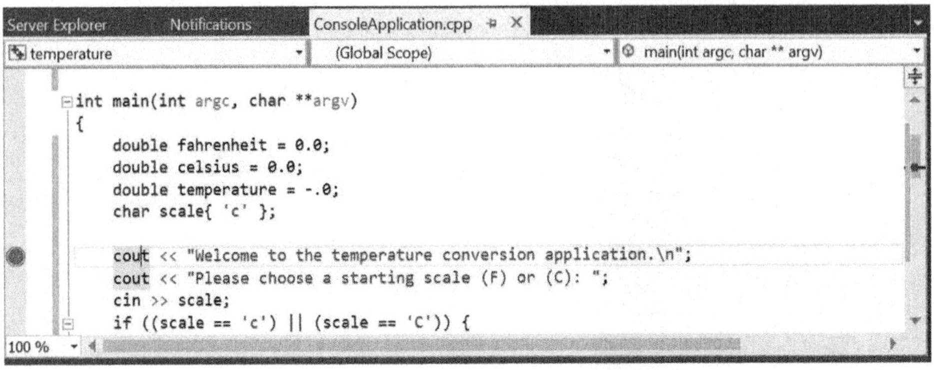

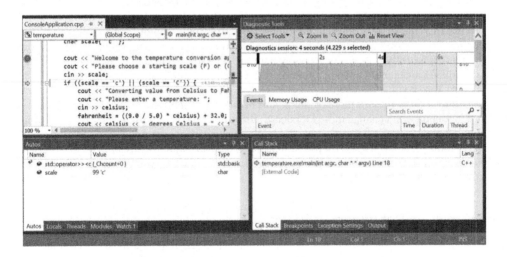

Figure 4-12. *Interactive debugger: setting breakpoints in the code editor*

To start the debugging session, press F5 or choose the **Debug ➤ Start Debugger** menu. This causes the application to start a new console window so that you can interact with the application. The debugger stops on the first line of code (where the breakpoint is set). You can then step through the code one line at a time by pressing F10 or by choosing the **Debug ➤ Step Into** menu.

When the debugger is running, several new windows open, including diagnostic tools that show memory and CPU usage, a tabbed window that allows you to watch variables (and more), and another tabbed window that allows you to see the call stack, breakpoints, and more. Figure 4-13 shows an excerpt of running the debugger with our sample project.

Figure 4-13. *Interactive debugger: inspection*

Go ahead and step through the application. Take some time to tour the various windows. Pay attention to the watch window and note how the variables change values. Clearly, there is far more that you can do with debugging Windows applications than what I have described here. Indeed, I consider the interactive debugger the most important feature of Visual Studio: one that requires some time and experience using to be able to fully utilize it. However, what I've shown here is the bare essentials for getting started.

■ **Tip**   To learn more about the interactive debugger, see `https://msdn.microsoft.com/en-us/library/dn986851.aspx`.

It is also possible to remotely debug an application running on your Windows 10 IoT device. I will demonstrate how to do this in a later chapter. Remote debugging allows you to run the application on the Windows 10 IoT device and monitor its progress with the interactive debugger on your PC. Due to the nature of console applications, if you were to try remote debugging this project, you would not get far. This is because the application prompts the user for input and there is no way to connect to the application to interact with it. Thus, remote debugging is best used for cases where you want to see how the code works on the device—checking variables, logic, and so forth, much like the interactive debugger.

Deploying a Debug Build

Now you're ready to deploy the application to your Raspberry Pi and start using it. In this section, I describe how to set up your project for deploying a debug build. Let's begin with the setting the project properties. Be sure to choose the build type (Debug) and platform (ARM) before opening the project properties.

Open the project properties by right-clicking the project and choose **Properties**. You see a dialog open that has a host of settings you can change. You want the debugging section. Click that label to open the debug settings panel. Figure 4-14 shows the completed dialog.

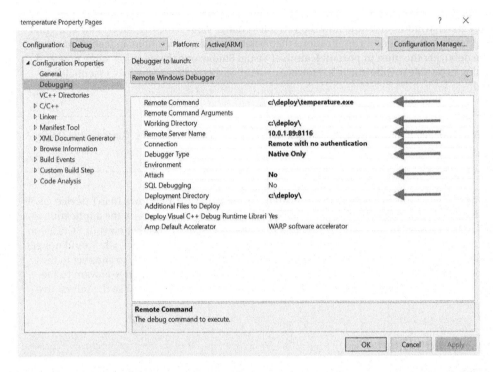

Figure 4-14. *Project properties (debug builds)*

The current configuration is at the top of the dialog. Make sure that everything corresponds to the debug build and ARM platform. There are seven settings you must change. I describe each next.

- *Remote Command*: Set this to the path and executable of your application. Use the working directory set below.

- *Working Directory*: Enter a working directory to deploy your application. The directory is created for you. Remember to add the trailing slash or else you may get a warning when you deploy the application.

- *Remote Server Name*: Set this to the IP address or hostname of your device along with port 8116 with .8116. For example, my Raspberry Pi was on 10.0.1.89 so I entered 10.0.1.89:8116. This is the most commonly misunderstood setting.

- *Connection*: Set this to **Remote with no authentication**.

- *Debugger Type*: Choose **Native Only**. You only want to deploy the native ARM application.

- *Attach*: Choose **No**. For a debug build deployment, you do not want to attach to the process once deployed.

- *Deployment Directory*: Set this to be the same as the working directory. You can choose a different directory, but if you do, make sure that the remote command reflects the difference in path.

Once these settings are entered, click **Apply** and then close the dialog. You are almost ready to deploy. But first, you must ensure that the remote debugger on the Raspberry Pi is turned on. Use your browser to browse to your device at port 8080 (e.g., 10.0.1.89:8080) or use the Device Portal from the Windows IoT Core dashboard. Click the Debug label and then click the Start button at the top of the page. This starts the remote debugger and readies the device for deployment. Figure 4-15 shows the result of start in the device portal. Note the port that the pop-up dialog presents. This is where you got the value for the debug settings.

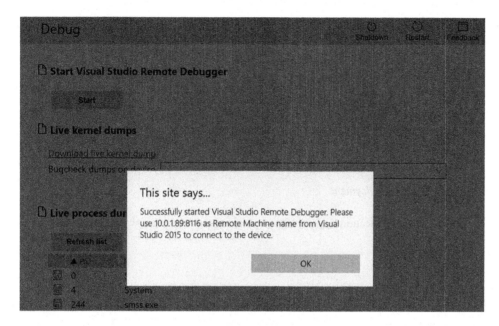

Figure 4-15. *Starting the Remote Debugger: Device Portal*

Now return to Visual Studio and choose **Build ➤ Deploy temperature** from the menu or click the **Remote Debugger** button on the toolbar. This initiates a build (if needed) and deploys the application to the Raspberry Pi. Once the deployment is complete, you see a message the output window.

■ **Note** The remote debugger is no longer installed by default in the 10586 release of Windows 10 IoT Core. This is because the deployment of the remote debugger was added in Visual Studio 2015 Update 1. If you try to deploy your debug build to your device and you get a message about MSVSMON.EXE missing or you get an error in the Device Portal when you try to start the remote debugger, you can try deploying a blank C# application to your device. This should install the remote debugger and allow you to deploy your application without errors. You can also try closing the MSVSMON.EXE application on your device. Sometimes the debugger can hang resulting in errors when deploying. Shutting down the remote debugger and restarting it can help solve deployment issues.

To test your application, use PuTTY to open a SSH session to log into your device. Once connected, you can change into the directory that you specified in the project properties. Figure 4-16 shows an example deployment. Notice the files that are deployed. Among those is the temperature.exe file, which you can use to run the application.

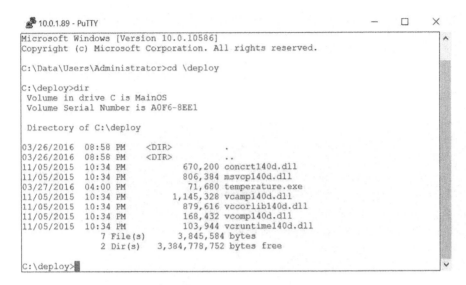

Figure 4-16. *Checking the debug deployment*

Now go ahead and run the application by entering temperature.exe. You can now interact with the application as I did in Figure 4-17.

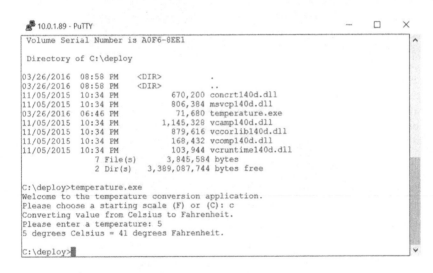

Figure 4-17. *Testing the application*

That's it! You've just deployed your first Windows 10 IoT Core application to your device and ran it. Very nice.

▪ **Tip** If you encountered problems during deployment, be sure the project settings are correct and that the build type and platform are set correctly. You can access the configuration management dialog to check these.

Deploying a Release Build

Deploying a release build is very similar to deploying a debug build. That is, you must set the project properties correctly. This may seem a bit strange, but you will use the same dialog that you did for deploying the debug build but with slightly different settings.

Like the debug build, you must set the build type to release and the platform to ARM. Then open the project properties and set the values, as shown in Figure 4-18.

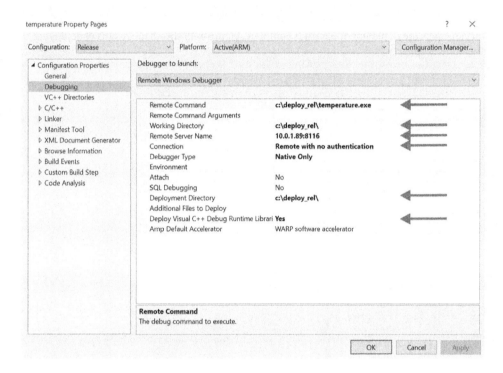

Figure 4-18. *Project properties (release builds)*

There are six settings that need to be set as described next. If this is your first time visiting the dialog after setting the project settings for the debug build, you may notice the values are back to the defaults. This is because each build type has its own settings and you must set them for each build you want to work with.

- *Remote Command*: Set this to the path and executable of your application. Use the working directory set below.

- *Working Directory*: Enter a working directory to deploy your application. The directory is created for you. Remember to add the trailing slash or else you may get a warning when you deploy the application.

- *Remote Server Name*: Set this to the IP address or hostname of your device along with port 8116 with :8116. For example, my Raspberry Pi was on 10.0.1.89, so I entered 10.0.1.89:8116. This is the most commonly misunderstood setting. Even though this is a release build, you still use the remote debugger to do the deployment.

- *Connection*: Set this to **Remote with no authentication**.

- *Debugger Type*: Choose **Native Only**. You only want to deploy the native ARM application. "Native" means that the code is compiled for the ARM platform; it does not include managed code.

- *Deployment Directory*: Set this to be the same as the working directory. You can choose a different directory but if you do make sure that the remote command reflects the difference in path.

I set the deployment directory to deploy_rel so that I can tell the difference. Once you've completed the settings, apply them, and close the dialog. Then, choose the **Build ➤ Deploy temperature** to deploy the application. You can check the deployment once it is complete. Figure 4-19 shows the results of deploying the release build to the Raspberry Pi.

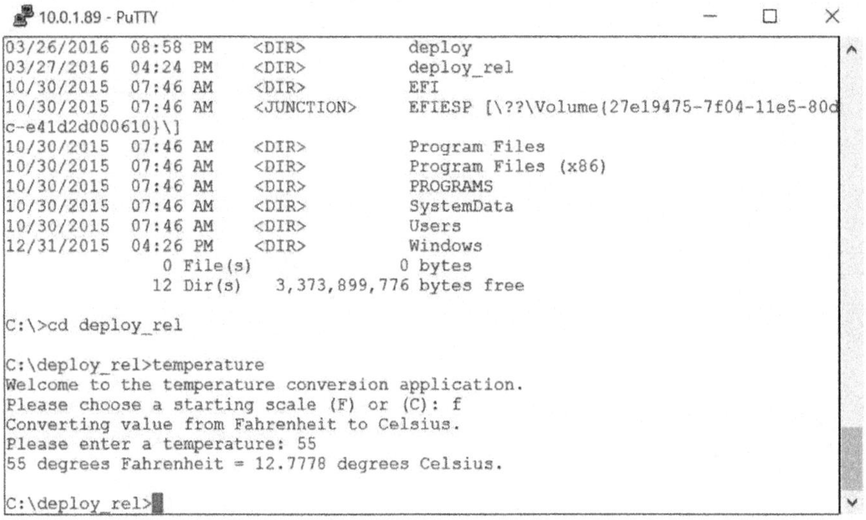

```
🖳 10.0.1.89 - PuTTY                                    —    □    ×
03/26/2016  08:58 PM    <DIR>          deploy
03/27/2016  04:24 PM    <DIR>          deploy_rel
10/30/2015  07:46 AM    <DIR>          EFI
10/30/2015  07:46 AM    <JUNCTION>     EFIESP [\??\Volume{27e19475-7f04-11e5-80d
c-e41d2d000610}\]
10/30/2015  07:46 AM    <DIR>          Program Files
10/30/2015  07:46 AM    <DIR>          Program Files (x86)
10/30/2015  07:46 AM    <DIR>          PROGRAMS
10/30/2015  07:46 AM    <DIR>          SystemData
10/30/2015  07:46 AM    <DIR>          Users
12/31/2015  04:26 PM    <DIR>          Windows
               0 File(s)              0 bytes
              12 Dir(s)   3,373,899,776 bytes free

C:\>cd deploy_rel

C:\deploy_rel>temperature
Welcome to the temperature conversion application.
Please choose a starting scale (F) or (C): f
Converting value from Fahrenheit to Celsius.
Please enter a temperature: 55
55 degrees Fahrenheit = 12.7778 degrees Celsius.

C:\deploy_rel>
```

Figure 4-19. *Checking the release deployment*

If you have followed along and succeeded in deploying both the debug and release builds, you're ready to learn more about the power and level of sophistication available for writing IoT solutions with Windows 10.

Summary

Developing applications for Windows 10 IoT Core requires the use of Visual Studio and the choice of several programming languages. Visual Studio is loaded with advanced features that meet the needs of even the most serious developers for the most complex solutions.

Fortunately for you, you do not need to master every nuance of the IDE. Indeed, you need only learn the most basics of getting around in the IDE, including starting new projects, coding, compiling, testing and debugging, and deploying your applications to your Windows 10 IoT Core devices.

In this chapter, you explored the Visual Studio 2015 interface, which included learning how the windows are laid out in the IDE and the sample project templates used in this book to write applications for your Windows 10 IoT Core devices. You learned the basic features that you will use for most of the book's projects. You also had a detailed overview of a basic Windows 10 IoT Core application using C++.

In the next chapter, you'll take a closer look at developing Windows 10 IoT Core applications by using C++. I present a short tutorial on the major language constructs and walk you through an application that blinks an LED.

CHAPTER 5

■ ■ ■

Windows 10 IoT Development with C++

Now that you have a basic understanding of how to use Visual Studio 2015, you can learn more about some of the languages you may encounter when developing your IoT solutions. One of those languages is C++—a very robust and powerful language that you can use to write very powerful applications. Mastering C++ is not a trivial task and indeed could take someone several years to be fully knowledgeable of all of its features.[1] However, you do not need to achieve a Zen-like harmony with C++ to be able to write applications for Windows 10 IoT Core. You saw this in action in the last chapter. In fact, if you are just getting started programming or know little about C++, all you need to get going is knowledge of the fundamentals of the language and how to use it in Visual Studio.

This chapter presents a crash course on the basics of C++ programming in Visual Studio-including an explanation about some of the most commonly used language features. As such, this chapter provides you with the skills you need to understand the growing number of IoT project examples available on the Internet. The chapter concludes with a walk-through of another C++ example project that shows you how to interact with hardware. Specifically, you will implement the LED project you saw in Chapter 3. Only this time, you'll be writing it as a Windows 10 IoT Core application. So let's get started!

■ **Tip** If you are not interested in using C++ in your IoT solutions, or you already know the basics of C++ programming, feel free to skim through this chapter. I recommend working through the example project at the end of the chapter, especially if you've not written IoT applications.

Getting Started

Microsoft's implementation of C++ is named Visual C++ and is often referred to as VC++ or MSVC. Visual C++ conforms to the latest standards, including C++11 and C++14.[2] Early versions of Visual C++ were sold as a separate product. Visual C++ today is integrated in the Visual Studio product. As you saw in Chapter 2, you must explicitly select Visual C++ during installation of Visual Studio because it is not installed by default. However, as you saw, it is very easy to add optional features in Visual Studio.

---

[1]But who has the time?
[2]For more about compatibility with C++ standards, see https://msdn.microsoft.com/en-us/library/hh567368.aspx.

While the formal product name is Visual C++, and it is designed to work with the Microsoft Windows operating system, learning the basics of the Visual C++ language is no different than learning C++ that is offered for other platforms. Thus, in this section and throughout the rest of the book, I use C++ as shorthand for Visual C++. Most of the material in the next section applies to C++ in general and is not specific to Visual C++. Indeed, except for the Microsoft frameworks and the components of Visual Studio, the basics of the C++ language apply universally. Let's learn more about the origins and merits of C++.

The C++ programming language has been around for over 35 years and is one of the most widely used programming languages among all platforms. There are many reasons for this. C++ was designed to permit programmers to express hardware, machine, and programming concepts as close to the hardware level as possible. The C++ language therefore is designed to be both very expressive with concepts close to actual hardware. Indeed, well-designed and well written C++ applications are typically faster and more efficient than those written in other languages.[3] Part of this is due to the fact that C++ can be used to program a wide array of solutions—from computer games to highly sophisticated scientific analysis applications to device drivers and even entire frameworks. The longevity of C++ is a testament to the resiliency of the language.

However, there has been some misunderstanding in the history of C++ for which Visual C++ was not immune. Early releases of the C++ language specification occurred during a time when the C language was popular. While it is true C++ was based on C and to this day you can use a number of C features (but not nearly as many now that C++ has evolved from those early days), C++ has become a completely different language.

The evolution included the addition of static type safety (every object, name, value, etc. must have a type known at compile time), full object-oriented programming features (classes, inheritance, etc.), and a host of improvements designed to make programming easier and faster (e.g., the standard template library). What started out as "C with classes" (circa 1979), which included the beginnings of object-oriented programming, the language was renamed C++ (circa 1983) to distinguish it from C.

Finally, C++ programs are compiled into executable files. More specifically, the code you write is translated into binary code for the platform chosen. This is another reason C++ is potentially faster than other languages because there is not interpreter involved. For example, the Python language is an interpreted language that, while there is a building stage, the code is turned into an intermediate form that is platform agnostic, which is then executed by the interpreter. The power of compiling the code directly into binary code with the gains in speed and efficiency are why many people choose C++ over other languages.

Should you require more in-depth knowledge of C++ (Visual C++), there are a number of excellent books on the topic. The following is a list of a few of my favorites.

- *Beginning C++* by Ivor Horton (Apress, 2014)

- *The C++ Programming Language* by Bjarne Stroustrup (Addison-Wesley Professional, 2013)[4]

- *C++ Recipes: A Problem-Solution Approach* by Bruce Sutherland (Apress, 2015)

Another excellent resource is Microsoft's documentation on MSDN. The following are some excellent resources for learning Visual C++.

- Visual C++ in Visual Studio 2015 (`https://msdn.microsoft.com/en-us/library/60k1461a.aspx`)

- C++ Language Reference (`https://msdn.microsoft.com/en-us/library/3bstk3k5.aspx`)

---

[3]However, I have seen the opposite happen for poorly written code. Language features can never overcome poor programming.
[4]Creator of the C++ language.

C++ Crash Course

Now let's learn some of the basic concepts of C++ programming. You begin with the building blocks of the language, such as comments, variables, and basic control structures, and then move into the more complex concepts of data structures and libraries.

While the material may seem to come at you in a rush (hence, the crash part), this crash course on C++ covers only the most fundamental knowledge of the language and how to use it in Visual Studio. It is intended to get you started writing C++ Windows 10 IoT Core applications. If you find you want to write more complex applications than the examples in this book, I encourage you to acquire one or more of the resources listed earlier to learn more about the intriguing power of C++ programming.

The following sections present many of the basic features of C++ programming that you need to know in order to understand example projects for Windows 10 IoT Core and vital to successfully implementing the C++ projects in this book.

The Basics

There are a number of basic concepts about the C++ programming language that you need to know in order to get started. In this section, I describe some of the fundamental concepts used in C++, including how the code is organized, how libraries are used, namespaces, and how to document your code. Before you begin, let's take a look at a slightly different version of the temperature application you saw in Chapter 4. Listing 5-1 shows the code rewritten slightly to use functions.

Listing 5-1. Temperature Code Example Rewrite

```
//
// Windows 10 for the IoT
//
// Example C++ console application rewrite.
//
// Created by Dr. Charles Bell
//
#include pch.h

using namespace std;

double convert_temp(char scale, double base_temp) {
        if ((scale == 'c') || (scale == 'C')) {
                return ((9.0 / 5.0) * base_temp) + 32.0;
        } else if ((scale == 'f') || (scale == 'F')) {
                return (5.0 / 9.0) * (base_temp - 32.0);
        }
        return 0.0;
}

int main(int argc, char **argv) {
        double temp_read = 0.0;
        char scale{'c'};
```

```
      cout << "Welcome to the temperature conversion application.\n";
      cout << "Please choose a starting scale (F) or (C): ";
      cin >> scale;
      cout << "Please enter a temperature: ";
      cin >> temp_read;
      if ((scale == 'c') || (scale == 'C')) {
              cout << "Converting value from Celsius to Fahrenheit.\n";
              cout << temp_read << " degrees Celsius = " <<
                  convert_temp(scale, temp_read) << " degrees Fahrenheit.\n";
      } else if ((scale == 'f') || (scale == 'F')) {
              cout << "Converting value from Fahrenheit to Celsius.\n";
              cout << temp_read << " degrees Fahrenheit = " <<
                  convert_temp(scale, temp_read) << " degrees Celsius.\n";
      } else {
              cout << "\nERROR: I'm sorry, I don't understand '" << scale << "'.";
              return -1;
      }
      return 0;
}
```

Wow, that's quite a change! While the functionality is exactly the same, the code looks very different. The following describe the C++ concepts I have implemented in this example.

Functions

Notice that I've added a new function named convert_temp() that converts the temperature based on the scale chosen. This effectively moves that logic out of the main() function thereby simplifying the code. This technique is a key technique you use when writing C++ applications. More specifically, in C++, applications are always built using functions.

Recall that the main() function is the starting or initial execution for the C++ console project. Traditional C++ applications (such as the console application) must have a main function.

Notice the main function again. Here you see the function name is preceded by a type (integer). This tells the C++ compiler that this method returns an integer value. On Windows GUI applications (not command-line interface applications), it is common practice to not return a value from main() but to be pedantic; you should do so as I have in the preceding example.

```
int main(int argc, char **argv)
```

Next, you see the name, main, followed by a list of parameters enclosed in parenthesis. For the main() function, the parameters are fixed and are used to store any command line arguments provided by the user. In this case, you have the number of arguments stored in argc and a list (array) of the arguments stored in argv (which is actually a double pointer—more on pointers later).

A function in C++ is used as an organizational mechanism to group functionality and make your programs easier to maintain (functions with hundreds of lines of code are very difficult to maintain), improves comprehensibility, and localize specialized operations in a single location thereby reducing duplication.

Functions are used in code to express the concepts of the functionality that they provide. Notice how I used the convert_temp() function. Here I declared it as a function that returned a double and takes a character and a double as input. As you can see, the body of the function (defined inside the curly braces) uses the character as the scale in the same was as you do in main and uses the double parameter as the target (or base) temperature to convert.

■ **Tip** Function parameters and values passed must match on type and order when called.

Notice also that I placed it in the line of code that prints the value to the screen. This is a very common practice in C++ (and other programming languages). That is, you use the function to perform some operation and rather than store the result in a variable, you use it directly in the statements (code).

Curly Braces

Notice that both methods are implemented with a pair of curly braces that define the body of the function. Curly braces in C++ are used to define a block of code or simply to express grouping of code. Curly braces are used to define the body of functions, structures, classes, and more. Notice that you use them everywhere, even in the conditional statements (see the if statements).

■ **Tip** Some C++ programmers prefer to place the starting curly brace on the same line as the line of code to which it belongs like I did in the example. However, others prefer the open curly brace placed on the next line. Neither preference matters to the compiler; rather, this is an example of code style. You should choose the style you like best.

Including Libraries

If you recall from earlier examples, there were some lines of code at the top of the source code that indicated something was to be included. These are called *preprocessor directives* and often look like the following. They are called *preprocessor directives* because they signal the compilation process to perform some tasks before the code is compiled.

```
#include "pch.h"
```

The directive does what it sounds like: it tells the compiler to include that file along with your source code. When the compiler encounters this directive, it "includes" that source file with your source code and compiles it.

The #include directive is one of the fundamental mechanisms that support modularity in C++. That is, you can create a library of source code that provides some functionality that resides in one or more separate source code files. Even if you do not create a new library, you can use modularity to split your source code into separate parts that form some high-level abstraction. More specifically, you would place like functionality together making the code easier to maintain or allow more than yourself to work on it at the same time. However, you would not use modularity to separate random sections of code—that would gain you nothing except confusion as to where the bits of code reside.

The file that the preceding example includes is called a *header file*. A header file is named .h and contains only the declaration of the code. You can think of it as a blueprint or pattern for the code. A header file often contains only the primitives of the code that you will use, hence making it possible for the compiler to resolve any references to the features in the header file. A separate companion file called the *source file* is named .cpp (you can also use .cc, but most prefer .cpp) and contains the actual code for the features. In the preceding example, there are two files: one named pch.h and the other pch.cpp.

Using Namespaces

Notice the line of code that begins with using in the example. This is another preprocessor directive. The using directive tells the compiler that you are using the namespace std. A namespace is a special organizational feature that allows you to group identifiers (names of variables, constants, etc.) under a group that is localized to the namespace. Using the namespace tells the compiler to look in the namespace for any identifier you've used in your code that is not found.

```
using namespace std;
```

For example, the following cout statement is included in the std namespace. Had I not added the using directive, the compiler would not know what cout was. Notice in the second line that I could have included the namespace std followed by two colons. This tells the compile to look in the std namespace for cout or more correctly use the cout in the std namespace.

```
cout << "Hello, World!\n";
std::cout << "Right back at you!";
```

Namespaces can also help you avoid duplication of identifiers. Had I wanted to, I could have created a new module (library, source file) that contains a new namespace and reused the cout identifier. However, care must be taken when reusing identifiers among namespaces because the compiler may not know which namespace I want to use. Creating namespaces is a bit advanced for a crash course, but I encourage you to consider using them for applications that grow beyond a single source code and header file.

Finally, namespaces associated with libraries of classes often form hierarchies that you can chain together. For example, if you wanted to use the namespace inside the Windows Foundations library named Collections, you would refer to it as follows. This is a very common occurrence in Windows C++ applications. In fact, you use several namespaces in our example project.

```
using namespace Windows::Foundation::Collections;
```

Comments

One of the most fundamental concepts in any programming language is the ability to annotate your source code with text that not only allows you to make notes among the lines of code does but also forms a way to document your source code.[5]

To add comments to your source code, use two slashes, // (no spaces between the slashes). Place them at the start of the line to create a comment for that line repeating the slashes for each subsequent line. This creates what is known as a *block comment*, as shown. Notice that I used a comment without any text to create whitespace. This helps with readability and is a common practice for block comments.

```
//
// Windows 10 for the IoT
//
// Example C++ console application rewrite.
//
// Created by Dr. Charles Bell
//
```

---

[5]If you ever hear someone claim, "My code is self-documenting," be cautious when using his or her code. There is no such thing. Sure, plenty of good programmers can write code that is easy to understand (read), but all fall short of that lofty claim.

You can also use the double slash to add a comment at the end of a line of code. That is, the compiler ignores whatever is written after the double slash to the end of the line. You see an example of this next. Notice that I used the comment symbol (double slash) to comment out a section of code. This can be really handy when testing and debugging, but generally discouraged for final code. That is, don't leave any commented out code in your deliverable (completed) source code. If it's commented out, it's not needed!

```
if (size < max_size) {
  size++;
} //else {
//  return -1;
//}
```

Writing good comments and indeed documenting your code well is a bit of an art form; one that I encourage you to practice regularly. Since it is an art rather than a science, keep in mind that your comments should be written to teach others what your code does or is intended to do. As such, you should use comments to describe any preconditions (or constraints) of using the code, limitations of use, errors handled, and a description of how the parameters are used and what data is altered or returned from the code (should it be a function or class member).

Variables and Types

No program would be very interesting if you did not use variables to store values for calculations. Variables are declared with a type and once defined with a specific type cannot be changed. Since C++ is strongly typed, the compiler ensures that anywhere you use the variable that it obeys its type. For example, that the operation on the variable is valid for the type. Thus, every variable must have a type assigned.

There are a number of simple types that the C++ language supports (often called *built-in types*). They are the basic building blocks for more complex types. Each type consumes a small segment of memory which defines not only how much space you have to store a value, but also the range of values possible.[6]

For example, an integer consumes 4 bytes and you can store values in the range –2,147,483,648 to 2,147,483,647. In this case, the integer variable is signed (the highest bit is used to indicate positive or negative values). An unsigned integer can store values in the range 0 to 4,294,967,295.

You can declare a variable by specifying its type first and then an identifier. The following shows a number of variables using a variety of types.

```
int num_fish = 0;          // number of fish caught
double max_length {0.0}; // length of the longest fish in feet
char fisherman[25];       // name of the fisherman
char rod_used[40];        // name or type of rod used
```

Notice also that I have demonstrated how to assign a value to the variable in the declaration. I demonstrate two widely used techniques: using a simple assignment and using the initialization mechanism available in C++11 (meaning it is the C++ standard adopted in 2011) and newer.

The assignment operator is the equal sign. All assignments must obey the type rules. That is, I cannot assign a floating-point number (e.g., 17.55) to an integer value. The C++ initialization mechanism uses curly braces (called an *initializer list*) that contain the value you want to assign. The following shows an example.

```
int x {14};
```

---

[6]For a complete list, see https://msdn.microsoft.com/en-us/library/s3f49ktz.aspx.

Note that you can include the assignment operator with the curly braces (the compile will not complain) but that is considered sloppy and discouraged. For example, the following code will compile, but it is considered bad form.

```
int y = {15};
```

Table 5-1 shows a list of the commonly used built-in types that you use in your applications.

Table 5-1. *Commonly Used Types in C++*

Symbol	Size in Bytes	Range
bool	1	false or true
char	1	–128 to 127 by default
signed char	1	–128 to 127
unsigned char	1	0 to 255
short	2	–32,768 to 32,767
unsigned short	2	0 to 65,535
int	4	–2,147,483,648 to 2,147,483,647
unsigned int	4	0 to 4,294,967,295
long	4	–2,147,483,648 to 2,147,483,647
unsigned long	4	0 to 4,294,967,295
float	4	3.4E +/- 38 (7 digits)
long long	8	–9,223,372,036,854,775,808 to 9,223,372,036,854,775,807
unsigned long long	8	0 to 18,446,744,073,709,551,615
double	8	1.7E +/- 308 (15 digits)

It is always a good practice to initialize your variables when you declare them. It can save you from some nasty surprises if you use the variable before it is given a value (although the compiler will complain about this). For example, it is possible for code to work correctly in debug mode (since variables may be initialized by the debugger) but fail in release mode.

There is also a convenient automatic type keyword (auto) that you can use to permit the compiler to choose the correct type. This is helpful for things like loops as follows but also helps with maintainability as well as permitting advanced reuse through templates.

```
auto j {3};    // integer is used here
for (auto i=0; i < 10; ++i) {
  cout << i + j << "\n";
}
```

Here you see you create the variable j with the auto keyword, which given the initialization is an integer results in j being an integer. Similarly, you use the auto keyword in the for loop counting variable. Since it was assigned an integer, the variable i will be an integer.

Arithmetic

You can perform a number of mathematical operations in C++, including the usual primitives but also logical operations and operations used to compare values. Rather than discuss these in detail, I provide a quick reference in Table 5-2 that shows the operation and example of how to use the operation.

Table 5-2. *Arithmetic, Logical, and Comparison Operators in C++*

Type	Operator	Description	Example
Arithmetic	+	Addition	int_var + 1
	-	Subtraction	int_var - 1
	*	Multiplication	int_var * 2
	/	Division	int_var / 3
	%	Modulus	int_var % 4
	-	Unary subtraction	-int_var
	+	Unary addition	+int_var
Logical	&	Bitwise and	var1&var2
	|	Bitwise or	var1|var2
	^	Bitwise exclusive	var1^var2
	~	Bitwise compliment	~var1
	&&	Logical and	var1&&var2
	||	Logical or	var1||var2
Comparison	==	Equal	expr1==expr2
	!=	Not equal	expr1!=expr2
	<	Less than	expr1<expr2
	>	Greater than	expr1>expr2
	<=	Less than or equal	expr1<=expr2
	>=	Greater than or equal	expr1>=expr2

Bitwise operations produce a result on the values performed on each bit. Logical operators (and, or) produce a value that is either true or false and are often used with expressions or conditions.

Finally, C++ has a concept called *constants*, where a value is set at compile time. There are two types of constants. One, signified by using the const keyword, creates a value (think variable) that will never be changed. The other, signified by using the constexpr keyword, creates a function whose body or functionality is evaluated at compile time. The following are examples of constants in C++.

```
const int fish_catch_limit {7};    // Creates a constant variable whose value cannot change
constexpr double square(double z) { return z*z; }  // A constant expression
```

Now that you understand variables and types, the operations permitted on them, and expressions, let's look at how you can use them in flow control statements.

Flow Control Statements

Flow control statements change the execution of the program. They can be conditionals that define gates using expressions that restrict execution to only those cases where the expression evaluates true (or negated), special constructs that allow you to repeat a block of code (loops), and the use of functions to switch context to perform some special operations. You've already seen how functions work so let's look at conditional and loop statements.

Conditionals

Conditional statements allow you to direct execution of your programs to sections (blocks) of code based on the evaluation of one or more expressions. There are two types of conditional statements in C++—the if statement and the switch statement.

You have seen the if statement in action in our example code. Notice in the example that you can have one or more (optional) else phrases that you execute once the expression for the if conditions evaluate to false. You can chain if/else statements to encompass multiple conditions where the code executed depends on the evaluation of several conditions. The following shows the general structure of the if statement.

```
if (expr1) {
  // execute only if expr1 is true
} else if ((expr2) || (expr3)) {
  // execute only if expr1 is false *and* either expr2 or expr3 is true
} else {
  // execute if both sets of if conditions evaluate to false
}
```

Although you can chain the statement as much as you want, use some care here because the more else/if sections you have, the harder it becomes to understand, maintain, and avoid logic errors in your expressions.

If you have a situation where you want to execute code based on one of several values for a variable or expression that returns a value (such as a function or calculation), you can use the switch statement. The following shows the structure of the switch statement.

```
switch (eval) {
  case <value1> :
    // do this if eval == value1
    break;
  case <value2> :
    // do this if eval == value2
    break;
  default :
    // do this if eval != any case value
    break;  // Not needed, but good form
}
```

The case values must match the type of the thing you are evaluating. That is, case values must be same type as eval. Notice the break statement. This is used to halt evaluation of the code once the case value is found. Otherwise, each successive case value will be compared. Finally, there is a default section for code that you want to execute, should eval fail to match any of the values.

■ **Tip**   Code style varies greatly in how to space/separate these statements. For example, some indent the case statements, some do not.

Loops

Loops are used to control the repetitive execution of a block of code. There are three forms of loops that have slightly different behavior. All loops use conditional statements to determine whether to repeat execution or not. That is, they repeat as long as the condition is true. The three types of loops are while, do, and for. I explain each with an example.

The while loop has its condition at the "top" or start of the block of code. Thus, while loops only execute the body if and only if the condition evaluates to true on the first pass. The following illustrates the syntax for a while loop. This form of loop is best used when you need to execute code only if some expression(s) evaluate to true. For example, iterating through a collection of things those number of elements is unknown (loop until you run out of things in the collection).

```
while (expression) {
  // do something here
 }
```

The do loop places the condition at the "bottom" of the statement, which permits the body of the loop to execute at least once. The following illustrates the do loop. This form of loop is handy for cases where you want to execute code that, depending on the results of that execution, may require repetition. For example, repeatedly asking the user for input that matches one or more known values repeating the question if the answer doesn't match.

```
do {
  // do something here - always done once
} while (expression);
```

for loops are sometimes called *counting loops* because of their unique form. The for loop allows you to define a counting variable, a condition to evaluate, and an operation on the counting variable. More specifically, for loops allow you to define stepping code for a precise number of operations. The following illustrates the structure of the for loop. This form of loop is best used for a number of iterations for a known number (either at run time or as a constant) and commonly used to step through memory, count things, and so forth.

```
for (<init> ; <expression> ; <increment>) {
// do someting
}
```

The <init> section or counting variable declaration is executed once and only once. The <expression> is evaluated on every pass. The <increment> code is executed every pass except the last. The following is an example for loop.

```
for (int i; i < 10; i++) {
  // do something here
}
```

Now let's look at some commonly used data structures.

Basic Data Structures

What you have learned so far about C++ allows you to create applications that do simple to moderately complex operations. However, when you start needing to operate on data (either from the user or from sensors and similar sources), you need a way to organize and store data and operations on the data in memory. The following introduces three data structures in order of complexity: arrays, structures, and classes.

Arrays allocate a contiguous area of memory for multiple storage of a specific type. That is, you can store several integers, characters, and so forth, set aside in memory. Arrays also provide an integer index that you can use to quickly access a specific element. The following illustrates how to create an array of integers and iterate through them with a for loop. Array indexes start at 0.

```
int num_array[10] {0,1,2,3,4,5,6,7,8,9};  // an array of 10 integers
for (int i; i < 10; ++i) {
  cout << "num_array[" << i << "] = " << num_array[i] << "\n";
}
```

Notice the ++i in the for loop. This is a shorthand for i = i + 1 and is very common in C++. You can also define multiple dimensional arrays (arrays of arrays). Arrays can be used with any type or data structure.

If you have a number of data items that you want to group together, you can use a special data structure called, amazingly, struct. A struct is formed as follows.

```
struct <name> {
  // one or more declarations go here
};
```

You can add whatever declarations you want inside the struct body (defined by the curly braces). The following shows a crude example. Notice that you can use the structure in an array.

```
struct address {
  char first_name[30];
  char last_name[30];
  int street_num;
  char street_name[40];
  char city[40];
  char state[2];
  char zip_code[12];
};
```

```
address address_book[100];
```

Arrays and structures can increase the power of your programs by allowing you to work with more complex data types. However, there is one data structure that is even more powerful: the class.

A class is more than a simple data structure. You use classes to create abstract data types and to model concepts that include data and operations on data. Like structures, you can name the class and use that name to allocate (instantiate) a variable of that type. Indeed, structs and classes are closely related.

You use classes to break your programs down into modules. More specifically, you place the definition of a class in a header file and the implementation in a source file. The following shows the header file (myclass.h) for a simple and yet trivial class to store an integer and provide operations on the integer.

```
class MyClass {
  public:
    MyClass();
    int get_num();
    void inc();
    void dec();
  private:
    int num;
};
```

Notice several things here. First, the class has a name (MyClass), a public section where anything in this area is visible (and usable) outside of the class. In this case, there are three functions. The function with the same name as the class is called a *constructor*, which is called whenever you instantiate a variable of the class (type). The private section is only usable from functions defined in the class (private or public).

The source code file (myclass.cpp) is where you implement the methods for the class as follows.

```
#include "myclass.h"

MyClass::MyClass() {
  num = 0;
}

int MyClass::get_num() {
  return num;
}

void MyClass::inc() {
  ++num;
}

void MyClass::dec() {
  --num;
}
```

Notice that you define the methods in this file prefixed with the name of the class and two colons (MyClass::). While missing in this example, you can also provide a destructor (noted as ~MyClass) that is executed when the class instantiation is deallocated. Finally, notice at the top is the #include preprocessor directive to include the header file so that the compiler knows how to compile this code (using the class header or declaration). You can then use the class in the program, as follows.

```
#include <iostream>
#include "myclass.h"

using namespace std;

int main(int argc, char **argv) {
  MyClass c = MyClass();
  c.inc();
  c.inc();
  cout << "contents of myclass: " << c.get_num() << "\n";
}
```

Notice how you use the class. This is actually allocating memory for the class—both data and operations. Thus, you can use classes to operate on things or provide functionality when you need it saving you time and making your programs more sophisticated. Classes are used to form libraries of functionality that can be reused. Indeed, you have entire suites of libraries built using classes.

As you may have surmised, classes are the building block for object-oriented programming and as you learn more about using classes, you can build complex libraries of your own.

Pointers

Pointers are one of the most difficult things for new programmers to understand. However, the following attempts to explain the basics of using pointers. There is a lot more that you can do with pointers, but this is the fundamental concept of simple pointers.

A pointer (also called a *pointer variable*) stores the memory address of variable or data. Thus, a pointer "points to" a section of memory. You declare by type and the * symbol. All pointers must be typed and any operation on what the pointer points to must obey the condition of that type. When you access the thing the pointer "points to", you call that dereferencing and use the * symbol to tell the compiler you want the value of the thing the pointer is "pointing to."[7] The following shows how you can declare a pointer and then dereference it. Note that you expect int_ptr to be assigned a value; otherwise, the code may not compile or exhibit side effects.

```
int *int_ptr; // pointer to an integer
int i = *int_ptr; // Store what int_ptr is pointing to
```

To store an address in a pointer variable, you use the & symbol (also called the *address of* operator). The following shows an example.

```
int *int_ptr = &i; // Store address of i in int_ptr
```

You can perform arithmetic and comparison on pointers. You can add or subtract an integer to change the address of the pointer (the actual value of the pointer variable, not the thing the pointer points to) by multiples of the size of the type. For example, adding 1 to an integer pointer advances (increases) the memory value by 4 bytes.

You can also compare pointers to determine equality and subtract one pointer from another to find distance (in bytes) between the pointers. This could be handy for calculating distance for contiguous memory segments. When performing arithmetic on pointers, you should use parenthesis to avoid nasty mistakes with operator precedence. For example, the following code is not equivalent. The second line increments the thing that the pointer points to, but the third line increments the value of the pointer variable (memory address). Be careful when performing math on pointers, because you could unexpectedly end up dereferencing portions of memory.

```
*int_ptr = 10;       // set the thing that the pointer points to = 10
i = *int_ptr + 1;    // add one to the thing that the pointer points to
i = *(int_ptr + 1);  // add 4 bytes to the pointer variable (size of integer) - ERROR?
points to nowhere!
```

Finally, always use nullptr to initialize a pointer variable when the address is not known, as follows.

```
int *int_ptr {nullptr};
```

---

[7] I think you get the point.

Now that you know how to declare pointers, dereference them to retrieve the value of the thing the pointer points to and to find the address to store in a pointer variable, let's see pointers in action. The following demonstrates how pointers are used. What follows is an overly simplified example where the memory locations use fictitious values but it illustrates the concept of pointers as memory addresses. You will follow several code statements as they execute.

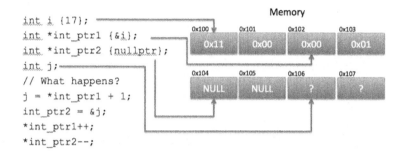

Figure 5-1. *Pointer Illustration: initial state*

■ **Note**   The values shown are in hexadecimal and shown using only two bytes for integers and pointers to integers. You also show data saved with the low byte first.

Take a few moments to study the drawing until you are confident you see how each value is stored. Here you see you have two integer variables and two pointers to integer variables. The drawing shows how each is allocated in memory. Notice that you see the variable i is stored in memory (at address 0x100) with the value 17, the first integer pointer stored with a value of i (stored in memory location 0x102), the second integer pointer stored with the value of nullptr (stored in memory location 0x104), and the integer variable j (at address 0x106) stored without an initial value.

Now let's see what happens when the first line of code is executed. Figure 5-2 shows the results in our fictional memory space.

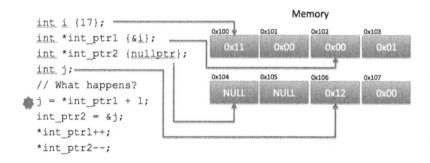

Figure 5-2. *Pointer illustration: step 1*

Here you see that the variable j was assigned the value dereferenced from int_ptr1 plus one. To execute this statement, the compiler dereferenced int_ptr1 by using its value (memory address 0x100) to get the value from that memory location (0x11) and then adding 1 (0x12) and storing it in memory.

Now let's see what happens when you execute the next statement. Figure 5-3 shows the results in our fictional memory space.

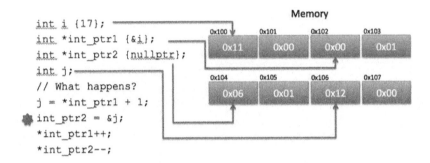

Figure 5-3. *Pointer illustration: step 2*

Here you see the address of the variable j (0x104) is stored in the pointer variable int_ptr2 (located at memory address 0x014).

If you are following along and the change to the drawing makes sense, you are seeing and comprehending how pointers work. If you are not entirely certain the drawing is correct, take a few moments and work through the figures and code again until you are convinced it is working correctly. If you are in this frame of mind, do not be disappointed (or frustrated) as learning how pointers work takes some time to get your mind around. These figures are designed to help you understand how they work. Just make sure you see how the values are changing before proceeding.

■ **Tip**   There are entire books written about pointers! If you want a more in-depth look at pointers or want to dive into the details of how to use pointers, see the book *Understanding and Using C Pointers* by Richard Reese (O'Reilly, 2013). The book was written for C programmers, and although some of the data is outdated, it is an excellent study on pointers.

Now let's see what happens when you execute the next statement. Figure 5-4 shows the results in our fictional memory space.

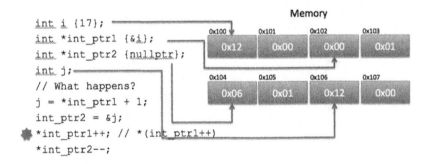

Figure 5-4. *Pointer illustration: step 3*

Here you see the statement executed increments the value that int_ptr1 points to by 1 (the ++ operator). In the drawing, you see the compiler dereferences int_ptr1 (which contains the memory address of 0x100) and added 1 to it. Thus, the value 17 (0x11) becomes 18 (0x12).

■ **Tip** There is one more pointer-related concept that you will encounter—the hat or caret symbol (^). Visual C++ as a special pointer handler that automatically destroys the allocated memory when it is no longer in use uses this symbol. For more information about the caret symbol and objects, see https://msdn.microsoft. com/en-us/library/yk97tc08.aspx.

Now let's see what happens when you execute the last statement. Figure 5-5 shows the results in our fictional memory space.

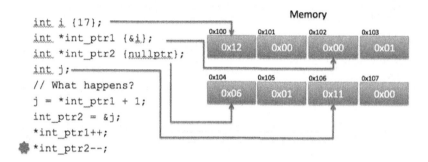

Figure 5-5. Pointer illustration: step 4

Here you see a very similar operation take place as the last statement only this time you decrement the value. You see the statement executed decrements the value that int_ptr2 points to by 1 (the -- operator). In the drawing, you see the compiler dereferences int_ptr2 (which contains the memory address of 0x106) and subtracted 1 from it. Thus, the value 18 (0x12) becomes 17 (0x11).

Wow! That was a wild ride, wasn't it? I hope that this short crash course in C++ has explained enough about the sample programs shown so far that you now know how they work. This crash course also forms the basis for understanding the other C++ examples in this book.

OK, now it's time to see some of these fundamental elements of C++ in action. Let's look at the blink an LED application you saw in Chapter 3 only this time you're going to write it for Windows 10 IoT Core!

Blink an LED, C++ Style

OK, let's write some C++ code! This project is the same concept as the project from Chapter 3 where you used Python to blink an LED on your Raspberry Pi. Rather than simply duplicate that project, you'll mix it up a bit and make this example a headed application (recall that a headed application has a user interface). The user interface presents the user with a greeting, a symbol that changes color in time with the LED, and a button to start and stop the blink timer.

Rather than build the entire application at once by presenting you a bunch of code, you will walk through this example in two phases. The first phase builds the basic user interface. The second phase adds the code for the GPIO. By using this approach, you can test the user interface on your PC, which is really convenient.

Recall that the PC does not support the GPIO libraries (there is no GPIO!) so if you built the entire application, you would have to test it on the device, which can be problematic if there are serious logic errors in your code. This way, you can ensure that the user interface is working correctly and therefore eliminate any possible issues in that code before you deploy it.

Before you get into the code for the user interface, let's see what components you will use and then set up the hardware.

Required Components

The following lists the components that you need. All of these are available in the Microsoft Internet of Things Pack for the Raspberry Pi from Adafruit. If you do not have that kit, you can find these components separately on the Adafruit web site (adafruit.com), from SparkFun (sparkfun.com), or any electronics store that carries electronic components.

- 560 ohm 5% 1/4W resistor (green, blue, brown stripes[8])

- Diffused 10mm red LED (or similar)

- Breadboard (mini, half, or full sized)

- (2) male-to-female jumper wiresYou may notice that this is the same set of components you used in Chapter 3.

Set up the Hardware

Begin by placing the breadboard next to your Raspberry Pi. Power off the Raspberry Pi, orienting it with the label facing you (GPIO pins in the upper left). Next, take one of the jumper wires and connect the female connector to pin 6 on the GPIO. The pins are numbered left-to-right starting with the lower-left pin. Thus, the left two pins are 1 and 2 with pin 1 below pin 2. Connect the other wire to pin 7 on the GPIO.

■ **Tip**   The only component that is polarized is the LED. The longer side is the positive side.

Next, plug the resistor into the breadboard with each pin on one side of the center groove. You can choose whichever area you want on the breadboard. Next, connect the LED so that the long leg is plugged into the same row as the resistor and the other pin on another row. Finally, connect the wire from pin 6 to the same row as the negative side of the LED and the wire from pin 7 to the row with the resistor. Figure 5-6 shows how all of the components are wired together. Be sure to study this drawing and double-check your connections prior to powering on your Raspberry Pi. Once you're satisfied that everything is connected correctly, you're ready to power on the Raspberry Pi and write the code.

---

[8]https://en.wikipedia.org/wiki/Electronic_color_code

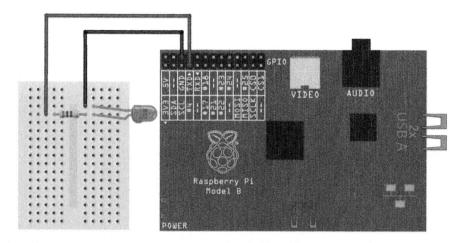

Figure 5-6. *Wiring the LED to a Raspberry Pi*

Since you are building a headed application, you'll also need a keyboard, mouse, and monitor connected to the Raspberry Pi.

OK, now that you have your hardware set up, it's time to start writing the code.

Write the Code: User Interface

Begin by opening a new project template. Choose **C++** ➤ **Windows** ➤ **Universal** in the tree and the **Blank Application** template in the list. This template creates a new solution with all of the source files and resources you need for a UWP headed application. Figure 5-7 shows the project template that you need. Use **BlinkCPPStyle** for the project name.

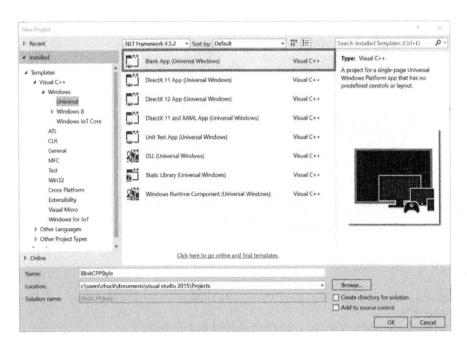

Figure 5-7. *New Project dialog: blank application*

A number of files have been created. First, add the XAML code in the MainPage.xaml file. Listing 5-2 shows the bare XAML code placed in the file by default. I've added a note that shows where add new code.

Listing 5-2. Bare XAML code (MainPage.xaml)

```
<Page
    x:Class="BlinkCPPStyle.MainPage"
    xmlns="http://schemas.microsoft.com/winfx/2006/xaml/presentation"
    xmlns:x="http://schemas.microsoft.com/winfx/2006/xaml"
    xmlns:local="using:BlinkCPPStyle"
    xmlns:d="http://schemas.microsoft.com/expression/blend/2008"
    xmlns:mc="http://schemas.openxmlformats.org/markup-compatibility/2006"
    mc:Ignorable="d">

    <Grid Background="{ThemeResource ApplicationPageBackgroundThemeBrush}">
    --> Our code goes here.
    </Grid>
</Page>
```

Recall that the XAML file is used to define a user interface in a platform independent way using an XML-like language. In this project, I demonstrate some of the more basic controls: a text box, a button, and an ellipse (circle) placed inside a special control called a *stacked panel*. The stacked panel allows you to arrange the controls in a vertical "stack," making it easier to position them. As you can see in the listing, you want to place your XAML user interface items in the <Grid></Grid> section.

In this example, you want a text box at the top, a circle (ellipse) to represent the LED that you use to turn on (change to green) and off (change to gray) to correspond with the hardware on/off code that you will add later. You also need a button to toggle the blink operation on and off. Finally, you'll add another text box to allow you to communicate with the user about the state of the GPIO code (that you'll add later).

Now let's add the code. Since the stacked panel is a container, all of the controls are placed inside it. Listing 5-3 shows the code that you want to add (shown in bold).

Listing 5-3. Adding XAML Code for the User Interface: MainPage.xaml

```
<Page
    x:Class="BlinkCPPStyle.MainPage"
    xmlns="http://schemas.microsoft.com/winfx/2006/xaml/presentation"
    xmlns:x="http://schemas.microsoft.com/winfx/2006/xaml"
    xmlns:local="using:BlinkCPPStyle"
    xmlns:d="http://schemas.microsoft.com/expression/blend/2008"
    xmlns:mc="http://schemas.openxmlformats.org/markup-compatibility/2006"
    mc:Ignorable="d">

    <Grid Background="{ThemeResource ApplicationPageBackgroundThemeBrush}">
        <StackPanel Width="400" Height="400">
            <TextBlock x:Name="title" Height="60" TextWrapping="NoWrap"
                       Text="Hello, Blinky C++ Style!" FontSize="28" Foreground="Blue"
                       Margin="10" HorizontalAlignment="Center"/>
            <Ellipse x:Name="led_indicator" Fill="LightGray" Stroke="Gray"  Width="75"
                       Height="75" Margin="10" HorizontalAlignment="Center"/>
            <Button x:Name="start_stop_button" Content="Start" Width="75" ClickMode="Press"
                       Click="start_stop_button_Click" Height="50" FontSize="24"
                       Margin="10" HorizontalAlignment="Center"/>
            <TextBlock x:Name="status" Height="60" TextWrapping="NoWrap"
                       Text="Status" FontSize="28" Foreground="Blue"
                       Margin="10" HorizontalAlignment="Center"/>
        </StackPanel>
    </Grid>
</Page>
```

Notice the button control. Here you have an event that you want to associate with the button named start_stop_button_Click, which you assigned via the Click attribute. That is, when the user clicks it, a method named start_stop_button_Click() is called.

XAML provides a great way to define a simple, easy user interface with the XML-like syntax. However, it also provides a mechanism to associate code with the controls. The code is placed in another file called a *source-behind file*, including a header and source file named MainPage.xaml.h and MainPage.xaml.cpp. Recall that you place declarations in the header file and the body of the code in the source file.

If you were typing this code in by hand, you notice a nifty feature of Visual Studio—context-sensitive help called IntelliSense that automatically completes the code you're typing and provides drop-down lists of choices. For example, when you type in the button control and then type **Click**=, a drop-down box appears, allowing you to create the event handler (a part of the code that connects to the XML). In fact, it creates the code in the MainPage.xaml.cpp (and MainPage.xaml.h) file for you. If you copy and pasted the code, you would not get this option and would have to type in the code manually. However, I will show you the code so that you can complete it yourself.

Let's see the code for the button control starting with the header file (MainPage.xaml.h). Listing 5-4 shows the code you need to add in bold. Notice that you place everything in the private section because it is only used by the BlinkCPPStyle (application) class.

Listing 5-4. Adding the Declarations: MainPage.xaml.h

```
//
// MainPage.xaml.h
// Declaration of the MainPage class.
//

#pragma once
#include "MainPage.g.h"

namespace BlinkCPPStyle
{
  /// <summary>
  /// An empty page that can be used on its own or navigated to within a Frame.
  /// </summary>

  public ref class MainPage sealed
  {
    public:
      MainPage();
    private:
      // Add references for color brushes to paint the led_indicator control
      Windows::UI::Xaml::Media::SolidColorBrush ^greenFill_ =
        ref new Windows::UI::Xaml::Media::SolidColorBrush(Windows::UI::Colors::Green);
      Windows::UI::Xaml::Media::SolidColorBrush ^grayFill_ =
      ref new Windows::UI::Xaml::Media::SolidColorBrush(Windows::UI::Colors::LightGray);

      // Add the start and stop button click event header
      void start_stop_button_Click(Platform::Object^ sender,
                                   Windows::UI::Xaml::RoutedEventArgs^ e);

      // Variables for blinking
      bool blinking{ false };
  };
}
```

OK, there are a few extra bits here that may not be very obvious why they're here. Recall that you want to paint the LED control green and gray for on and off. To do that, you need a reference to the green and gray brush resources. Thus, I create a new object (using the caret for cleanup) from the Windows user interface colors namespace. This is a common way to express brushes for painting controls (but not the only way).

You also add the header for the button click event—start_stop_button_Click() as well as a boolean member variable that you use to trigger the LED timer.

Let's see the source code file (MainPage.xaml.cpp) where you use these variables and fill in the code for the event. Again, only the code in bold is new. The project template provided the rest of the code.

Listing 5-5. Adding Code for the Event: MainPage.xaml.cpp

```
//
// MainPage.xaml.cpp
// Implementation of the MainPage class.
//
```

```cpp
#include "pch.h"
#include "MainPage.xaml.h"

using namespace BlinkCPPStyle;
using namespace Platform;
using namespace Windows::Foundation;
using namespace Windows::Foundation::Collections;
using namespace Windows::UI::Xaml;
using namespace Windows::UI::Xaml::Controls;
using namespace Windows::UI::Xaml::Controls::Primitives;
using namespace Windows::UI::Xaml::Data;
using namespace Windows::UI::Xaml::Input;
using namespace Windows::UI::Xaml::Media;
using namespace Windows::UI::Xaml::Navigation;

// The Blank Page item template is documented at http://go.microsoft.com/fwlink/?LinkId=402
352&clcid=0x409

MainPage::MainPage()
{
        InitializeComponent();
}
void BlinkCPPStyle::MainPage::start_stop_button_Click(Platform::Object^ sender,
Windows::UI::Xaml::RoutedEventArgs^ e)
{
        blinking = !blinking;
        if (blinking) {
                led_indicator->Fill = greenFill_;
                start_stop_button->Content = "Stop";
        }
        else {
                led_indicator->Fill = grayFill_;
                start_stop_button->Content = "Start";
        }
}
```

Notice that you added code that inverts the blinking variable (toggles between false and true) and depending on the value, you turn the led indicator control green (meaning the LED is on) or gray (meaning the LED is off). You also change the label of the button to correspond with the operation. That is, if the button is labeled Start, the LED indicator is off and when clicked, the label changes to Stop and the LED indicator is turned on.

That's it! You've finished the user interface. Go ahead and build the solution, correcting any errors that may appear. Once compiled, you're ready to test it.

Test and Execute: User Interface Only

That was easy, wasn't it? Better still, since this is a Universal app, you can run this code on your PC. To do so, choose debug and x86 (or x64 for 64-bit machines) from the platform box and press **Ctrl+F5**. Figure 5-8 shows an excerpt of the output (just the controls itself).

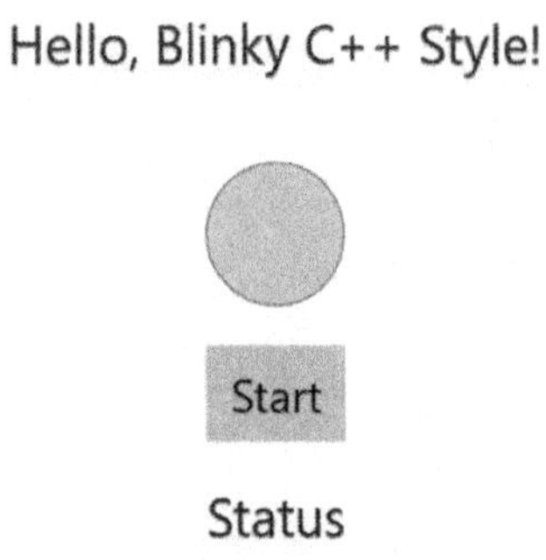

Hello, Blinky C++ Style!

Status

Figure 5-8. *The user interface: timer off*

Figure 5-8 shows what happens when you click the button. Cool, eh?

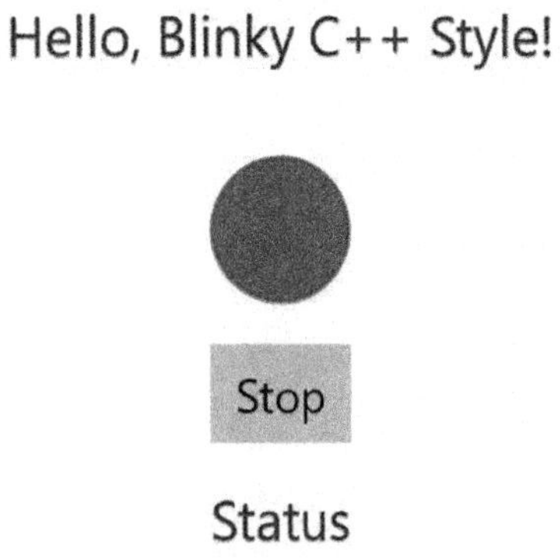

Hello, Blinky C++ Style!

Status

Figure 5-9. *The user interface: timer on*

You may be wondering where the blink part is. Well, you haven't implemented it yet. You will do that in the next phase.

Add the GPIO Code

Now, let's add the code to work with the GPIO header. For this phase, you cannot run the code on your PC because the GPIO header doesn't exist, but you can add code to check the GPIO header status—hence, the extra text box in the interface.

■ **Note** The following is a bit more complicated and requires changes to the header and source files. Thus, I will walk through the code changes one part at a time.

Let's start with adding the resource you need to access the GPIO header. Right-click the References in the project and choose **Add Reference….** When dialog opens, choose **Universal Windows ➤ Extensions** from the tree view and **Windows IoT Extensions for the UWP** from the list. This allows you to include additional namespaces for the GPIO. Figure 5-10 shows the resources dialog with the item selected.

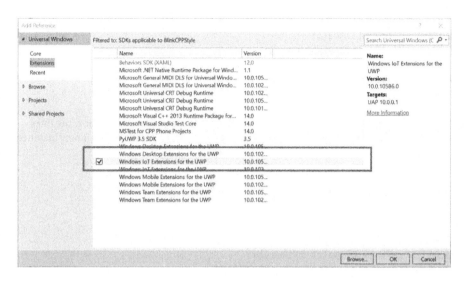

Figure 5-10. *Adding a new resource*

Notice on my system there are two entries: one for each version of the libraries I have loaded. You may only see the one entry. Click **OK** once you have the item selected.

■ **Tip**   Henceforth, for brevity, I present excerpts of the files that you will be editing.

Now, let's add the headers you need in the header file (`MainPage.xaml.h`). The following shows the code you need to add. Here you add a new method to initialize the GPIO: `InitGPIO()`, a new event named `OnTick()`, which you use with the timer object, a reference to the timer object, a variable to store the pin value, and finally, a constant set to GPIO 4 (hardware pin #7) and a pointer handler to a pin variable. Notice that the use of underscore for class member names. This is an optional but normal practice.

```cpp
// Add a constructor for the InitGPIO class
void InitGPIO();

// Add an event for the timer object
void OnTick(Platform::Object ^sender, Platform::Object ^args);

// Add references for the timer and a variable to store the GPIO pin value
Windows::UI::Xaml::DispatcherTimer ^timer_;
Windows::Devices::Gpio::GpioPinValue pinValue_ = Windows::Devices::Gpio::GpioPinValue::High;

// Variables for blinking
bool blinking{ false };
const int LED_PIN = 4;   // physical pin#7
Windows::Devices::Gpio::GpioPin^ pin_;
```

Now let's see the code for the new event (in `MainPage.xaml.cpp`). You begin by adding some namespaces, as shown follows. The new ones are in bold. Here you added namespaces for enumeration, the GPIO, and concurrency (for the timer).

```
...
using namespace Windows::UI::Xaml::Media;
using namespace Windows::UI::Xaml::Navigation;
using namespace Windows::Devices::Enumeration;    // Add this
using namespace Windows::Devices::Gpio;           // Add this
using namespace concurrency;                      // Add this
```

Next, you need to add some code to the constructor for the MainPage class as follows. I show the lines you add in bold. Notice that you have a new syntax, namely the use of ref new, which creates a new instance of the DispatcherTimer class.

```
MainPage::MainPage()
{
        InitializeComponent();
        InitGPIO();
        if (pin_ != nullptr) {
                timer_ = ref new DispatcherTimer();
                TimeSpan interval;
                interval.Duration = 500 * 1000 * 20;
                timer_->Interval = interval;
                timer_->Tick += ref new EventHandler<Object ^>(this, &MainPage::OnTick);
        }
}
```

You have added a bit of code here to set up the GPIO header calling the new method (you'll add that shortly), and code that checks to see if the pin variable is allocated (has a value), you instantiate the DispatcherTimer class, set an interval, and add the event handler named OnTick.

Next, let's add the new InitGPIO() method. You can place this after the constructor or at the end of the file as follows. You use a new syntax, -> (arrow), that dereferences a pointer to access a method or attribute.

```
void MainPage::InitGPIO()
{
        auto gpio = GpioController::GetDefault();
        if (gpio == nullptr) {
                pin_ = nullptr;
                status->Text = "No GPIO Controller!";
                return;
        }
        pin_ = gpio->OpenPin(LED_PIN);
        pin_->Write(pinValue_);
        pin_->SetDriveMode(GpioPinDriveMode::Output);
        status->Text = "You're good to go!";
}
```

OK, there's some stuff going on here. Like the constructor, you check the status of the GPIO but this time if the GPIO is null (nullptr), you set the text of the status text box with an error. This shows how easy it is to add code to affect the XAML controls. You also see code to open the GPIO pin, set a value (in this case, high or positive voltage), set the mode, and then update the status text box with a success message.

Next, you need to complete the OnTick() method code as follows.

```
void MainPage::OnTick(Object ^sender, Object ^args)
{
        if (pinValue_ == Windows::Devices::Gpio::GpioPinValue::High) {
                pinValue_ = Windows::Devices::Gpio::GpioPinValue::Low;
                pin_->Write(pinValue_);
                led_indicator->Fill = grayFill_;
        } else {
                pinValue_ = Windows::Devices::Gpio::GpioPinValue::High;
                pin_->Write(pinValue_);
                led_indicator->Fill = greenFill_;
        }
}
```

Here is where the real operation of the code happens. In this code, if the pin is set to high (on), you set it to low (off), and paint the LED control gray. Otherwise, you set the pin to high (on) and paint the LED control green.

■ **Note**   You could change this color to match the color of your LED if you wanted. Just remember to change the brush accordingly in the header file.

Finally, you need to add code to the start_stop_button_Click() method to start and stop the timer. The changes are in bold.

```
void BlinkCPPStyle::MainPage::start_stop_button_Click(Platform::Object^ sender, Windows::UI:
:Xaml::RoutedEventArgs^ e)
{
        blinking = !blinking;
        if (blinking) {
                timer_->Start();
                led_indicator->Fill = greenFill_;
                start_stop_button->Content = "Stop";
        } else {
                timer_->Stop();
                led_indicator->Fill = grayFill_;
                start_stop_button->Content = "Start";
        }
}
```

That's it! Now, let's build the solution and check for errors. You should see something like the following in the output window.

```
1>------ Build started: Project: BlinkCPPStyle, Configuration: Debug Win32 ------
1>  pch.cpp
1>  App.xaml.cpp
1>  MainPage.xaml.cpp
1>  XamlTypeInfo.Impl.g.cpp
1>  XamlTypeInfo.g.cpp
```

```
1>     Creating library c:\users\chuck\documents\visual studio 2015\Projects\BlinkCPPStyle\
Debug\BlinkCPPStyle\BlinkCPPStyle.lib and object c:\users\chuck\documents\visual studio
2015\Projects\BlinkCPPStyle\Debug\BlinkCPPStyle\BlinkCPPStyle.exp
1>  BlinkCPPStyle.vcxproj -> c:\users\chuck\documents\visual studio 2015\Projects\
BlinkCPPStyle\Debug\BlinkCPPStyle\BlinkCPPStyle.exe
1>  BlinkCPPStyle.vcxproj -> c:\users\chuck\documents\visual studio 2015\Projects\
BlinkCPPStyle\Debug\BlinkCPPStyle\BlinkCPPStyle.pdb (Partial PDB)
========== Build: 1 succeeded, 0 failed, 0 up-to-date, 0 skipped ==========
```

OK, now you're ready to deploy the application to your device. Go ahead—set up everything and power on your device.

Deploy and Execute: Completed Application

Once you're code compiles, you're ready to deploy the application to your Raspberry Pi (or other device). Recall from Chapter 4, you have to set up the debug settings to specify the IP address of your Raspberry Pi. Fortunately, unlike the console application, you only have to change two items as indicated in Figure 5-11. Remember to choose ARM for the platform. Click **Apply** and then **OK** to close the dialog.

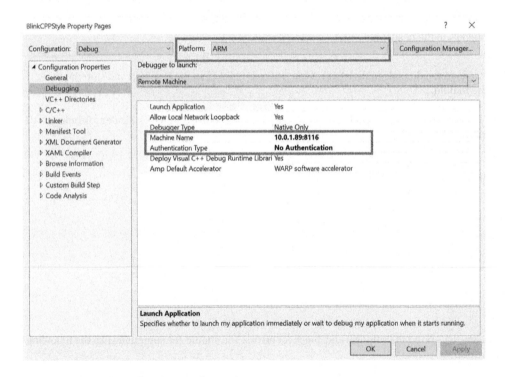

Figure 5-11. *Setting Debug Settings for Deployment*

Next, you should also name your application so that you can recognize it on the device. Double-click the Package.appxmanifest file. You see a new tabbed window open in the IDE. Click the **Packaging** tab and change the name of the application, as shown in Figure 5-12. You do this so that the application shows up in the application list on the device. Otherwise, you would get a strange name that is not readily recognizable.

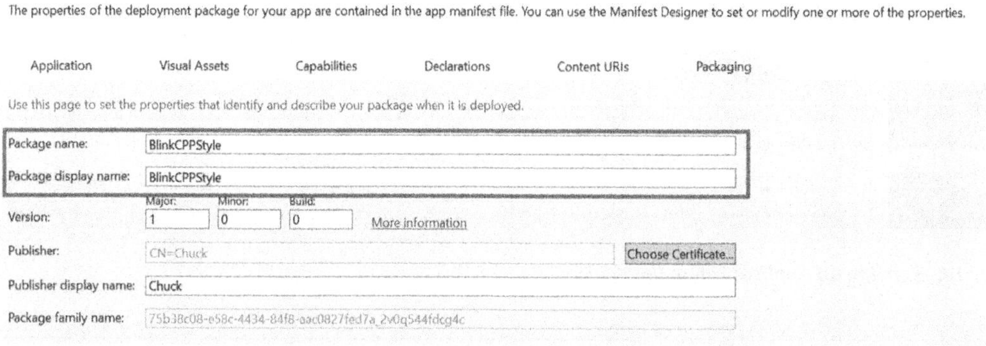

The properties of the deployment package for your app are contained in the app manifest file. You can use the Manifest Designer to set or modify one or more of the properties.

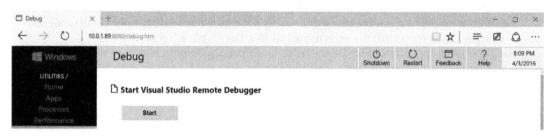

Figure 5-12. *Setting Package Name*

Once you have these set, you can power on your Raspberry Pi, and once it is booted, go to the Windows Device Portal and turn on the remote debugger, as shown in Figure 5-13.

Figure 5-13. *Turning on the Remote Debugger*

Now you can deploy the application from the Build menu. When complete, you'll get messages like the following from the Output window.

```
1>------ Deploy started: Project: BlinkCPPStyle, Configuration: Debug ARM ------
1>Creating a new clean layout...
1>Copying files: Total 2 mb to layout...
1>Checking whether required frameworks are installed...
1>Registering the application to run from layout...
1>Deployment complete (25198ms). Full package name: "BlinkCPPStyle_1.0.0.0_
arm__2v0q544fdcg4c"
========== Build: 0 succeeded, 0 failed, 1 up-to-date, 0 skipped ==========
========== Deploy: 1 succeeded, 0 failed, 0 skipped ==========
```

Notice the name that you were shown. This is the name that appears on your device. Go back to the Device Portal, click **Apps**, and then choose the application in the drop-down list and click **Run**. Figure 5-14 shows the settings.

Figure 5-14. *Starting an application on the device*

■ **Note** If the app deployed successfully but doesn't show in the drop-down list, try disconnecting and reconnecting. If that doesn't work, try rebooting your device.

You should soon see the application start on your device. Go ahead—click the start/stop button and observe the application running. The LED on the breadboard should blink in unison with the LED represented in the user interface. Play with it until the thrill dissipates.

If the LED is not blinking, double-check your wiring and ensure that you chose pin 4 in the code and pin 7 on the pin header (recall that the pin is named GPIO 4, but it is pin #7 on the header).

You can stop the application in the Device Portal by clicking the **Remove** button, as shown in the Figure 5-14. And that's it! Congratulations, you've just written your first C++ application that uses the GPIO header to power some electronics!

■ **Tip** As a challenge, you can modify this project to add a Close or an Exit button to stop the application.

Summary

If you are learning how to work with Windows 10 IoT Core and don't know how to program with C++, learning C++ can be a daunting challenge. While there are many examples on the Internet you can use, very few are documented in such a way as to provide enough information for someone new to C++ to understand or much less get started or even compile and deploy the sample!

This chapter has provided a crash course in Visual C++ that covers the basics of the things you encounter when examining most of the smaller example projects. You discovered the basic syntax and constructs of a Visual C++ application, including a walk-through of building a real C++ application that blinks an LED. Through the course of that example, you learned a little about XAML, including how to wire events to controls, and even a little about how to use the dispatcher timer.

In the next chapter, you discover another programming language called C#. You implement the same example project you saw in this chapter, so if you want to see how to do it in C#—read on!

■ ■ ■

## CHAPTER 6

Windows 10 IoT Development with C#

Now that you have a basic understanding of how to use Visual Studio 2015, you can learn more about some of the languages you may encounter when developing your IoT solutions. One of those languages is C# (pronounced "see sharp"[1])—a very robust and powerful object-oriented language that you can use to write managed Windows .NET and UWP applications. Mastering C# is not a trivial task, but it is not quite as challenging as other programming languages.

WHAT IS .NET?

In short, the .NET Framework is a huge library designed to provide a layer above the operating system for building Windows applications. The .NET Framework supports a number of languages, including C# as well as a number of platforms. As you will see when you deploy your IoT application to your device, Visual Studio includes a subset of the framework for use on Windows 10 IoT Core. Some of the classes within the namespaces that we will use derive from the .NET Framework.

If you are used to using C++ or Java, you may find C# to be familiar, but a bit more verbose. That is, it may seem like you're typing a lot more or adding more lines of code. Both are true to some extent. The libraries we will use have more verbose naming conventions (names or identifiers are longer) and we will use a few more lines of code in the process. However, as you will see, C# source code reads easier than some other languages making it easier to understand and modify in the future. It also helps when debugging your code. You don't have to guess what a library class and method may do because the name is more descriptive (in general).

Some find learning C# easier than other programming languages because, if you are familiar with C++ or especially Java, some of what you will learn is similar. Essentially, C# is an improvement on languages like C++ and Java. In fact, you may only need a little knowledge of the fundamentals of the language and how to use it in Visual Studio to become proficient in creating Windows 10 IoT Core applications.

This chapter presents a crash course on the basics of C# programming in Visual Studio including an explanation about some of the most commonly used language features. As such, this chapter provides you with the skills you need to understand the growing number of IoT project examples available on the Internet. The chapter concludes with a walk-through of a C# example project that shows you how to interact with hardware. Specifically, you will implement the LED project you saw in Chapter 3. Only this time, you'll be writing it as a C# Windows 10 IoT Core application. So let's get started!

---

[1]Not "see-hash" or worse, "see-hashtag"—both of which may show ignorance, so don't do that.

© Charles Bell 2016
C. Bell, *Windows 10 for the Internet of Things*, DOI 10.1007/978-1-4842-2108-2_6

■ **Tip** If you are not interested in using C# in your IoT solutions, or you already know the basics of C# programming, feel free to skim through this chapter. I recommend working through the example project at the end of the chapter, especially if you've not written IoT applications.

Getting Started

The C# language has been around since the first introduction of the .NET Framework (pronounced "dot net"). In fact, C# was specifically designed to be the object-oriented programming language of choice for writing .NET applications. C# was released in 2000 with the release of the .NET Framework. The latest version of C# is version 6.0 and is sometimes called Visual C# (but Microsoft seems to prefer C#).

You may be thinking that C# and .NET may restrict the types of applications you can write, but that also is not true. You can use C# to write a host of applications from Windows 10 IoT Core to desktop to web applications and beyond. As you will see, you can also write C# console applications like you did in Chapter 3 with C++.

C# was heavily influenced by Java and C++. Indeed, if you have programmed in Java or C++, C# will seem familiar to you. What sets C# apart is it is a purely object-oriented language. That is, you must write all of your programs as an object using a class. In fact, with an easy to understand class syntax,[2] which makes developing your object-oriented programming easier. This gives C# a powerful advantage over C++ and other languages that have more complex syntax where objects are largely optional.

While some may be tempted to think C# is another flavor of C++, there are some serious differences. The most important is the use of an execution manager to keep the C# applications in a protected area (called a *managed application*). This protected area ensures that the C# code has access only to portions that the execution manager permits—those defined in the code. One of the great side effects of this arrangement is automatic garbage collection—freeing of allocated memory that is no longer used (or in scope). While you can choose to make a C# application an unmanaged application, the practice is discouraged (and largely unnecessary for the vast majority of use cases).

■ **Tip** For more information about the differences between C++ and C#, see www.differencebetween. info/difference-between-cplusplus-and-csharp.

Another very important difference between C# and C++ is how they are made into an executable (compiled). C++ is compiled and linked to form an executable program with object code native to the platform, but C# is compiled in two steps: the first is a platform-independent intermediate object code called the *common intermediate language* (CIL) and the second when that is converted to native object code by the just-in-time compiler (JIT). The JIT compiler operates in the background and as the name implies prepares the code for execution when needed. In fact, you cannot tell the JIT compiler is running.

When a CIL is built, the compiler includes all of the references needed to execute the application. This may include framework files, dynamic libraries, and metadata needed by the system. This is called an *assembly*. You need not think of this other than an executable but technically it is a small repository (hence, assembly).

This mechanism allows C# applications to execute within a cordoned off area of memory (called a *managed application*) that the .NET common language runtime (CLR) can monitor and protect other applications from harm. A side effect allows C# applications to have garbage collection (by the CLR) freeing memory automatically based on scope and use. Figure 6-1 shows a pictorial example of the way C# applications are compiled and executed in phases.

---

[2]As opposed to other languages like C++ with differing notation based on how an object is instantiated.

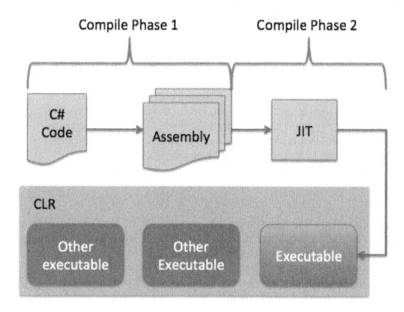

Figure 6-1. *How C# applications are compiled and executed*

Here you see the two phases of compilation. Phase one occurs when you compile the application in Visual Studio. Phase two occurs when you execute the application placing the executable in the CLR for execution. Notice that I depict other applications running in the same CLR, each protected from the other by the CLR's managed features.

■ **Note**   While C# is technically compiled in two phases, the rest of the chapter focuses on compilation as executed from Visual Studio.

Should you require more in-depth knowledge of C#, there are a number of excellent books on the topic. Here is a list of a few of my favorites.

- *Beginning C# 6.0 Programming with Visual Studio 2015* by Benjamin Perkins and Jacob Vibe Hammer (Wrox, 2015)

- *The C# Player's Guide* by R. B. Whitaker (Starbound Software, 2015)

- *Microsoft Visual C# Step by Step (Developer Reference)* by John Sharp (Microsoft Press, 2015)

Another excellent resource is Microsoft's documentation on MSDN. The following are some excellent resources for learning C#.

- Getting Started with C# (https://msdn.microsoft.com/en-us/library/a72418yk.aspx)

- C# Programming Guide (https://msdn.microsoft.com/en-us/library/67ef8sbd.aspx)

Now that you know some of the origins and unique features of C# and the .NET CLR, let's learn about the syntax and basic language features for creating applications.

C# Crash Course

Now let's learn some of the basic concepts of C# programming. Let's begin with the building blocks of the language, such as classes, methods, variables, and basic control structures, and then move into the more complex concepts of data structures and libraries.

While the material may seem to come at you in a rush (hence the crash part), this crash course on C# covers only the most fundamental knowledge of the language and how to use it in Visual Studio. It is intended to get you started writing C# Windows 10 IoT Core applications. If you find you want to write more complex applications than the examples in this book, I encourage you to acquire one or more of the resources listed earlier to learn more about the intriguing power of C# programming.

C# Fundamentals

There are a number of basic concepts about the C# programming language that you need to know in order to get started. In this section, I describe some of the fundamental concepts used in C#, including how the code is organized, how libraries are used, namespaces, and how to document your code.

C# is a case sensitive language so you must take care when typing the names of methods or classes in libraries. Fortunately, Visual Studio's IntelliSense feature recognizes mistyped case letters, which allows you to choose the correct spelling from a drop-down list as you write your code. Once you get used to this feature, it is very hard to live without it.

Namespaces

The first thing you may notice is that C# is an object-oriented language and that every program you write is written as a class. Applications are implemented with a namespace that has the same name. A namespace is a special organizational feature that allows you to group identifiers (names of variables, constants, etc.) under a group that is localized to the namespace. Using the namespace tells the compiler to look in the namespace for any identifier you've used in your code that is not found.

You can also create namespaces yourself, as you see in the upcoming example source code. Namespaces may contain any number of classes and may extend to other source files. That is, you can define a namespace so that it spans several source files.

■ **Tip**   Source files in C# have a file extension of .cs.

Classes

The next thing you may notice is a class definition. A class is more than a simple data structure. You use classes to model concepts that include data and operations on the data. A class can contain private and public definitions (called *members*) and any number of operations (called *methods*) that operate on the data and give the class meaning.

You can use classes to break your programs down into more manageable chunks. That is, you can place a class you've implemented in its own .cs file and refer to it in any of the code provided there aren't namespace issues and even then you simply use the namespace you want.

Let's look at a simple class named Vector implemented in C#. This class manages a list of double variables hiding the data from the caller while providing rudimentary operations for using the class. Listing 6-1 shows how such a class could be constructed.

Listing 6-1. Vector Class in C#

```csharp
class Vector
{
    private double[] elem;
    private int sz;

    public Vector(int s)
    {
        elem = new double[s];
        sz = s;
    }

    public ~Vector() { /* destructor body */ }

    public int size() { return sz; }

    public double this[int i]
    {
        get { return elem[i]; }
        set { elem[i] = value; }
    }
}
```

This is a basic class that is declared with the keyword `class` followed by a name. The name can be any identifier you want to use but convention is to use an initial capital letter for the name. All classes are defined with a set of curly braces that define the body or structure of the class.

By convention, you list the member variables (also called *attributes*) indented from the outer curly braces. In this example, you see two member variables that are declared as private. You make them private to hide information from the caller. That is, only member methods inside the class itself can access and modify private member variables. Note that derivatives (classes built from other classes—sometimes called a *child class*—can also access protected member variables.

Next, you see a special method that has the same name as the class. This is called the *constructor*. The constructor is a method that is called when the class is instantiated (used). You call the code that defines the attributes and methods a class and when executed, you call the resulting object an instance of the class.

In this case, the constructor takes a single integer parameter that is used to define the size of the private member variable that stores the array of double values. Notice how this code is used to dynamically define that array. More specifically, you use the `new` command to allocate memory for the array.

Following the constructor is another special method called the *destructor*. This method is called when the object is destroyed. Thus, you can place any cleanup code that you want to occur when the object is destroyed. For example, you can add code to close files or remove temporary storage. However, since C# runs on .NET with a garbage collector (a special feature that automatically frees allocated memory when no longer in scope), you do not have to worry about freeing (deleting) any memory you've allocated.

Next are two public methods, which users (or callers of the instance) can call. The first method, `size()`, looks as you would expect and in this case simply returns the value of the private member variable `sz`. The next method is a special form of method called an *operator* (or get/set) method. Notice that there are two sections or cases: get and set.

The method allows you to use the class instance (object) as if it were an array. The method is best understood by way of an example. The following shows how this operator method is used. Notice the lines in bold.

```
Vector v = new Vector(10);
for (int i=0; i < v.size(); ++i)
{
    v[i] = (i * 3);
}
Console.Write("Values of v: ");
for (int j=0; j < v.size(); ++j)
{
    Console.Write(v[j]);
    Console.Write(" ");
}
Console.WriteLine();
```

The code instantiates an instance of the Vector class requesting storage for 10 double values. Next, you iterate over the values in the instance using the operator to set the value of each of the 10 elements. This results in the set portion of the operator method executing. Next, the code iterates over the values again this time requesting (getting) the value for each. This results in the get portion of the operator method executing. Neat, eh?

■ **Note**   Some programming language books on C# use the term *function* and *method* interchangeably while other books make a distinction between the two.[3] However, Microsoft uses the term *method*.

I built this example as a single code file but had I wanted to use modularization, I would have placed the code for the Vector class in its own source (.cs) file. The name of the source file is not required to be the same as the class it contains. Indeed, a source file may contain multiple classes. Still, you would likely choose a meaningful name. To add a new source file to a Visual Studio C# solution, simply right-click the project name in Solution Explorer, and then choose **Add ➤ Add new item** and choose **C#** in the tree view and finally **C# Code file**. At the bottom of the dialog you can name the file. When ready, click the **Add** button. You can then create any classes you want (or move classes) in the file. You can also use the same namespace to keep all of your classes in the same namespace. However, if you create a new namespace, you must use the using command to use the new namespace.

As you may have surmised, classes are the building block for object-oriented programming and as you learn more about using classes, you can build complex libraries of your own.

■ **Tip**   Visual Studio provides a tool called the Class View window that you can use to explore the libraries and classes used in your application.

Curly Braces

Notice that both methods are implemented with a pair of curly braces {} that define the body of the method. Curly braces in C# are used to define a block of code or simply to express grouping of code. Curly braces are used to define the body of methods, structures, classes, and more. Notice that they are used everywhere, even in the conditional statements (see the if statements).

---

[3]A function returns a value, whereas a method does not.

■ **Tip** Some C# programmers prefer to place the starting curly brace on the same line as the line of code to which it belongs like I did in the example. However, others prefer the open curly brace placed on the next line. You should choose the style you like best.

Comments

One of the most fundamental concepts in any programming language is the ability to annotate your source code with text that not only allows you to make notes among the lines of code but also forms a way to document your source code.[4]

To add comments to your source code, use two slashes, // (no spaces between the slashes). Place them at the start of the line to create a comment for that line repeating the slashes for each subsequent line. This creates what is known as a *block comment*, as shown. Notice that I used a comment without any text to create whitespace. This helps with readability and is a common practice for block comments.

```
//
// Windows 10 for the IoT
//
// Example C# console application rewrite.
//
// Created by Dr. Charles Bell
//
```

You can also use the double slash to add a comment at the end of a line of code. That is, the compiler ignores whatever is written after the double slash to the end of the line. You see an example of this next. Notice that I used the comment symbol (double slash) to comment out a section of code. This can be really handy when testing and debugging, but generally discouraged for final code. That is, don't leave any commented out code in your deliverable (completed) source code. If it's commented out, it's not needed!

```
if (size < max_size) {
  size++;  /* increment the size */
} //else {
// return -1;
//}
```

Notice that you also see the use of /* */, which is an alternative C-like mechanism for writing comments. Anything that is included between the symbols becomes a comment and can include multiple lines of code. However, convention seems to favor the // symbol but as you can see you can mix and match however you like. I recommend choosing one or the other with consistency over variety.

Writing good comments and indeed documenting your code well is a bit of an art form; one that I encourage you to practice regularly. Since it is an art rather than a science, keep in mind that your comments should be written to teach others what your code does or is intended to do. As such, you should use comments to describe any preconditions (or constraints) of using the code, limitations of use, errors handled, and a description of how the parameters are used and what data is altered or returned from the code (should it be a method or class member).

---

[4]If you ever hear someone claim, "My code is self-documenting," be cautious when using his or her code. There is no such thing. Sure, plenty of good programmers can write code that is easy to understand (read), but all fall short of that lofty claim.

How C# Programs Are Structured

Now let's look at how C# programs are structured by examining a slightly different version of the temperature application you saw in Chapter 4. Listing 6-2 shows the code rewritten for C#.

Listing 6-2. Temperature Code Example Rewrite

```
//
// Windows 10 for the IoT
//
// Example C# console application rewrite
//
// Created by Dr. Charles Bell
//
using System;
using System.Collections.Generic;
using System.Linq;
using System.Text;
using System.Threading.Tasks;

namespace temperature_csharp
{
    class Program
    {
        static double convert_temp(char scale, double base_temp)
        {
            if ((scale == 'c') || (scale == 'C'))
            {
                return ((9.0/5.0) * base_temp) + 32.0;
            }
            else
            {
                return (5.0 / 9.0) * (base_temp - 32.0);
            }
        }

        static void Main(string[] args)
        {
            double temp_read = 0.0;
            char scale = 'c';

            Console.WriteLine("Welcome to the temperature conversion application.");
            Console.Write("Please choose a starting scale (F) or (C): ");
            scale = Console.ReadKey().KeyChar;
            Console.WriteLine();
            Console.Write("Please enter a temperature: ");
            temp_read = Convert.ToDouble(Console.ReadLine());
```

```
        if ((scale == 'c') || (scale == 'C'))
        {
            Console.WriteLine("Converting value from Celsius to Fahrenheit.");
            Console.Write(temp_read);
            Console.Write(" degrees Celsius = ");
            Console.Write(convert_temp(scale, temp_read));
            Console.WriteLine(" degrees Fahrenheit.");
        }
        else if ((scale == 'f') || (scale == 'F'))
        {
            Console.WriteLine("Converting value from Fahrenheit to Celsius.");
            Console.Write(temp_read);
            Console.Write(" degrees Fahrenheit = ");
            Console.Write(convert_temp(scale, temp_read));
            Console.WriteLine(" degrees Celsius.");
        }
        else
        {
            Console.Write("ERROR: I'm sorry, I don't understand '");
            Console.Write(scale);
            Console.WriteLine("'.");
        }
    }
  }
}
```

In the example, the only methods created are convert_temp() and MainPage() but this is because you are implementing a very simple solution. Had you wanted to model (create a separate class for) temperature, you would still have the one class named Program with the MainPage() method (which is the starting method for the application), but would have added a new class named Temperature, which would contain its own methods for working with temperature. Indeed, this is how one should think when writing C# code—model each distinct concept as a class. Here you see the sample application named temperature_csharp was implemented with a class with the name Program.

Wow, that's quite a change from the code in the last chapter! While the functionality is exactly the same, the code looks very different from the C++ version. The following describe the C# concepts I have implemented in this example.

The using Keyword

First, you notice a number of lines that begin with using. These are pre-processor directives that tell the compiler you want to "use" a classes or a class hierarchy that exists in a particular namespace. The using directive tells the compiler that you are using the namespace System.

```
using System;
```

In the other lines I have included additional namespaces with multiple names separated by a period. This is how you tell the compiler to use a specific namespace located in libraries of classes often form hierarchies that you can chain together. For example, if you wanted to use the namespace inside the Windows Foundations library named Collections, you would refer to it as follows.

```
using System.Threading.Tasks;
```

This is a very common occurrence in Windows C# applications. In fact, you will use several namespaces in our example project. The following is an example of using the Tasks namespace located in the Threading sub-class namespace of the System namespace.

The MainPage() Method

The method named MainPage(). The MainPage() method is the starting or initial execution for the C# console project. Here you see the name is preceded by the keyword static (which means its value with not change and indeed cannot change during runtime) followed by a type (in this case void). This tells the C# compiler that this method will not return any value (but you can make methods that return a value).

```
static void Main(string[] args)
```

Next, you see the name, main, followed by a list of parameters enclosed in parenthesis. For the MainPage() method, the parameters are fixed and are used to store any command line arguments provided by the user. In this case, you have the arguments stored in args, which is an array of strings.

A method in C# is used as an organizational mechanism to group functionality and make your programs easier to maintain (methods with hundreds of lines of code are very difficult to maintain), improves comprehensibility, and localize specialized operations in a single location thereby reducing duplication.

Methods therefore are used in your code to express the concepts of the functionality they provide. Notice how I used the convert_temp() method. Here I declared it as a method that returned a double and takes a character and a double as input. As you can see, the body of the method (defined inside the curly braces) uses the character as the scale in the same was as you do in main and uses the double parameter as the target (or base) temperature to convert. Since I made the parameters generic, I can use only the one variable.

■ **Tip** Method parameters and values passed must match on type and order when called.

Notice also that I placed it in the line of code that prints the value to the screen. This is a very common practice in C# (and other programming languages). That is, you use the method to perform some operation and rather than store the result in a variable, you use it directly in the statements (code) .

Variables and Types

No program would be very interesting if you did not use variables to store values for calculations. As you saw earlier, variables are declared with a type and once defined with a specific type cannot be changed. Since C# is strongly typed, the compiler ensures that anywhere you use the variable that it obeys its type. For example, that the operation on the variable is valid for the type. Thus, every variable must have a type assigned.

There are a number of simple types that the C# language supports. They are the basic building blocks for more complex types. Each type consumes a small segment of memory which defines not only how much space you have to store a value, but also the range of values possible.[5]

For example, an integer consumes 4 bytes and you can store values in the range –2,147,483,648 to 2,147,483,647. In this case, the integer variable is signed (the highest bit is used to indicate positive or negative values). An unsigned integer can store values in the range 0 to 4,294,967,295.

---

[5]For a complete list, see https://msdn.microsoft.com/en-us/library/s3f49ktz.aspx.

You can declare a variable by specifying its type first and then an identifier. The following shows a number of variables using a variety of types.

```
int num_fish = 0;              // number of fish caught
double max_length = 0.0;        // length of the longest fish in feet
char[] fisherman = new char[25]; // name of the fisherman
```

Notice also that I have demonstrated how to assign a value to the variable in the declaration. The assignment operator is the equal sign. All assignments must obey the type rules. That is, I cannot assign a floating-point number (e.g., 17.55) to an integer value. Table 6-1 shows a list of the commonly used built-in types you will use in your applications.

Table 6-1. *Commonly Used Types in C#*

Symbol	Size in bytes	Range
bool	1	false or true
char	1	–128 to 127
string	User-defined	–128 to 127 per character
sbyte	1	–128 to 127
byte	1	0-255
short	2	–32,768 to 32,767
ushort	2	0 to 65,535
int	4	–2,147,483,648 to 2,147,483,647
uint	4	0 to 4,294,967,295
long	4	–2,147,483,648 to 2,147,483,647
ulong	4	0 to 4,294,967,295
float	4	3.4E +/- 38 (7 digits)
decimal	8	(-7.9 x 1028 to 7.9 x 1028) / (100 to 28)
double	8	1.7E +/- 308 (15 digits)

It is always a good practice to initialize your variables when you declare them. It can save you from some nasty surprises if you use the variable before it is given a value (although the compiler will complain about this).

Arithmetic

You can perform a number of mathematical operations in C#, including the usual primitives but also logical operations and operations used to compare values. Rather than discuss these in detail, I provide a quick reference in Table 6-2 that shows the operation and example of how to use the operation.

Table 6-2. *Arithmetic, Logical, and Comparison Operators in C#*

Type	Operator	Description	Example
Arithmetic	+	Addition	int_var + 1
	-	Subtraction	int_var - 1
	*	Multiplication	int_var * 2
	/	Division	int_var / 3
	%	Modulus	int_var % 4
	-	Unary subtraction	-int_var
	+	Unary addition	+int_var
Logical	&	Bitwise and	var1&var2
	|	Bitwise or	var1|var2
	^	Bitwise exclusive	var1^var2
	~	Bitwise compliment	~var1
	&&	Logical and	var1&&var2
	||	Logical or	var1||var2
Comparison	==	Equal	expr1==expr2
	!=	Not equal	expr1!=expr2
	<	Less than	expr1<expr2
	>	Greater than	expr1>expr2
	<=	Less than or equal	expr1<=expr2
	>=	Greater than or equal	expr1>=expr2

Bitwise operations produce a result on the values performed on each bit. Logical operators (and, or) produce a value that is either true or false and are often used with expressions or conditions.

Now that you understand variables and types, the operations permitted on them, and expressions, let's look at how you can use them in flow control statements.

Flow Control Statements

Flow control statements change the execution of the program. They can be conditionals that use expressions that restrict execution to only those cases where the expression evaluates true (or negated), special constructs that allow you to repeat a block of code (loops), and the use of methods to switch context to perform some special operations. You've already seen how methods work so let's look at conditional and loop statements.

Conditionals

Conditional statements allow you to direct execution of your programs to sections (blocks) of code based on the evaluation of one or more expressions. There are two types of conditional statements in C#—the if statement and the switch statement.

You have seen the if statement in action in our example code. In the example, you can have one or more (optional) else phrases that you execute once the expression for the if conditions evaluate to false. You can chain if/else statements to encompass multiple conditions where the code executed depends on the evaluation of several conditions. The following shows the general structure of the if statement.

```
if (expr1) {
  // execute only if expr1 is true
} else if ((expr2) || (expr3)) {
  // execute only if expr1 is false *and* either expr2 or expr3 is true
} else {
  // execute if both sets of if conditions evaluate to false
}
```

While you can chain the statement as much as you want, use some care here because the more else/if sections you have, the harder it becomes to understand, maintain, and avoid logic errors in your expressions.

If you have a situation where you want to execute code based on one of several values for a variable or expression that returns a value (such as a method or calculation), you can use the switch statement. The following shows the structure of the switch statement.

```
switch (eval) {
  case <value1> :
    // do this if eval == value1
    break;
  case <value2> :
    // do this if eval == value2
    break;
  default :
    // do this if eval != any case value
    break;  // Not needed, but good form
}
```

The case values must match the type of the thing you are evaluating. That is, case values must be same type as eval. Notice the break statement. This is used to halt evaluation of the code once the case value is found. Otherwise, each successive case value will be compared. Finally, there is a default section for code you want to execute should eval fail to match any of the values.

■ **Tip**   Code style varies greatly in how to space/separate these statements. For example, some indent the case statements, some do not.

Loops

Loops are used to control the repetitive execution of a block of code. There are three forms of loops that have slightly different behavior. All loops use conditional statements to determine whether to repeat execution or not. That is, they repeat as long as the condition is true. The three types of loops are while, do, and for. I explain each with an example.

The while loop has its condition at the "top" or start of the block of code. Thus, while loops only execute the body if and only if the condition evaluates to true on the first pass. The following illustrates the syntax for a while loop. This form of loop is best used when you need to execute code only if some expression(s) evaluate to true. For example, iterating through a collection of things those number of elements is unknown (loop until you run out of things in the collection).

```
while (expression) {
  // do something here
 }
```

The do loop places the condition at the "bottom" of the statement which permits the body of the loop to execute at least once. The following illustrates the do loop. This form of loop is handy for cases where you want to execute code that, depending on the results of that execution, may require repetition. For example, repeatedly asking the user for input that matches one or more known values repeating the question if the answer doesn't match.

```
do {
  // do something here - always done once
} while (expression);
```

for loops are sometimes called *counting loops* because of their unique form. for loops allow you to define a counting variable, a condition to evaluate, and an operation on the counting variable. More specifically, for loops allow you to define stepping code for a precise number of operations. The following illustrates the structure of the for loop. This form of loop is best used for a number of iterations for a known number (either at run time or as a constant) and commonly used to step through memory, count, and so forth.

```
for (<init> ; <expression> ; <increment>) {
// do someting
}
```

The <init> section or counting variable declaration is executed once and only once. The <expression> is evaluated on every pass. The <increment> code is executed every pass except the last. The following is an example for loop.

```
for (int i; i < 10; i++) {
  // do something here
}
```

Now let's look at some commonly used data structures.

Basic Data Structures

What you have learned so far about C# will allow you to create applications that do simple to moderately complex operations. However, when you start needing to operate on data—either from the user or from sensors and similar sources—you need a way to organize and store data and operations on the data in memory. The following introduces three data structures in order of complexity: arrays, structures, and classes.

Arrays allocate a contiguous area of memory for multiple storage of a type. That is, you can store several integers, characters, and so forth, set aside in memory. Arrays also provide an integer index that you can use to quickly access a specific element. The following illustrates how to create an array of integers and iterate through them with a for loop. Array indexes start at 0.

```
Int[] num_array = {0,1,2,3,4,5,6,7,8,9};  // an array of 10 integers
for (int i=0; i < 10; ++i) {
  Console.Write(num_array[i]);
  Console.Write(" ");
}
Console.Writeline();
```

You can also define multiple dimensional arrays (arrays of arrays). Arrays can be used with any type or data structure.

If you have a number of data items that you want to group together, you can use a special data structure called, amazingly, struct. A struct is formed as follows.

```
struct <name> {
  // one or more declarations go here
};
```

You can add whatever declarations you want inside the struct body (defined by the curly braces). The following shows a crude example. Notice that you can use the structure in an array.

```
struct address {
  int street_num;
  string street_name;
  string city;
  string state;
  string zip;
};

address[] address_book = new address[100];
```

Arrays and structures can increase the power of your programs by allowing you to work with more complex data types.

Wow! That was a wild ride, wasn't it? I hope that this short crash course in C# has explained enough about the sample programs shown so far that you now know how they work. This crash course also forms the basis for understanding the other C# examples in this book.

OK, now it's time to see some of these fundamental elements of C# in action. Let's look at the blink an LED application you saw in Chapter 3 only this time you're going to write it for Windows 10 IoT Core!

Blink an LED, C# Style

OK, let's write some C# code! This project is the same concept as the project from Chapter 3 where you used Python to blink an LED on your Raspberry Pi. Rather than simply duplicate that project, you'll mix it up a bit and make this example a headed application (recall a headed application has a user interface). The user interface presents the user with a greeting, a symbol that changes color in time with the LED, and a button to start and stop the blink timer.

Rather than build the entire application at once by presenting you a bunch of code, we walk through this example in two phases. The first phase builds the basic user interface built. The code for the GPIO is added in the second phase. By using this approach, you can test the user interface on your PC, which is really convenient.

Recall that the PC does not support the GPIO libraries (there is no GPIO!) so if you built the entire application, you would have to test it on the device, which can be problematic if there are serious logic errors in your code. This way, you can ensure that the user interface is working correctly and therefore eliminate any possible issues in that code before you deploy it.

Before you get into the code for the user interface, let's look at the components that you will use, and then set up the hardware.

Required Components

The following lists the components that you need. All of these are available in the Microsoft Internet of Things Pack for the Raspberry Pi from Adafruit. If you do not have that kit, you can find these components separately on the Adafruit web site (www.adafruit.com), from SparkFun (www.sparkfun.com), or any electronics store that carries electronic components.

- 560 ohm 5% 1/4W resistor (green, blue, brown stripes[6])

- Diffused 10mm red LED (or similar)

- Breadboard (mini, half, or full sized)

- (2) male-to-female jumper wiresYou may notice that this is the same set of components you used in Chapter 3.

Set up the Hardware

Begin by placing the breadboard next to your Raspberry Pi and power the Raspberry Pi off orienting the Raspberry Pi with the label facing you (GPIO pins in the upper-left). Next, take one of the jumper wires and connect the female connector to pin 6 on the GPIO. The pins are numbered left-to-right starting with the lower left pin. Thus, the left two pins are 1 and 2 with pin 1 below pin 2. Connect the other wire to pin 7 on the GPIO.

■ **Tip**   The only component that is polarized is the LED. This longer side is the positive side.

Next, plug the resistor into the breadboard with each pin on one side of the center groove. You can choose whichever area you want on the breadboard. Next, connect the LED so that the long leg is plugged into the same row as the resistor and the other pin on another row. Finally, connect the wire from pin 6 to the same row as the negative side of the LED and the wire from pin 7 to the row with the resistor. Figure 6-2 shows how all of the components are wired together. Be sure to study this drawing and double-check your connections prior to powering on your Raspberry Pi. Once you're satisfied everything is connected correctly, you're ready to power on the Raspberry Pi and write the code.

---

[6]https://en.wikipedia.org/wiki/Electronic_color_code

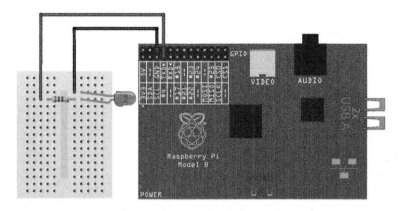

Figure 6-2. *Wiring the LED to a Raspberry Pi*

Since you are building a headed application, you'll also need a keyboard, mouse, and monitor connected to the Raspberry Pi.

OK, now that the hardware is set up, it's time to start writing the code.

Write the Code: User Interface

Begin by opening a new project template. Choose **C# ➤ Windows ➤ Universal** in the tree and the Blank App (Universal Windows) template in the list. This template creates a new solution with all of the source files and resources you need for a UWP headed application. Figure 6-3 shows the project template you need. Use the project name BlinkCSharpStyle.

Figure 6-3. *New Project dialog: blank application*

171

Notice that there are a number of files created. You first add the XAML code in the `MainPage.xaml` file. Listing 6-3 shows the bare XAML code placed in the file by default. I've added a note that shows where to add new code.

Listing 6-3. Bare XAML code (MainPage.xaml)

```xaml
<Page
    x:Class="BlinkCSharpStyle.MainPage"
    xmlns="http://schemas.microsoft.com/winfx/2006/xaml/presentation"
    xmlns:x="http://schemas.microsoft.com/winfx/2006/xaml"
    xmlns:local="using:BlinkCSharpStyle"
    xmlns:d="http://schemas.microsoft.com/expression/blend/2008"
    xmlns:mc="http://schemas.openxmlformats.org/markup-compatibility/2006"
    mc:Ignorable="d">
    <Grid Background="{ThemeResource ApplicationPageBackgroundThemeBrush}">
        --> Our code goes here.
    </Grid>
</Page>
```

Recall that the XAML file is used to define a user interface in a platform independent way using an XML-like language. In this project, I demonstrate the more basic controls: a text box, a button, and an ellipse (circle) placed inside a special controlled called a *stacked panel*. The stacked panel allows you to arrange the controls in a vertical "stack," making it easier to position them. As you can see in the listing, you want to place your XAML user interface items in the `<Grid></Grid>` section.

In this example, you want a text box at the top, a circle (ellipse) to represent the LED that you will use to turn on (change to green) and off (change to gray) to correspond with the hardware on/off code that you will add later. You also need a button to toggle the blink operation on and off. Finally, you'll add another text box to allow you to communicate with the user about the state of the GPIO code (that you'll add later).

Now let's add the code. Since the stacked panel is a container, all of the controls are placed inside it. Listing 6-4 shows the code you want to add (shown in bold).

Listing 6-4. Adding XAML Code for the User Interface: MainPage.xaml

```xaml
<Page
    x:Class="BlinkCSharpStyle.MainPage"
    xmlns="http://schemas.microsoft.com/winfx/2006/xaml/presentation"
    xmlns:x="http://schemas.microsoft.com/winfx/2006/xaml"
    xmlns:local="using:BlinkCSharpStyle"
    xmlns:d="http://schemas.microsoft.com/expression/blend/2008"
    xmlns:mc="http://schemas.openxmlformats.org/markup-compatibility/2006"
    mc:Ignorable="d">
    <Grid Background="{ThemeResource ApplicationPageBackgroundThemeBrush}">
        <StackPanel Width="400" Height="400">
            <TextBlock x:Name="title" Height="60" TextWrapping="NoWrap"
                    Text="Hello, Blinky C# Style!" FontSize="28" Foreground="Blue"
                    Margin="10" HorizontalAlignment="Center"/>
            <Ellipse x:Name="led_indicator" Fill="LightGray" Stroke="Gray"  Width="75"
                    Height="75" Margin="10" HorizontalAlignment="Center"/>
            <Button x:Name="start_stop_button" Content="Start" Width="75" ClickMode="Press"
                    Click="start_stop_button_Click" Height="50" FontSize="24"
                    Margin="10" HorizontalAlignment="Center"/>
```

```
        <TextBlock x:Name="status" Height="60" TextWrapping="NoWrap"
                Text="Status" FontSize="28" Foreground="Blue"
                Margin="10" HorizontalAlignment="Center"/>
    </StackPanel>
  </Grid>
</Page>
```

Notice the button control. Here you have an event that you want to associate with the button named start_stop_button_Click, which you assigned via the Click attribute. That is, when the user clicks it, a method named start_stop_button_Click() will be called.

XAML provides a great way to define a simple, easy user interface with the XML-like syntax. However, it also provides a mechanism to associate code with the controls. The code is placed in another file called a *source-behind file* named MainPage.xaml.cs. You will place all of the source code for the application in this file.

If you were typing this code in by hand, you will notice a nifty feature of Visual Studio—a context-sensitive help called IntelliSense that automatically completes the code you're typing and provides drop-down lists of choices. For example, when you type in the button control and type **Click**=. A drop-down box will appear, allowing you to create the event handler (a part of the code that connects to the XML). In fact, it creates the code in the MainPage.xaml.cs file for you. If you copy and pasted the code, you would not get this option and would have to type in the code manually.

Let's look at the code for the button control implemented in the class that you created in the source code file (MainPage.xaml.cs). Listing 6-5 shows the code you need to add in bold. You place everything in the class named BlinkCPPStyle (the application).

Listing 6-5. Adding the Code: MainPage.xaml.cs

```
using System;
using Windows.UI.Xaml;
using Windows.UI.Xaml.Controls;
using Windows.UI.Xaml.Media;
using Windows.UI.Xaml.Navigation;
// Add using clause for GPIO
using Windows.Devices.Gpio;

namespace BlinkCSharpStyle
{
public sealed partial class MainPage : Page
    {
        // Create brushes for painting contols
        private SolidColorBrush greenBrush = new SolidColorBrush(Windows.UI.Colors.Green);
        private SolidColorBrush grayBrush = new SolidColorBrush(Windows.UI.Colors.Gray);

        // Add a variable to control button
        private Boolean blinking = false;

        public MainPage()
        {
            this.InitializeComponent();
            // Add code to initialize the controls
            this.led_indicator.Fill = grayBrush;
        }
```

```
    private void start_stop_button_Click(object sender, RoutedEventArgs e)
    {
        this.blinking = !this.blinking;
        if (this.blinking)
        {
            this.start_stop_button.Content = "Stop";
            this.led_indicator.Fill = greenBrush;
        }
        else
        {
            this.start_stop_button.Content = "Start";
            this.led_indicator.Fill = grayBrush;
        }
    }
}
}
```

OK, there are a few extra bits here that may not be very obvious why they're here. You want to paint the LED control green and gray for on and off. To do that, you need a reference to the green and gray brush resources. Thus, I create a new object from the Windows user interface colors namespace.

You also add the code for the button click event, start_stop_button_Click(), as well as a boolean member variable that you use to trigger the LED timer. You add code that inverts the blinking variable (toggles between false and true) and depending on the value, you turn the led indicator control green (meaning the LED is on) or gray (meaning the LED is off). You also change the label of the button to correspond with the operation. That is, if the button is labeled Start, the LED indicator is off, and when clicked, the label changes to Stop and the LED indicator is turned on.

That's it! You've finished the user interface. Go ahead and build the solution correcting any errors that may appear. Once compiled, you're ready to test it. The following shows an example of what you should see in the output window.

```
Restoring NuGet packages...
To prevent NuGet from restoring packages during build, open the Visual Studio Options
dialog, click on the Package Manager node and uncheck 'Allow NuGet to download missing
packages during build.'
1>------ Rebuild All started: Project: BlinkCSharpStyle, Configuration: Debug ARM ------
1>   BlinkCSharpStyle -> C:\Users\Chuck\Documents\Visual Studio 2015\Projects\
BlinkCSharpStyle\bin\ARM\Debug\BlinkCSharpStyle.exe
========== Rebuild All: 1 succeeded, 0 failed, 0 skipped ==========
```

Test and Execute: User Interface Only

That was easy, wasn't it? Better still, since this is a Universal app, you can run this code on your PC. To do so, choose debug and x86 (or x64) from the platform box and press **Ctrl+F5**. Figure 6-4 shows an excerpt of the output (just the controls itself).

Hello, Blinky C# Style!

Status

Figure 6-4. *The user interface: timer off*

Figure 6-5 shows what happens when you click the button. Cool, eh?

Hello, Blinky C# Style!

Status

Figure 6-5. *The user interface: timer on*

You may be wondering where the blink part is. Well, you haven't implemented it yet. You will do that in the next phase.

Add the GPIO Code

Now, let's add the code to work with the GPIO header. For this phase, you cannot run the code on your PC because the GPIO header doesn't exist, but you can add code to check the GPIO header status—hence the extra text box in the interface.

■ **Note**   The following is a bit more complicated and requires additional objects. Thus, I walk through the code changes one part at a time.

Let's start with adding the resource you need to access the GPIO header. Right-click References in the project and choose **Add Reference...**. When dialog opens, choose **Universal Windows ➤ Extensions** from the tree view and **Windows IoT Extensions for the UWP** from the list. This allows you to include additional namespaces for the GPIO. Figure 6-6 shows the resources dialog with the item selected.

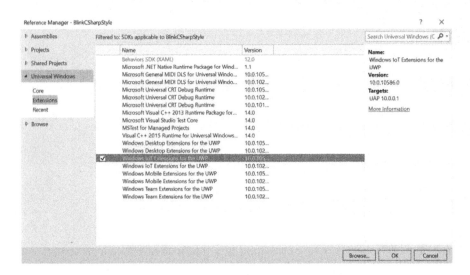

Figure 6-6. Adding a new resource

Notice on my system there are two entries—one for each version of the libraries I have loaded. You may only see the one entry. Click **OK** once you have the item selected.

■ **Tip**   Henceforth, for brevity, I present excerpts of the files that we will be editing.

Now, let's add the variables you need for the timer in the source file (`MainPage.xaml.cs`). The following shows the code you need to add. Here you add an instance of the DispatchTimer as well as several private variables for working with the pin in the GPIO library. I show the changes in context with the new lines in bold.

```
...
        private SolidColorBrush greenBrush = new SolidColorBrush(Windows.UI.Colors.Green);
        private SolidColorBrush grayBrush = new SolidColorBrush(Windows.UI.Colors.Gray);

        // Add a Dispatch Timer
        private DispatcherTimer blinkTimer;

        // Add a variable to control button
        private Boolean blinking = false;

        // Add variables for the GPIO
        private const int LED_PIN = 4;
        private GpioPin pin;
        private GpioPinValue pinValue;
...
```

The private variables store the pin value, a constant set to GPIO 4 (hardware pin #7), and a variable o store the pin variable result.

Next, you need a new method to initialize the GPIO—InitGPIO(). You add a new private method to the class and complete it with code to control the GPIO. As you can see, there is a lot going on here.

```csharp
private void InitGPIO()
{
    var gpio_ctrl = GpioController.GetDefault();

    // Check GPIO state
    if (gpio_ctrl == null)
    {
        this.pin = null;
        this.status.Text = "ERROR: No GPIO controller found!";
        return;
    }
    // Setup the GPIO pin
    this.pin = gpio_ctrl.OpenPin(LED_PIN);
    // Check to see that pin is Ok
    if (pin == null)
    {
        this.status.Text = "ERROR: Can't get pin!";
        return;
    }
    this.pin.SetDriveMode(GpioPinDriveMode.Output);
    this.pinValue = GpioPinValue.Low; // turn off
    this.pin.Write(this.pinValue);
    this.status.Text = "Good to go!";
}
```

The code first creates an instance of the default GPIO controller class. Next, you check to see if that instance is null, which indicates the GPIO header cannot be initiated and if so you change the label of the status text and return. Otherwise, you open the GPIO pin defined earlier. If that value is null, you print the message that you cannot get the pin. Otherwise, you set up the pin for output mode, and then turn off the bin and state all is well in the status label text.

Next, you add a new method to handle the event fired from the DispatchTimer. The timer fires (or call) this method on the interval you specify (see the following changes to the MainPage() method).

```csharp
private void BlinkTimer_Tick(object sender, object e)
{
    // If pin is on, turn it off
    if (this.pinValue == GpioPinValue.High)
    {
        this.led_indicator.Fill = grayBrush;
        this.pinValue = GpioPinValue.Low;
    }
    // else turn it on
```

```
    else
    {
        this.led_indicator.Fill = greenBrush;
        this.pinValue = GpioPinValue.High;
    }
    this.pin.Write(this.pinValue);
}
```

In this method, you check the value of the pin. Here is where the real operation of the code happens. In this code, if the pin is set to high (on), you set it to low (off), and paint the LED control gray. Otherwise, you set the pin to high (on) and paint the LED control green. Cool, eh?

■ **Note**   You could change this color to match the color of your LED if you wanted. Just remember to change the brush accordingly in the source file.

Next, you need to add code to the start_stop_button_Click() method to start and stop the timer. The changes are in bold.

```
private void start_stop_button_Click(object sender, RoutedEventArgs e)
{
    this.blinking = !this.blinking;
    if (this.blinking)
    {
        this.start_stop_button.Content = "Stop";
        this.led_indicator.Fill = greenBrush;
        this.blinkTimer.Start();
    }
    else
    {
        this.start_stop_button.Content = "Start";
        this.led_indicator.Fill = grayBrush;
        this.blinkTimer.Stop();
        this.pinValue = GpioPinValue.Low;
        this.pin.Write(this.pinValue);
    }
```

You see here a few things going on. First, notice that you invert the blinking variable to toggle blinking on and off. You then add a call to the blinkTimer instance of the DispatchTimer to start the timer if the user presses the button when it is labeled Start. Notice you also set the label of the button to Stop so that, when clicked again, the code turns off the timer and set the pin value to low (off). This is an extra measure to ensure that if the button is clicked when the timer is between tick events, the LED is turned off. Try removing it and you'll see.

Finally, you must add a few lines of code to the MainPage() method to get everything started when you launch the application. The following shows the code with changes in bold.

```
public MainPage()
{
    this.InitializeComponent();
    // Add code to initialize the controls
    this.led_indicator.Fill = grayBrush;
    // Add code to setup timer
    this.blinkTimer = new DispatcherTimer();
    this.blinkTimer.Interval = TimeSpan.FromMilliseconds(1000);
    this.blinkTimer.Tick += BlinkTimer_Tick;
    this.blinkTimer.Stop();
    // Initalize GPIO
    InitGPIO();
}
```

Notice you add code to create a new instance of the DispatchTimer class, set the interval for the tick event to 1 second (1000 milliseconds), add the BlinkTimer_Tick() method to the new instance (this is how you assign a method reference to an existing event handle). Next, you stop the timer and finally call the method that you wrote to initialize the GPIO.

That's it! Now, let's build the solution and check for errors. You should see something like the following in the output window.

```
Restoring NuGet packages...

To prevent NuGet from restoring packages during build, open the Visual Studio Options
dialog, click on the Package Manager node and uncheck 'Allow NuGet to download missing
packages during build.'
1>------ Build started: Project: BlinkCSharpStyle, Configuration: Debug x86 ------
1> BlinkCSharpStyle -> c:\users\chuck\documents\visual studio 2015\Projects\
BlinkCSharpStyle\bin\x86\Debug\BlinkCSharpStyle.exe
========== Build: 1 succeeded, 0 failed, 0 up-to-date, 0 skipped ==========
```

OK, now you're ready to deploy the application to your device. Go ahead and set up everything and power on your device.

Deploy and Execute: Completed Application

Once your code compiles, you're ready to deploy the application to your Raspberry Pi (or other device). Recall from Chapter 4, you have to set up the debug settings to specify the IP address of your Raspberry Pi. Fortunately, you only have to change three things: IP address, authentication, and uninstall/install check box, as indicated in Figure 6-7. Remember to choose ARM for the platform. Click **Apply** and then **OK** to close the dialog.

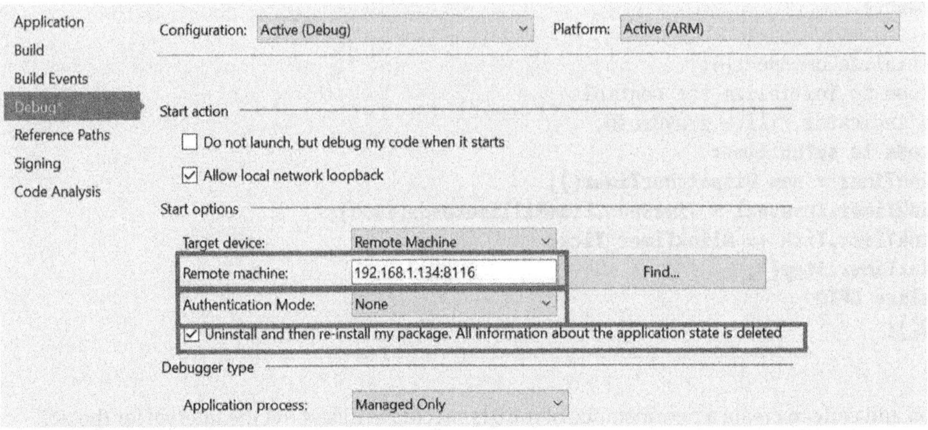

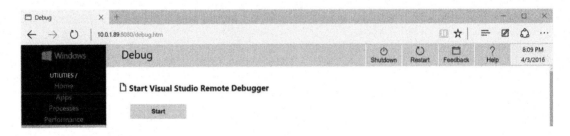

Figure 6-7. Debug Settings for Deployment

Next, you should also name your application so that you can recognize it on the device. Double-click the Package.appxmanifest file. You see a new tabbed window open in the IDE. Click the **Packaging** tab and change the name of the application, as shown in Figure 6-8. You do this so that the application shows up in the application list on the device. Otherwise, you would get a strange name that is not readily recognizable.

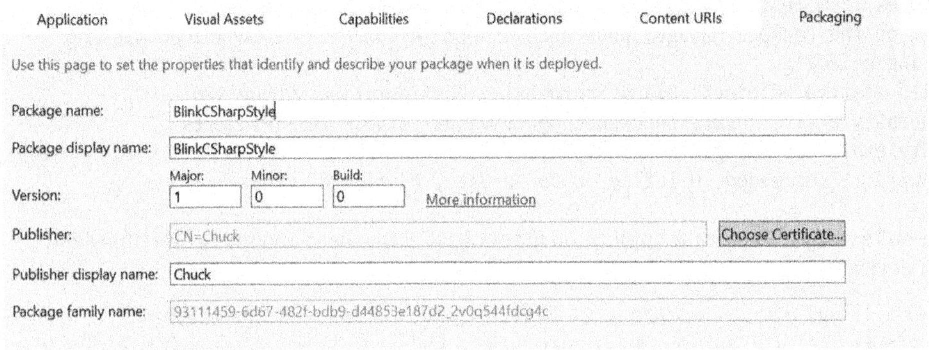

Figure 6-8. Setting Package Name

Once you have these set, you can power on your Raspberry Pi, and once it is booted, go to the Device Portal and turn on the remote debugger, as shown in Figure 6-9.

Figure 6-9. Turning on the remote debugger

Now you can deploy the application from the Build menu. When complete, you'll get messages like the following from the Output window.

```
1>------ Build started: Project: BlinkCSharpStyle, Configuration: Debug ARM ------
1>  BlinkCSharpStyle -> c:\users\chuck\documents\visual studio 2015\Projects\
BlinkCSharpStyle\bin\ARM\Debug\BlinkCSharpStyle.exe
2>------ Deploy started: Project: BlinkCSharpStyle, Configuration: Debug ARM ------
2>Creating a new clean layout...
2>Copying files: Total 16 mb to layout...
2>Checking whether required frameworks are installed...
2>Framework: Microsoft.NET.CoreRuntime.1.0/ARM, app package version 1.0.23430.0 is not
currently installed.
2>Installing missing frameworks...
2>Registering the application to run from layout...
2>Deployment complete (54476ms). Full package name: "BlinkCSharpStyle_1.0.0.0_
arm__2v0q544fdcg4c"
========== Build: 1 succeeded, 0 failed, 0 up-to-date, 0 skipped ==========
========== Deploy: 1 succeeded, 0 failed, 0 skipped ==========
```

Notice the name that you were shown. This is the name that appears on your device. Go back to the Device Portal and click **Apps**, and then choose the application in the drop-down list and click **Run**. Figure 6-10 shows the settings.

Figure 6-10. *Starting an application on the device*

■ **Note** If the app deployed successfully but doesn't show in the drop-down list, try disconnecting and reconnecting. If that doesn't work, try rebooting your device.

You should soon see the application start on your device. Go ahead and click the start/stop button and observe the application running. The LED on the breadboard should blink in unison with the LED represented in the user interface. Play with it until the thrill dissipates.

If the LED is not blinking, double-check your wiring and ensure that you chose pin 4 in the code and pin 7 on the pin header (recall that the pin is named GPIO 4 but it is pin #7 on the header).

You can stop the application in the Device Portal by clicking the **Remove** button, as shown in Figure 6-10. And that's it! Congratulations, you've just written your first C# application that uses the GPIO header to power some electronics!

■ **Tip** As a challenge, you can modify this project to add a Close or an Exit button to stop the application.

Summary

If you are learning how to work with Windows 10 IoT Core and don't know how to program with C#, learning C# can be a challenge. While there are many examples on the Internet you can use, very few are documented in such a way as to provide enough information for someone new to C# to understand or much less get started or even compile and deploy the sample!

This chapter has provided a crash course in C# that covers the basics of the things that you encounter when examining most of the smaller example projects. You discovered the basic syntax and constructs of a Visual C# application, including a walk-through of building a real C# application that blinks an LED. Through the course of that example, you learned a little about XAML, including how to wire events to controls, and even a little about how to use the dispatcher timer.

In the next chapter, you'll discover another programming language named Python. You implement the same example project you saw in this chapter but without the user interface. If you want to see how to work with the GPIO in Python, read on!

■ ■ ■

## CHAPTER 7

Windows 10 IoT Development with Python

Now that you have a basic understanding of how to use Visual Studio 2015, you can learn more about some of the languages you may encounter when developing your IoT solutions. One of those languages is Python—a very robust and powerful language that you can use to write very powerful applications. Mastering is very easy and some may suggest it doesn't require any formal training to use. This is largely true and thus you should be able to write applications for Windows 10 IoT Core with only a little bit of knowledge about Python.

Thus, this chapter presents a crash course on the basics of Python programming in Visual Studio, including an explanation about some of the most commonly used language features. As such, this chapter provides you with the skills you need to understand the growing number of IoT project examples available on the Internet. The chapter concludes with a walk-through of an example project that shows you how to interact with hardware. Specifically, you will implement the LED project you saw in Chapter 3. Only this time, you'll be writing it as a Windows 10 IoT Core application in Python. So let's get started!

■ **Tip** If you are not interested in using Python in your IoT solutions, or you already know the basics of Python programming, feel free to skim through this chapter. I recommend working through the example project at the end of the chapter, especially if you've not written IoT applications.

I briefly discussed Python in Chapter 3, but you will see much more of the language in the following sections. That is, I explain the code that you will see in the sample application at the end of the chapter.

Getting Started

Python is a high-level, interpreted, object-oriented scripting language. One of the biggest tenants of Python is to have a clear, easy to understand syntax that reads as close to English as possible. That is, you should be able to read a Python script and understand it even if you haven't learned Python. Python also has less punctuation (special symbols) and fewer syntactical machinations than other languages.

■ **Tip** Although you can terminate a statement with the semicolon (;), it is not needed and considered bad form.

Here are a few of the key features of Python.

- An interpreter processes python at runtime. No compiler is used.

- Python supports object-oriented programming constructs by way of a class.

- Python is a great language for the beginner-level programmers and supports the development of a wide range of applications.

- Python is a scripting language but can be used for a wide range of applications.

- Python is very popular and used throughout the world giving it a huge support base.

- Python has few keywords, simple structure, and a clearly defined syntax. This allows the student to pick up the language quickly.

- Python code is easy to read and understand.

Python was developed by Guido van Rossum from the late 1980s to early 1990s at the National Research Institute for Mathematics and Computer Science in the Netherlands and maintained by a core development team at the institute. It was derived from and influenced by many languages, including Modula-3, C, C++, and even Unix shell scripting languages.

A fascinating fact about Python is it was named after the BBC show *Monty Python's Flying Circus* and has nothing to do with the reptile by the same name.[1] Quoting Monty Python in source code documentation (and even a humorous diversion for error messages) is very common and while some professional developers may cringe at the insinuation, it's considered by Pythonistas[2] as showing your Python street cred.

Python is available for download (www.python.org/downloads) for just about every platform that you may encounter or use—even Windows! However, the Python additions to Visual Studio and especially Windows 10 IoT are limited. In fact, you can only create background (headless) applications in Python for Windows 10 IoT Core.[3] While this may seem very limited, there is a very good reason why I've included Python in this book.

Why Do I Need to Learn Python?

Recall that I stated that the volume of Python examples for the Raspberry Pi and other low-cost computing boards is vast. So much so that you can find Python code snippets to do just about anything you want when learning to work with the GPIO.

It would be very sad if Windows 10 IoT enthusiasts could not take advantage of the wealth of information available for programming the Raspberry Pi with Python. Fortunately, this is not the case! Indeed, many of the Python examples can be adapted for use on Windows 10 IoT Core. You may need to change the modules included (I'll explain this later), and you may be limited to working with GPIO, but by and large you can reuse that Python code in your own solutions! Thus, learning Python is a powerful resource for developing applications for Windows 10 IoT Core.

Furthermore, you can often develop your Python scripts using the interpreter on the Raspberry Pi prior to building your Windows 10 solution. This is why I covered the topic in Chapter 3. Using the interpreter with a basic script on the Raspberry Pi is faster and less error-prone than developing it in Visual Studio and deploying it. This is because since Python is interpreted, the Visual Studio tools cannot catch all of the problems or errors in your code before you deploy and run it. Using the interpreter (either on Windows or Raspberry Pi if you're using GPIO) allows you to correct any syntax errors before wasting effort in remote debugging (which can fail for simple syntax errors).

---

[1]Monty Python refers to a group of comedians and not a single individual. However, their comedy is undeniably brilliant. https://en.wikipedia.org/wiki/Monty_Python

[2]Pythonistas are expert Python developers and advocates for all things Python.

[3]Currently. Hopefully, more project types will be available in the future.

■ **Tip** You can use the Windows Python interpreter to debug basic, non-GPIO code outside of Visual Studio.

Installing Python on Windows 10

How can you use it to write Windows 10 IoT applications? Fortunately, a Python interpreter is available for Windows 10 and indeed for the Windows 10 IoT Core. In fact, in order to use Python for your IoT projects, you need to download and install the following tools and components (if you haven't already).

- Python 3.5 for Windows: http://www.python.org/downloads
- Python Tools for Visual Studio (PTVS): https://github.com/microsoft/ptvs/releases
- Python UWP SDK (pyuwpsdk.vsix): https://github.com/ms-iot/python/releases
- Python Windows Devices (pywindevices.zip) modules: https://github.com/ms-iot/python/releases

First and foremost, you must install Python 3.5 on your PC. If you have an older version, such as 2.6, 2.7, or even 3.4, you should uninstall them (if you don't need them). This is because the PTVS requires Python 3.5 and ignore the older versions. If you attempt to work with Python solutions with an older version of Python, you may get some strange build or deployment errors or your solution will fail to run on the IoT device. Follow the instructions on the link to download and install Python 3.5.

Next, you need the Python Tools for Visual Studio. This is a third-party component that you can download and install but must do so in order to see the Python projects in the new project dialog tree. Follow the instructions on the web site to download and install PTVS. Note that this installation could take a while depending on your Internet connection and processor speed.

Next, you need to install the SDK for Python UWP applications. This is a Visual Studio plugin that you can install by double-clicking the file or by installing it via Visual Studio 2015.

Finally, you need the Python modules for the GPIO and other libraries. These are found in the pywindevices.zip file. Simply download this file and extract the files in a common place that you can find when building your solutions. For example, I placed mine in a folder (pywindevices) in the root of the Visual Studio project directory.

OK, once these prerequisites are complete, you're ready to start wiring Python Windows 10 IoT solutions!

■ **Note** You may encounter blogs that claim Python does not work on Windows 10 or that the projects cannot load or run. This is largely due to improper setup of the PC or project. If you follow the example in this chapter, your projects will work correctly.

Should you require more in-depth knowledge of Python, there are a number of excellent books on the topic. The following lists a few of my favorites. A great resource is the documentation on the Python site (www.python.org/doc/).

- *Pro Python,* 2nd Edition, by J. Burton Browning and Marty Alchin (Apress, 2014)
- *Learning Python,* 5th Edition, by Mark Lutz (O'Reilly Media, 2013)
- *Automate the Boring Stuff with Python: Practical Programming for Total Beginners* by Al Sweigart (No Starch Press, 2015)

Python Crash Course

Now let's learn some of the basic concepts of Python programming. Let's begin with the building blocks of the language, such as variables, modules, and basic statements, and then move into the more complex concepts of flow control and data structures.

While the material may seem to come at you in a rush (hence the crash part), this crash course on Python covers only the most fundamental knowledge of the language and how to use it in Visual Studio. It is intended to get you started writing Python Windows 10 IoT Core applications. If you find that you want to write more complex applications than the examples in this book, I encourage you to acquire one or more of the resources listed earlier to learn more about Python programming.

The following sections present many of the basic features of Python programming that you need to know in order to understand example projects for Windows 10 IoT Core and vital to successfully implementing the Python projects in this book.

The Basics

Python is a very easy language to learn with very few constructs that are even mildly difficult to learn. Rather than toss out a sample application, let's approach learning the basics of Python in a Python-like way: one step at a time.

The first thing you should learn is that Python does not use a code block demarcated with symbols like other languages. More specifically, code that is local to a construct, such as a function or conditional or loop is designated using indentation. Thus, the lines below that are indented (by spaces or, gag, tabs[4]) so that the starting characters align for the code body of the construct. The following shows this concept in action.

```
if (expr1):
    print("inside expr1")
    print("still inside expr1")
else:
    print("inside else")
    print("still inside else")

print("in outer level")
```

Here you see a conditional, or if, statement. Notice that the function call print() (a common way to display output to the console) is indented. This signals the interpreter that the lines belong to the construct above it. For example, the two print statements that mention expr1 form the code block for the if condition (expression evaluates to true). Similarly, the next two print statements form the code block for the else condition. Finally, the non-indented lines are not part of the conditional and thus are executed after either the if or else depending on the expression evaluation.

As you can see, indentation is a key concept to learn when writing Python. Even though it is very simple, making mistakes in indentation can result in code executing that you did not expect or worse errors from the interpreter.

■ **Note**   I use *program* and application interchangeably with *script* when discussing Python. While technically, Python code is a script, you often use it in contexts where *program* or *application* is more appropriate.

---

[4]In my experience and travels through geekdom, few programmers prefer tabs over spaces. Indeed, many coding guidelines prohibit the use of tabs because they can interfere with certain offline code analysis tools. Plus, they can scramble the format of the code if your editor is set differently than others or if tabs are used inconsistently.

There is one special symbol that you will encounter frequently. Notice the use of the colon (:) in the preceding code. This symbol is used to terminate a construct and signals the interpreter that the declaration is complete and the body of the code block follows.

Now let's look at how you can use variables in your programs (scripts).

Variables

Python does not have a formal type specification mechanism like other languages. However, you can still define variables to store anything you want. In fact, Python permits you to create and use variables based on context. However, you can use initialization to "set" the data type for the variable. The following show several examples.

```
# Numbers
float_value = 9.75
integer_value = 5

# Strings
my_string = "He says, he's already got one."

print("Floating number: {0}".format(float_value))
print("Integer number: {0}".format(integer_value))
print(my_string)
```

For situations where you need to convert types or want to be sure values are typed a certain way, there are many functions for converting data. Table 7-1 shows a few of the more commonly used type conversion functions. I discuss some of the data structures in a later section.

Table 7-1. *Type Conversion in Python*

Function	Description
int(x [,base])	Converts x to an integer. Base is optional (e.g., 16 for hex).
long(x [,base])	Converts x to a long integer.
float(x)	Converts x to a floating-point.
str(x)	Converts object x to a string.
tuple(t)	Converts t to a tuple.
list(l)	Converts l to a list.
set(s)	Converts s to a set.
dict(d)	Creates a dictionary.
chr(x)	Converts an integer to a character.
hex(x)	Converts an integer to a hexadecimal string.
oct(x)	Converts an integer to an octal string.

However, you should use these conversion functions with care to avoid data loss or rounding. For example, converting a float to an integer can result in truncation.

Including Modules

Python applications can be built from reusable libraries that are provided by the Python environment. They can also be built from custom modules or libraries that you create yourself or download from a third party. When you want to use a library (function, class, etc.) that is part of a module, you use the import keyword and list the name of the module. The following shows some examples.

```
import os
import sys
import mysql.utilities.common.server
import _wingpio as gpio
```

The first two lines demonstrate how to import a base or common module provided by Python. In this case, you are using or importing modules for the os and sys modules (operating system and Python system functions).

Next, you see a special dotted notation in use. The dotted notation is realized by using folders. In this case, you're using a module from MySQL Utilities (a set of MySQL scripts and library written in Python; see http://dev.mysql.com/doc/mysql-utilities/1.6/en/). Specifically, you're using the module located in the utilities/common folder named server. Indeed, if you were to locate that module, you would find a file named server.py. The starting point of the module dotted notation reference is any path in the PYTHONPATH environment variable.

Finally, the last example shows how you can use the as keyword to rename the module (give it a different name or namespace label) to simplify your code by allowing you to use shorter names or names that make more sense in context.

■ **Tip**   It is customary (but not required) to list your imports in alphabetical order, with built-in modules first, and then third-party modules, and finally, your own modules.

Comments

One of the most fundamental concepts in any programming language is the ability to annotate your source code with text that not only allows you to make notes among the lines of code, but also forms a way to document your source code.

To add comments to your source code, use the pound sign (#).[5] Place it at the start of the line to create a comment for that line; repeat the # symbol for each subsequent line. This creates what is known as a *block comment*, as shown next. Notice that I used a comment without any text to create white space. This helps with readability and is common practice with block comments.

```
#
# Windows 10 for the IoT
#
# Example Python application.
#
# Created by Dr. Charles Bell
#
```

---

[5]Sometimes called "hash" but I personally think that term is over used, especially by those that want to sound intelligent when speaking about social media.

You can also place comments on the same line as the source code. The compiler will ignore anything from the pound sign to the end of the line. For example, the following shows a common method for documenting variables.

```
zip = 35012            # Zip or postal code
address1= "123 Main St."  # Store the street address
```

Arithmetic

You can perform a number of mathematical operations in Python, including the usual primitives but also logical operations and operations used to compare values. Rather than discuss these in detail, I provide a quick reference in Table 7-2 that shows the operation and example of how to use the operation.

Table 7-2. *Arithmetic, Logical, and Comparison Operators in Python*

Type	Operator	Description	Example
Arithmetic	+	Addition	int_var + 1
	-	Subtraction	int_var - 1
	*	Multiplication	int_var * 2
	/	Division	int_var / 3
	%	Modulus	int_var % 4
	-	Unary subtraction	-int_var
	+	Unary addition	+int_var
Logical	&	Bitwise and	var1&var2
	|	Bitwise or	var1|var2
	^	Bitwise exclusive	var1^var2
	~	Bitwise compliment	~var1
	&&	Logical and	var1&&var2
	||	Logical or	var1||var2
Comparison	==	Equal	expr1==expr2
	!=	Not equal	expr1!=expr2
	<	Less than	expr1<expr2
	>	Greater than	expr1>expr2
	<=	Less than or equal	expr1<=expr2
	>=	Greater than or equal	expr1>=expr2

Bitwise operations produce a result on the values performed on each bit. Logical operators (and, or) produce a value that is either true or false and are often used with expressions or conditions.

Now that you understand variables and types, the operations permitted on them, and expressions, let's look at how to use them in flow control statements.

Flow Control Statements

Flow control statements change the execution of the program. They can be conditionals that use expressions that restrict execution to only those cases where the expression evaluates true (or negated), special constructs that allow you to repeat a block of code (loops), and the use of functions to switch context to perform some special operations. You've already seen how functions work so let's look at conditional and loop statements.

Conditionals

Conditional statements allow us to direct execution of your programs to sections (blocks) of code based on the evaluation of one or more expressions. There are two types of conditional statements in Python—the if statement and the switch statement.

You have seen the if statement in action in our example code. Notice in the example, you can have one or more (optional) else phrases that you execute once the expression for the if conditions evaluate to false. You can chain if / else statements to encompass multiple conditions, where the code executed depends on the evaluation of several conditions. The following shows the general structure of the if statement.

```
if (expr1):
    # execute only if expr1 is true
elif ((expr2) || (expr3)):
    # execute only if expr1 is false *and* either expr2 or expr3 is true
else:
    # execute if both sets of if conditions evaluate to false
```

While you can chain the statement as much as you want, use some care here because the more elif sections you have, the harder it becomes to understand, maintain, and avoid logic errors in your expressions.

Loops

Loops are used to control the repetitive execution of a block of code. There are three forms of loops that have slightly different behavior. All loops use conditional statements to determine whether to repeat execution or not. That is, they repeat as long as the condition is true. The three types of loops are while, do, and for. I explain each with an example.

The while loop has its condition at the "top" or start of the block of code. Thus, while loops only execute the body if and only if the condition evaluates to true on the first pass. The following illustrates the syntax for a while loop. This form of loop is best used when you need to execute code only if some expression(s) evaluate to true. For example, iterating through a collection of things those number of elements is unknown (loop until you run out of things in the collection).

```
while (expression):
    # do something here
```

For loops are sometimes called counting loops because of their unique form. For loops allow you to define a counting variable and a range or list to iterate over. The following illustrates the structure of the for loop. This form of loop is best used for performing an operation in a collection. In this case, Python automatically places each item in the collection in the variable for each pass of the loop until no more items are available.

```
for variable_name in list:
  # do something here
```

You can also do range loops or counting loops. This uses a special function called `range()` that takes up to three parameters, `range([start], stop[, step])`, where start is the starting number (an integer), stop is the last number in the series, and step is the increment. So you can count by 1, 2, 3, and so forth through a range of numbers. The following shows a simple example.

```
for i in range(2,9):
  # do something here
```

There are other uses for `range()` that you may encounter. See the documentation on this function and other built-in functions at `https://docs.python.org/3/library/functions.html` for more information.

Python also provides a mechanism for controlling the flow of the loop (e.g., duration or termination) using a few special keywords as follows.

- *break*: Exit the loop body immediately

- *continue*: Skip to next iteration of the loop

- *else*: Execute code when loop ends

There are some uses for these keywords, particularly break, but it is not the preferred method of terminating and controlling loops. That is, professionals believe the conditional expression or error handling code should behave well enough to not need these options.

Functions

Python allows you to use modularization in your code. While it supports object-oriented programming by way of classes (a more advanced feature that you are unlikely to encounter for most Python GPIO examples), on a more fundamental level you can break your code into smaller chunks using functions.

Functions use a special keyword construct (rare in Python) to define a function. You simply use `def` followed by a name and a parameter list in parenthesis. The colon is used to terminate the declaration. The following shows an example.

```
def print_dictionary(the_dictionary):
    for key, value in the_dictionary.items():
      print("'{0}': {1}".format(key, value))
```

You may be wondering what this strange code does. Note that the loop is assigning two values from the result of the `items()` function. This is a special function available from the dictionary object. (Yes, dictionaries are objects! So are tuples, lists, and many other data structures.) The `items()` function returns the key, value pairs; hence, the names of the variables.

The next line prints out the values. The use of formatting strings where the curly braces define the parameter number starting at 0 is common for Python 3.x applications. See the Python documentation for more information about formatting strings.

Here the body of the function is indented. All statements indented under this function declaration belong to the function and are executed when the function is called. You can call functions by name providing any parameters as follows.

```
print_dictionary(my_dictionary)
```

This example, when executed, generates the following output as shown in Figure 7-1. I generated this by writing the function in the Python interpreter on my PC and executing it. To run the interpreter, search for Python and select the Python command window. Thus, this demonstrates how to use the Python interpreter to execute Python code on the fly (as you type each line and press Enter).

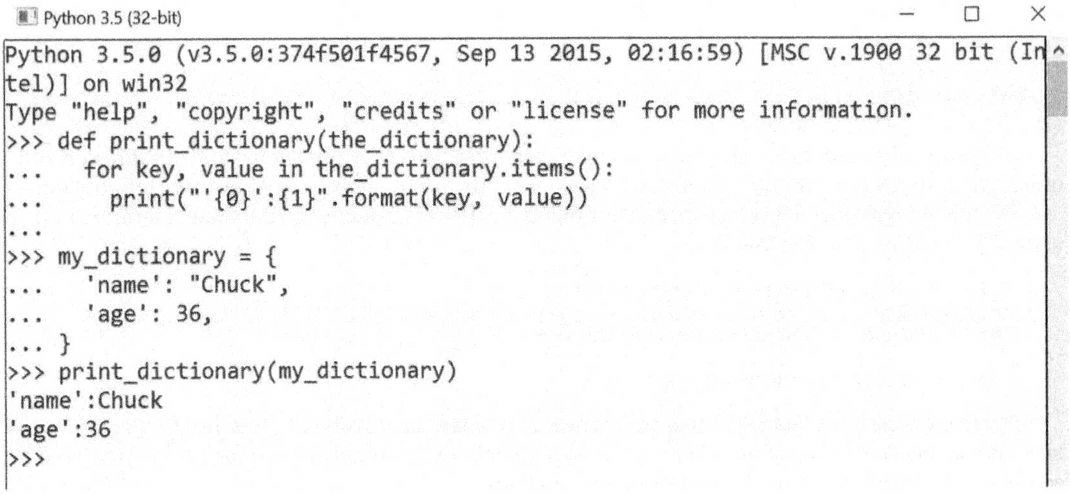

```
Python 3.5 (32-bit)                                                  —    □    ✕
Python 3.5.0 (v3.5.0:374f501f4567, Sep 13 2015, 02:16:59) [MSC v.1900 32 bit (In ^
tel)] on win32
Type "help", "copyright", "credits" or "license" for more information.
>>> def print_dictionary(the_dictionary):
...     for key, value in the_dictionary.items():
...         print("'{0}':{1}".format(key, value))
...
>>> my_dictionary = {
...     'name': "Chuck",
...     'age': 36,
... }
>>> print_dictionary(my_dictionary)
'name':Chuck
'age':36
>>>
```

Figure 7-1. *Using the Python interpreter*

■ **Tip**   Functions (methods) provided by objects (classes) can be called using the syntax `object_name.method_name()`.

Now let's look at some commonly used data structures, including this strange thing called a dictionary.

Basic Data Structures

What you have learned so far about Python is enough to write the most basic programs and indeed more than enough to tackle the example project later in this chapter. However, when you start needing to operate on data (either from the user or from sensors and similar sources), you need a way to organize and store data and operations on the data in memory. The following introduces three data structures in order of complexity: lists, tuples, and so forth. Next, I demonstrate the three that you encounter most frequently.

Lists

Lists are a way to organize data in Python. It is a free-form way to build a collection. That is, the items (or elements) need not be the same data type. Lists also allow you to do some interesting operations, such as adding things at the end, beginning, or at a special index. The following demonstrates how to create a list.

```
# List
my_list = ["abacab", 575, "rex, the wonder dog", 24, 5, 6]
my_list.append("end")
my_list.insert(0,"begin")
for item in my_list:
  print("{0}".format(item))
```

Here you see I created the list using square brackets ([ ]). The items in the list definition are separated by commas. Note that you can create an empty list simply be setting a variable equal to [ ]. Since lists, like other data structures, are objects, there are a number of operations available for lists, such as the following.

- *append(x)*: Add x to the end of the list

- *extend(l)*: Add all items to the end of the list

- *insert(pos,item)*: Insert item at a position pos

- *remove(value)*: Remove the first item that matches (==) the value

- *pop([i])*: Remove the item at position i or end of list

- *index(value)*: Return index of first item that matches

- *count(value)*: Count occurrences of value

- *sort()*: Sort the list (ascending)

- *reverse()*: Reverse sort the list

Lists are like arrays in other languages and very useful for building dynamic collections of data.

Tuples

Tuples on the other hand, are a more restrictive type of collection. That is, they are built from a specific set of data and do not allow manipulation like a list. The following shows an example of a tuple and how to use it.

```
# Tuple
my_tuple = (0,1,2,3,4,5,6,7,8,"nine")
for item in my_tuple:
  print("{0}".format(item))
if 7 in my_tuple:
  print("7 is in the list")
```

Here you see I created the tuple using parenthesis ( ). The items in the tuple definition are separated by commas. Note that you can create an empty tuple simply be setting a variable equal to ( ). Since tuples, like other data structures, are objects, there are a number of operations available, such as the following, including operations for sequences such as inclusion, location, and so forth.

- *x in t*: Determine if t contains x

- *x not in t*: Determine if t does not contain x

- *s + t*: Concatenate tuples

- *s[i]*: Get element i

- *len(t)*: Length of *t* (number of elements)

- *min(t)*: Minimal (smallest value)

- *max(t)*: Maximal (largest value)

If you want even more structure with storing data in memory, you can use a special construct (object) called a dictionary.

Dictionaries

A dictionary is a data structure that allows you to store key/value pairs where the data is assessed via the keys. Dictionaries are a very structured way of working with data and they are the most logical form you want to use when collecting complex data. The following shows an example of a dictionary.

```
# Dictionary
my_dictionary = {
  'first_name': vChuck",
  'last_name': "Bell",
  'age': 36,
  'my_ip': (192,168,1,225),
  42: "What is the meaning of life?",
}
# Access the keys:
print(my_dictionary.keys())
# Access the items (key, value) pairs
print(my_dictionary.items())
# Access the values
print(my_dictionary.values())
# Create a list of dictionaries
my_addresses = [my_dictionary]
```

There is a lot going on here! You see a basic dictionary declaration that uses curly braces to create a dictionary. Inside that, you can create as many key/value pairs you want separated by commas. Keys are defined using strings (I use single quotes by convention but double quotes will work) or integers, values can be any data type you want. For the my_ip attribute, you are also storing a tuple.

Following the dictionary, you see a number of operations performed on the dictionary from printing the keys, printing all of the values, and printing only the values. The following shows the output of executing this code snippet from the Python interpreter.

```
[42, 'first_name', 'last_name', 'age', 'my_ip']
[(42, 'what is the meaning of life?'), ('first_name', 'Chuck'), ('last_name', 'Bell'),
('age', 36), ('my_ip', (192, 168, 1, 225))]
['what is the meaning of life?', 'Chuck', 'Bell', 36, (192, 168, 1, 225)]
'42': what is the meaning of life?
'first_name': Chuck
'last_name': Bell
'age': 36
'my_ip': (192, 168, 1, 225)
```

As you have seen in this example, there are a number of operations (methods) available for dictionaries, including the following. Together this list of operations makes dictionaries a very powerful programming tool.

- *len(d)*: Number of items in *d*

- *d[k]*: Item of *d* with key *k*

- *d[k] = x*: Assign key *k* with value *x*

- *del d[k]*: Delete item with key *k*

- *k in d*: Determine if *d* has an item with key *k*

- *d.items()*: Return a list of the (key, value) pairs in *d*

- *d.keys()*: Return a list of the keys in *d*

- *d.values()*: Return a list of the values in *d*

Best of all, objects can be placed inside other objects. For example, you can create a list of dictionaries like I did, a dictionary that contains lists and tuples, and any combination you need. Thus, using lists, tuples, and dictionaries are a powerful way to manage data for your program.

Wow! That was a wild ride, wasn't it? I hope that this short crash course in Python has explained enough about the sample programs shown so far that you now know how they work. This crash course also forms the basis for understanding the other Python examples in this book.

OK, now it's time to see some of these fundamental elements of Python in action. Let's look at Let's look at the blink an LED application you saw in Chapter 3 only this time you're going to write it for Windows 10 IoT Core!

Blink an LED, Python Style

OK, let's write some Python code! This project is the same concept as the project from Chapter 3 where you used Python to blink an LED on the Raspberry Pi. Since we are limited to creating only a background application in Windows 10 IoT Core (recall that a headless application has no user interface), the code will be less and much simplified from the C++ and C# examples since there is no user interface code. In fact, you aren't producing any output other than the blinking LED.

Also, the mechanisms you will use to deploy, start, and stop the application are different from headed applications. I explain how to do this in detail in the following sections. Before we get into the code, let's look at the components that you will be using and set up the hardware.

Required Components

The following lists the components that you need. All of these are available in the Microsoft Internet of Things Pack for the Raspberry Pi from Adafruit. If you do not have that kit, you can find these components separately on the Adafruit web site (`www.adafruit.com`), from SparkFun (`www.sparkfun.com`), or any electronics store that carries electronic components.

- 560 ohm 5% 1/4W resistor (green, blue, brown stripes[6])

- Diffused 10mm red LED (or similar)

- Breadboard (mini, half, or full sized)

- (2) male-to-female jumper wires

You may notice that this is the same set of components that you used in Chapter 3.

---

[6]See `https://en.wikipedia.org/wiki/Electronic_color_code`.

Set up the Hardware

Begin by placing the breadboard next to your Raspberry Pi and power the Raspberry Pi off orienting the Raspberry Pi with the label facing you (GPIO pins in the upper-left). Next, take one of the jumper wires and connect the female connector to pin 6 on the GPIO. The pins are numbered left-to-right starting with the lower left pin. Thus, the left two pins are 1 and 2 with pin 1 below pin 2. Connect the other wire to pin 7 on the GPIO.

■ **Tip** The only component that is polarized is the LED. This longer side is the positive side.

Next, plug the resistor into the breadboard with each pin on one side of the center groove. You can choose whichever area you want on the breadboard. Next, connect the LED so that the long leg is plugged into the same row as the resistor and the other pin on another row. Finally, connect the wire from pin 6 to the same row as the negative side of the LED and the wire from pin 7 to the row with the resistor. Figure 7-2 shows how all of the components are wired together. Be sure to study this drawing and double-check your connections prior to powering on your Raspberry Pi. Once you're satisfied everything is connected correctly, you're ready to power on the Raspberry Pi and write the code.

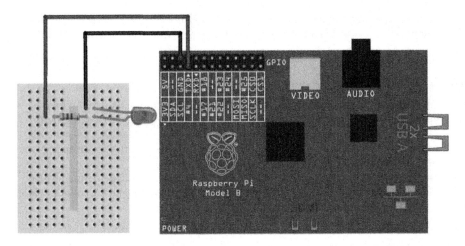

Figure 7-2. *Wiring the LED to a Raspberry Pi*

Since you are building a headed application, you'll also need a keyboard, mouse, and monitor connected to the Raspberry Pi.

OK, now you have the hardware set up, so it's time to start writing the code.

Write the Code

Begin by opening a new project template. Choose **Python ➤ Windows IoT Core** in the tree and the **Background Application (IoT)** template in the list. This template creates a new solution with all of the source files and resources you need for a UWP headed application. Figure 7-3 shows the project template that you need. Use BlinkPythonStyle for the project name.

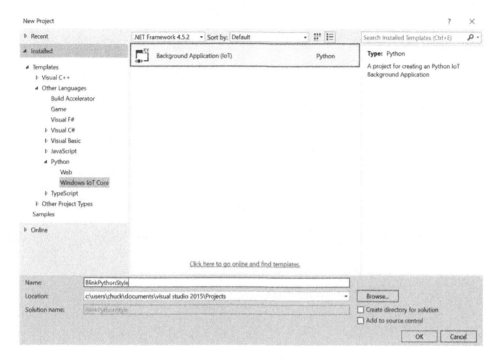

Figure 7-3. *New Project dialog—Background Application (IoT)*

■ **Tip** If you attempt to create this example but do not see Python in the tree view, you may need to install the Python prerequisites. See the "Getting Started" section of this chapter for information on what you need to install.

Once the project is open, you see a number of files like any other project. However, Python projects only have the one source code file named StartupTask.py. You can rename this file by right-clicking it, but it isn't necessary for our example. Go ahead and open that file now (just double-click it in the tree). Notice that the file is empty save for a single comment line (you can overwrite that).

The first thing you need to do is add a reference to the GPIO Python module. Remember that this is contained in the pywindevices.zip file from CPython UWP SDK. If you haven't already done so, download the file and extract the files in a directory you can remember.

To add the reference, Right-click the References item in the project and choose **Add Reference...**. Once the dialog opens, click **Browse** and then navigate to the folder where you extracted the files. In that folder are subfolders for ARM or x86 builds. Choose the folder that matches your board (e.g., ARM for Raspberry Pi). Once selected, you see all of the available modules.

The module you want is named _wingpio. There are two versions of the reference; one for release builds (_wingpio.pyd) and one for debug builds (_wingpio_d.pyd). I recommend using the debug version until your application runs correctly. Figure 7-4 shows the dialog you can use to add the reference.

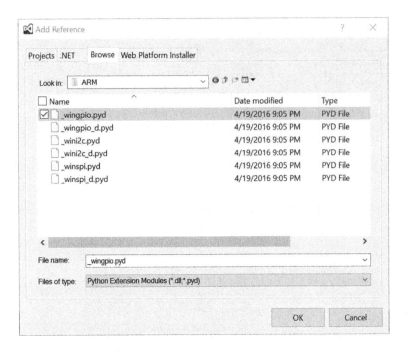

Figure 7-4. Adding the Python module references

Now you can return to the code file and add the modules you want to use. You need two: the GPIO module and the time module. You can add them as follows.

```
import _wingpio as gpio # The GPIO library
import time              # Time functions
```

Next, you can write the code to create some variables that you need and to initialize the GPIO module. The code to do this is as follows.

```
BLINK_TIME = 30.0       # blink for 30 seconds
GPIO_PIN = 4            # GPIO4 (not pin #4)
pin_val = gpio.HIGH     # Store pin value (high/low)
elapsed_time = 0        # Store time in seconds

# Setup the
gpio.setup(GPIO_PIN, gpio.OUT, gpio.PUD_OFF, gpio.HIGH)
```

Here you have a variable to store how long you want to blink the LED, a variable to store the GPIO pin number (GPIO4), a variable to store the GPIO value (high/low), and a variable to store how much time has passed.

As you may have anticipated, you are going to implement the code to turn the LED on and off in a loop. The following shows the loop that you will use.

```
# Run the blink series for
while (elapsed_time < BLINK_TIME):
    # If LED is on, turn it off
    if pin_val == gpio.HIGH:
        pin_val = gpio.LOW
        gpio.output(led_pin, pin_val)
    # else, turn it on
    else:
        pin_val = gpio.HIGH
        gpio.output(led_pin, pin_val)
    # Sleep for 1/2 second
    time.sleep(1)
    elapsed_time = elapsed_time + 1
```

Notice that you simply turn the GPIO pin on (high) if it is off or off (low) if it is on. At the end of the loop, you sleep (pause) for one second and sum the time. Thus, the loop simply toggles from one state (on) to the other (off) thereby blinking the LED.

Finally, you clean up the GPIO object when the loop ends.

```
# Close down and cleanup the GPIO
gpio.cleanup()
```

That's it! You've finished the code. Listing 7-1 shows the complete code.

Listing 7-1. Python Blink LED Script

```
import _wingpio as gpio # The GPIO library
import time              # Time functions

BLINK_TIME = 30.0       # blink for 30 seconds
GPIO_PIN = 4            # GPIO4 (not pin #4)
pin_val = gpio.HIGH     # Store pin value (high/low)
elapsed_time = 0        # Store time in seconds

# Setup the GPIO
gpio.setup(GPIO_PIN, gpio.OUT, gpio.PUD_OFF, gpio.HIGH)

# Run the blink series for
while (elapsed_time < BLINK_TIME):
    # If LED is on, turn it off
    if pin_val == gpio.HIGH:
        pin_val = gpio.LOW
        gpio.output(GPIO_PIN, pin_val)
    # else, turn it on
    else:
        pin_val = gpio.HIGH
        gpio.output(GPIO_PIN, pin_val)
```

```
# Sleep for 1/2 second
time.sleep(1)
elapsed_time = elapsed_time + 1

# Close down and cleanup the GPIO
gpio.cleanup()
```

As you can see, the code is much smaller than the C++ or C# examples from the previous chapters. You do not need to build the code (Python is interpreted). Indeed, building the code only performs some very high-level checks and does not check the validity of the code.

■ **Caution**   Compiling or building the code performs only the most basic of tests. It does not verify that the code is correct.

OK, now you're ready to deploy the application to your device. Go ahead and set everything up and power on your device.

Deploy and Set as Startup

Once the Raspberry Pi has booted, you're ready to deploy the application to it (or other device). Recall from Chapter 4, that you have to set up the debug settings to specify the IP address of your Raspberry Pi. Fortunately, you only have to change two items as indicated in Figure 7-5. Note that you may not need to set the remote port (5678 is the default). Remember to choose ARM for the platform. Click **Apply** and then **OK** to close the dialog.

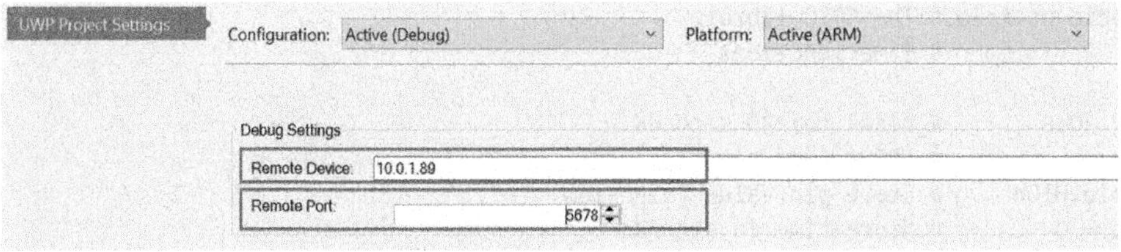

Figure 7-5. Setting Debug Settings for deployment

Notice that you use the IP address for the device. You can also use the computer name if you know that. Also, unlike the C++ example, you do not specify a port here. Instead, you must select port 5678. This is a special port that the Python debugger responds to when you deploy your application.

Once these are set, you can deploy the application from the Build menu. When complete, you'll get messages like the following from the Output window.

```
------ Deploy started: Project: BlinkPythonStyle, Configuration: Debug ARM ------
Creating a new clean layout...
Copying files: Total 16 mb to layout...
Checking whether required frameworks are installed...
Framework: Microsoft.VCLibs.140.00.Debug/ARM, app package version 14.0.22929.0 is not
currently installed.
```

```
Installing missing frameworks...
Registering the application to run from layout...
Deployment complete (33566ms). Full package name: "BlinkPythonStyle_1.0.0.0_
arm__2v0q544fdcg4c"
========== Build: 0 succeeded, 0 failed, 1 up-to-date, 0 skipped ==========
========== Deploy: 1 succeeded, 0 failed, 0 skipped ==========
```

If your deployment succeeded, you may expect to see something happen on the Raspberry Pi, but you won't see anything. That's normal. A headless application must be set as a startup application in order to execute. You can set the startup application via the Windows Device Portal, but it is much easier to use a command-line tool on the device. Plus, while you can set the startup application with the Device Portal, you cannot start the application. This is an unfortunate limitation of the Device Portal currently. It can only start headed applications. As a workaround, in the Processes page, you can use Run command to execute this command to start a background application (by specifying it as one of the startup apps).

Recall that you use Putty (or a similar application) to connect to the device via SSH. The command is iostartup. Listing 7-2 shows the help output for the command, including all of the options to work with startup applications. As you can see, it is very comprehensive.

Listing 7-2. Help Output for iostartup Command

```
C:\Data\Users\administrator>iotstartup
Usage:
  IotStartup [list|add|remove|startup] ([headed|headless]) (std::regex regular expression
with implied ^ prepended)
  IotStartup [help|-?|-h|--help]

Examples:
  IotStartup list                     // list installed applications
  IotStartup list headed              // list installed headed applications
  IotStartup list headless            // list installed headless applications
  IotStartup list MyApp               // list installed applications that match pattern MyApp
  IotStartup add headed MyApp         // add headed application that matches pattern MyApp.
                                      Pattern must match only one application.
  IotStartup add headless Task1       // add headless applications that match pattern Task1
  IotStartup remove headless Task1    // remove headless applications that match pattern Task1
  IotStartup startup                  // list headed and headless applications registered for
                                      startup
  IotStartup startup MyApp            // list headed and headless applications registered for
                                      startup that match pattern MyApp
  IotStartup startup headed MyApp     // list headed applications registered for startup that
                                      match MyApp
  IotStartup startup headless Task1 // list headless applications registered for startup
                                      that match Task1
```

You can see what applications are set for startup (they execute when the device is booted) using the iostartup startup command, as shown next. This command lists all of the applications that start on boot, grouped by headless or headed applications. Notice that you do not have to use the complete name for the application. You can just use the first few characters, or to be safer, the base name you used in Visual Studio.

```
C:\Data\Users\administrator>iotstartup startup
Headed   : IoTCoreDefaultApp_1w720vyc4ccym!App
No headless applications registered
```

To add an application that has been deployed to the device as a startup application, simply add the name of the application to the command, as shown next. Here you add the application and then use the command again without a name to verify that it has started.

```
C:\Data\Users\administrator>iotstartup add headless BlinkPythonStyle
Added Headless: BlinkPythonStyle_1.0.0.0_arm__2v0q544fdcg4c

C:\Data\Users\administrator>iotstartup startup
Headed    : IoTCoreDefaultApp_1w720vyc4ccym!App
Headless  : BlinkPythonStyle_1.0.0.0_arm__2v0q544fdcg4c
```

■ **Caution**   This command configures Windows 10 IoT Core to ensure that this background application is always running. If it crashes and you restart the device, the background application is started. If the crash is severe enough, it can cause Windows 10 to be unavailable, because the app will start and then immediately crash and force a reboot. The only way to break this cycle (sadly) is to reimage your SD card. Never attempt to set an untested application as a startup application.

The application should start running at this point. If it doesn't start, try rebooting the device. You can reboot the device via the command line with the following command. For information about other command-line commands, see https://ms-iot.github.io/content/en-US/win10/tools/CommandLineUtils.htm.

```
shutdown /r /t 0
```

To remove your application from the startup list, execute the following command.

```
iotstartup remove headless BlinkPythonStyle
```

If the LED is still not blinking after the reboot, double-check your wiring and ensure that you chose pin 4 in the code and pin 7 on the pin header (recall that the pin is named GPIO 4 but it is pin #7 on the header).

Once you get it working, you see the LED blink for 30 seconds and then the application stops. If this happens as expected, congratulations, you've just written your first Python application that uses the GPIO header to power some electronics!

Summary

If you are learning how to work with Windows 10 IoT Core and don't know how to program with Python, learning Python can be fun given its easy to understand syntax. While there are many examples on the Internet that you can use, very few are documented in such a way as to provide enough information for someone new to Python to understand—much less get started or compile and deploy the sample!

This chapter has provided a crash course in Python that covers the basics of the things you encounter when examining most of the smaller example projects. You discovered the basic syntax and constructs of a Python application, including a walk-through of building a real Python application that blinks an LED. Through that example, you learned how to work with headless applications, including how to manage a startup background application.

The next chapter takes a short detour in your exploration of Windows 10 IoT Core projects. You are introduced to the basics of working with electronics. Like the programming crash course, the chapter provides a short introduction to working with electronics. A mastery of electronics in general is not required for the projects you explore, but if you've never worked with electronic components before, the next chapter will prepare you for the more advanced projects in Chapters 10–15.

CHAPTER 8

Electronics for Beginners

If you're new to the IoT or have never worked with electronics, you may be wondering how you're going to get your ideas for an IoT solution realized. The projects in this book walk you through how to connect the various components used and thus you can complete them without a lot of additional information or specialized skills.

However, if something goes wrong or you want to create projects on your own, you may need a bit more information than "plug this end in here." More specifically, you need to know enough about how the components work in order to successfully complete your project—whether that is completing the examples in this book or examples found elsewhere on the Internet.

Rather than attempt to present a comprehensive tutorial on electronics, which would take several volumes, this chapter presents an overview of electronics for those who want to work with the types of electronic components commonly found in IoT projects. I include an overview of some of the basics, descriptions of common components, and a look at sensors. If you are new to electronics, this chapter gives you the extra boost you need to understand the components used in the projects in this book.

However, if you have experience with electronics either at the hobbyist or enthusiast level or have experience or formal training in electronics, you may want to skim this chapter or read the sections with topics that interest you.

SELF-PACED ELECTRONICS TRAINING

If you find you need or want to learn more about electronics, especially the types of electronics you need for an IoT solution, check out the set of electronics books by Charles Platt. I've found these books to be very well written opening the door for many to learn electronics without having to spend years learning the tedious (but no less important) theory and mathematics of electronics. I recommend the following books for anyone wanting to learn more about electronics.

- *Make: Electronics* by Charles Platt (O'Reilly, 2015)

- *Make: More Electronics* by Charles Platt (O'Reilly, 2014)

- *Encyclopedia of Electronic Components* by Charles Platt (O'Reilly, 2012)

Maker Shed (www.makershed.com/collections/electronics) sells companion kits that contain all of the parts you need to complete the experiments in the *Make: Electronics* and *Make: More Electronics* books. The books together with the kits make for an excellent self-paced learning experience.

© Charles Bell 2016

C. Bell, *Windows 10 for the Internet of Things*, DOI 10.1007/978-1-4842-2108-2_8

Let's begin with a look at the basics of electronics. Once again, this is in no way a tutorial that covers all there is to know, but it gets you to the point where the projects make sense in how they connect and use components.

The Basics

This section presents a short overview of some of the most common tools and techniques you need to use when working with electronics. As you will see, you only need the most basic of tools and the skills or techniques are not difficult to learn. However, before you get into those, let's discuss the most fundamental concept you must understand when working with electronics—power!

Powering Your Electronics

Electricity[1] is briefly defined as the flow of electric charge and when used, provides power for your electronics—a common lightbulb, a ceiling fan, a high-definition television, a tablet. Whether you are powering your electronics with batteries or a power supply, you are initiating a circuit where electrons flow in specific patterns. There are two forms (or kinds) of power that you will use. Your home is powered by alternating current and your electronics are powered by direct current.

The term alternating current (AC) is used to describe the flow of charged particles that changes direction periodically at a specific rate (or cycle) reversing the voltage along with the current. Thus, AC systems are designed to work with a specific range of cycles as well as voltage. Typically, AC systems use higher voltages than direct current systems.

The term direct current (DC) is used to describe the flow of charged particles that do not change direction and thus always flow in a specific "direction." Most electronics systems are powered with DC voltages and are typically at lower voltages than AC systems; for example, IoT projects typically run on lower direct current (DC) voltages in the range 3.3V–24V.

■ **Tip** For more information about AC and DC current and the differences, see `https://learn.sparkfun.com/tutorials/alternating-current-ac-vs-direct-current-dc`.

Since DC flows in a single direction, components that operate on DC have a positive side and a negative side, where current flows from positive to negative. The orientation of these sides—one to positive and one to negative—is called *polarity*. Some components, such as resistors, can operate in either "direction" but you should always be sure to connect your components according to its polarity. Most components are clearly marked but those that are not have a well-known arrangement. (For example, the positive pole (side) of an LED is the longer of the two legs; it is called an *anode*. The negative and shorter leg is called the *cathode*).

Despite the lower voltages, you mustn't think that they are completely harmless or safe. Incorrectly wiring electronics (reversing polarity) or shorting (connecting positive and negative together) can damage your electronics and in some cases cause overheating, which, in extreme cases, cause electronics to catch fire.

■ **Caution** Don't be tempted to think working with 3.3 or 5.5 volts is "safe." Even a small amount of voltage improperly connected can lead to potentially devastating results. Don't assume low DC voltage is harmless.

---

[1]See `https://learn.sparkfun.com/tutorials/what-is-electricity`.

I had a lesson in just how real this scenario can be a couple of years ago. I was changing the batteries in my smoke detectors. I took the old batteries out and placed them in my pocket. I had forgotten I had a small penknife in the same pocket. One of the 9V batteries shorted on the knife and within about ten minutes, the battery heated to an alarming temperature. It wasn't enough to burn but had I left something like that unattended, it could have been bad.

That's a scary thought, isn't it? Consider it an admonishment as well as a warning; you should never relax your safe handling practices even for lower voltage projects.

Finally, DC components are often rated for a specific voltage range. Recall from the discussion on the various low-cost computing boards and GPIO headers, some boards operate at 5V, whereas others operate at 3.3V (or less). Fortunately, there are several ways you can adapt components that work at different voltages— by using other components!

■ **Note** I have deliberately kept the discussion on power simple. There is far more to electrical current— even DC—than what I've described here. As long as you understand these basics, you'll be able to work with the projects in this book and more.

Now let's take a look at some of the tools you need to work on your IoT projects.

Tools

The vast majority of tools you need to construct your IoT projects are common hand tools (screwdrivers, small wrenches, pliers, etc.). For larger projects or for creating enclosures you may need additional tools, such as power tools, but I concentrate only on those tools needed for building the projects. The following is a list of tools that I recommend to get you started.

- Breadboard
- Breadboard wires (also called jumpers)
- Electrostatic discharge (ESD) safe tweezers
- Helping hands or printed circuit board (PCB) holder
- Multimeter
- Needle-nose pliers
- Screw drivers: assorted sizes (micro, small)
- Solder
- Soldering iron
- Solder remover (solder sucker)
- Tool case, roll, or box for storage
- Wire strippers

However, you cannot go wrong if you prefer to buy a complete electronics toolset, such as those from SparkFun (www.sparkfun.com/categories/47) or Adafruit (www.adafruit.com/categories/83). You can often find electronics kits at major brand electronics stores and home improvement centers. If you are fortunate enough to live near a Fry's Electronics, you can find just about any electronics tool made. Most electronics kits have all the hand tools that you need. Some even come with a multimeter, but more often you have to buy it separately.

Most of the tools in the list do not need any explanation except to say you should purchase the best tools that your budget permits. The following paragraphs describe some of the tools that are used for special tasks, such as stripping wires, soldering, and measuring voltage and current.

Multimeter

A *multimeter* is one of those tools that you need when building IoT solutions. You also need it to do almost any electrical repair on your circuits. There are many different multimeters available with prices ranging from inexpensive, basic units to complex, feature-rich, incredibly expensive units. For most IoT projects, including most IoT kits, a basic unit is all that you need. However, if you plan to build more than one IoT solution or want to assemble your own electronics, you may want to invest a bit more in a more sophisticated multimeter. Figure 8-1 shows a basic digital multimeter (costing about $10) on the left and a professional multimeter from BK Precision on the right.

Figure 8-1. *Digital multimeters*

Notice that the better meter has more granular settings and more features. Again, you probably won't need more than the basic unit. You need to measure voltage, current, and resistance at a minimum. Whichever meter you buy, make sure that it has modes for measuring AC and DC voltage, continuity testing (with an audible alert), and checking resistance. I explain how to use a multimeter in a later section.

■ **Tip** Most multimeters including the inexpensive ones come with a small instruction booklet that shows you how to measure voltage, resistance, and other functions of the unit.

Soldering Iron

A *soldering iron* is not required for any of the projects in this book because you use a breadboard to lay out and connect the components. However, if you plan to build a simple IoT solution where you need to solder wires, or maybe a few connectors, a basic soldering iron from an electronics store such as Radio

Shack is all you need. On the other hand, if you plan to assemble your own electronics, you may want to consider getting a good, professional soldering iron, such as a Hakko. The professional models include features that allow you to set the temperature of the wand, have a wider array of tips available, and tend to last a lot longer. Figure 8-2 shows a well-used entry-level Radio Shack. Figure 8-3 shows a professional model Hakko soldering iron.

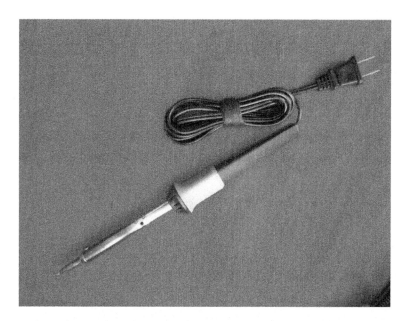

Figure 8-2. *Entry-level soldering iron*

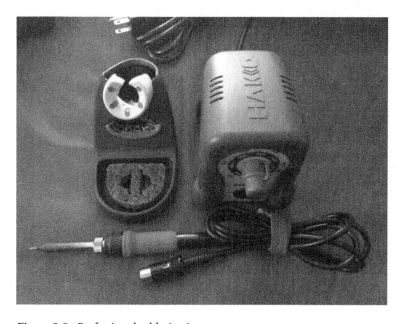

Figure 8-3. *Professional soldering iron*

■ **Tip** For best results, choose a solder with a low lead content in the 37%–40% range. If you use a professional soldering iron, adjust the temperature to match the melting point of the solder (listed on the label).

DO I NEED TO LEARN TO SOLDER?

If you do not know how to solder or it has been a while since you've used a soldering iron, you may want to check out the book *Learn to Solder* by Brian Jepson, Tyler Moskowite, and Gregory Hayes (O'Reilly Media, 2012) or Google how-to videos on soldering. Or you could buy the Getting Started Soldering Kit from Maker Shed (www.makershed.com/products/make-getting-started-kit-soldering), which comes with a soldering iron, wire cutters, supplies, and more—everything you need to learn how to solder. Cool.

Wire Strippers

There are several types of *wire strippers*. In fact, there are probably a dozen or more designs out there. But there really are two kinds: ones that only grip and cut the insulation as you pull it off the wire, and those that grip, cut, and remove the insulation. The first type is more common and, with some practice, does just fine for most small jobs (like repairing a broken wire); but the second type makes a larger job—such as wiring electronics from bare wire (no prefab connectors)—much faster. As you can imagine, the first type is considerably cheaper. Figure 8-4 shows both types of wire strippers. Either is a good choice.

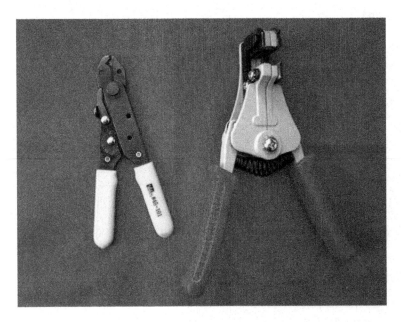

Figure 8-4. Wire strippers

Helping Hands

There is one other tool that you may want to get, especially if you need to do any soldering; it's called *helping hands* or a *third-hand tool*. Most have a pair of alligator clips to hold wires, printed circuit boards, or components while you solder. Figure 8-5 shows an excellent example from Adafruit (www.adafruit.com/products/291).

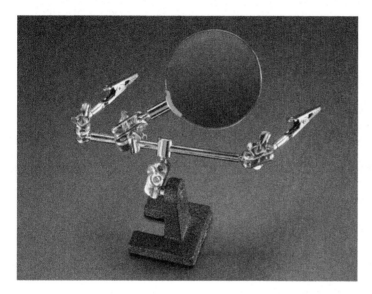

Figure 8-5. *Helping hands tool (courtesy of adafruit.com)*

Now let's take a look at some of the skills you are likely to need when working with advanced IoT projects.

ESD IS THE ENEMY

You should take care to make sure that your body, your workspace, and your project is grounded to avoid electrostatic discharge (ESD). ESD can damage your electronics—permanently. The best way to avoid this is to use a grounding strap that loops around your wrist and attaches to an anti-static mat like those available from https://www.uline.com/BL_7403/Anti-Static-Table-Mats.

Using a Multimeter

The electrical skills needed for IoT projects can vary from plugging in wires on a breadboard—as you saw with the projects so far—to needing to solder components together or to printed circuit boards (PCBs). Regardless of whether you need to solder the electronics, you need to be able to use a basic multimeter to measure resistance and check voltage and current.

A multimeter is a very useful and essential tool for any electronics hobbyist and downright required for any enthusiast of worth. A typical multimeter has a digital display[2] (typically an LCD or similar numeric display), a dial, and two or more posts or ports for plugging in test leads with probe ends. Most multimeters have ports for lower current (that you will use most) and ports for higher current. Test leads use red for positive and black for negative (ground). The ground port is where you plug in the black test lead and is often marked either with a dash or COM for common. Which of the other ports you use depends on what you are testing.

One thing to note on the dial is that there are many settings (with some values repeated) or those that look similar. For example, you see a set of values (sometimes called a *scale*) for ohms; one or two sets of values for amperage; and one or two sets of values for volts. The DC voltage is indicated by a V with a solid and dashed line over it; whereas the AC voltage is indicated by a V with a wavy line over it. Amperage ranges are marked in the same manner. Figure 8-6 shows a close-up of a multimeter dial labeled with the sets of values that I mentioned.

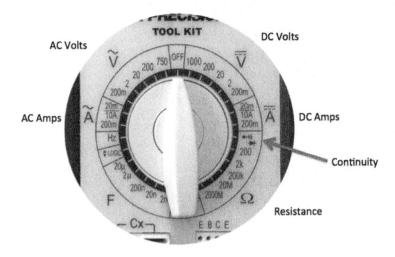

Figure 8-6. Multimeter dial (typical)

■ **Tip**   When not in use, be sure to turn your multimeter dial to off or one of the voltage ranges if it has a separate off button.

There is a lot you can do with a multimeter. You can check voltage, measure resistance, and even check continuity. Most basic multimeters do all of these functions; however, some multimeters have a great many more features, such as testing capacitors, and the ability to test AC and DC.

Let's see how you can use a multimeter to perform the most common tasks you need for IoT projects: testing continuity, measuring voltage in a DC circuit, measuring resistance, and measuring current.

Testing Continuity

You test for continuity to determine if there is a path for the charged particles to flow. That is, your wires and components are connected properly; for example, you may want to check to ensure that a wire has been spliced correctly.

---

[2]Older multimeters have an analog gauge. You can still find them if you want a bit of old school feel.

To test for continuity, turn your multimeter dial to the position marked with an audible symbol, bell, or triangle with an arrow through it. Plug the black test lead into the *COM* port and the red test lead in the port marked with Hz VΩ or similar. Now you can touch the probe end of the test leads together to hear an audible tone or beep. Some multimeters don't have an audible tone but instead may display "1" or similar to indicate continuity. Check your manual for how your multimeter indicates continuity. Figure 8-7 shows how to set a multimeter to check for continuity including which ports to plug in the test leads.

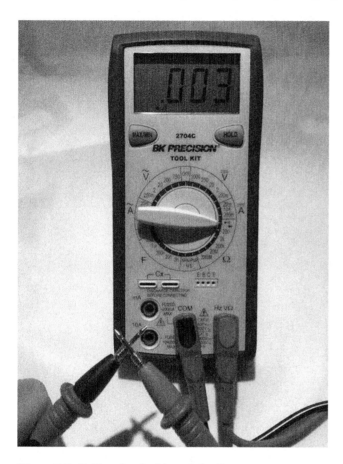

Figure 8-7. *Settings for checking continuity*

In Figure 8-7, I simply touched the probes together to demonstrate how to check for continuity. I like to do this just to ensure that my multimeter is turned on and in the correct setting.[3]

Another excellent use for the continuity test is when diagnosing or discovering how cables are wired. For example, you can use the continuity test to discover which connector is connected on each end of the cable (sometimes called *wire sorting* or *ringing out*, from the old telephone days).

---

[3]Yes, a bit of OCD there. Check, double check, check again.

Measuring Voltage

Our IoT projects use DC. To measure voltage in the circuit, you use the DC range on the multimeter. The DC range has several stops. This is a scale selection. Choose the scale that closely matches the voltage range you want to test. For example, for our IoT projects, you often measure 3.3V–12V, so you choose 20 on the dial. Next, plug the black test lead into the COM port and the red test lead into the port labeled **Hz VΩ**.

Now you need something to measure! Take any battery you have in the house and touch the black probe to the negative side and the red probe to the positive side. You should see a value appear on the display that is close to the range for the battery; for example, if you used a 1.5V battery, you should see close to 1.5V. It may not be exactly 1.5V–1.6V if the battery is depleted. So now you know how to test batteries for freshness! Figure 8-8 shows how to measure voltage of a battery.

Figure 8-8. *Measuring voltage of a battery*

The readout displays 1.50, which is the correct voltage for this AA battery. If I had reversed the probes— the red one on negative and the black on positive, the display would have read -1.50. This is OK because it shows the current is flowing in the opposite direction of how the probes are oriented.

■ **Note** If you use the wrong probe when measuring voltage in a DC circuit, most multimeters display the voltage as a negative number. Try that with your battery. It won't hurt the multimeter (or the battery)!

You can use this technique to measure voltage in your projects. Just be careful to place the probes on the appropriate positions and try not to cross or short by touching more than one component at a time with a single probe tip.

Measuring Current

Current is measured as amperage (actually milliamps (mA)). Thus, you use the range marked with an A with a straight and dashed line (not the wavy one, that's AC). You measure current in series. That is, you must place the multimeter in the circuit. This can be a little tricky because you must interrupt the flow of current and put the meter inline.

Let's set up an experiment to measure current. Get your breadboard, an LED, a resistor, and two jumper wires you used in the blink project. Wire everything up the same way except don't complete the circuit for the GPIO4 pin. Instead, you use the multimeter to complete the circuit by touching one probe to the positive 5V pin on the GPIO and the other probe on the resistor. Figure 8-9 shows how to set up the circuit with the multimeter inline.

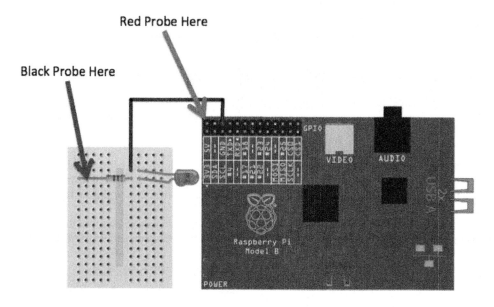

Figure 8-9. Measuring current

Before powering on your Raspberry Pi, plug the black test lead into the COM port and the other test leads into the port labeled mA. Some multimeters use the same port for measuring voltage as well as current. Turn the dial on the multimeter to the 200mA setting. Then power on the Raspberry Pi and touch the leads to the places indicated. Be careful to touch only the 5V pin on the header. If you want to err on the side of caution, use the remaining jumper wire and connect it to the 5V pin, and then touch the probe to the other end of the jumper wire. Once the Raspberry Pi is powered on, you should see a value on the multimeter. Figure 8-10 shows how to use a multimeter to measure current in a circuit.

Figure 8-10. *Measuring current*

In Figure 8-10, I am using a breadboard with a breadboard power supply instead of the Raspberry Pi. Whereas the value you see on the multimeter may differ, the demonstration accomplishes the same goal. In this case, I touch the red probe to the positive pole on the power supply and the black probe on the resistor in the same manner as I described.

There is one other tricky thing about measuring current. If you attempt to measure current that is greater than the maximum for the port (for example, the meter in the photo has a maximum of 20mA on the one port. If I exceeded that by say 5A, I would likely blow a fuse in the multimeter. This is not desirable but at least there is a fuse that you can replace should you make a mistake and choose the wrong port.

Measuring Resistance

Resistance is measured in ohms (Ω). A *resistor* is the most common component that you use to introduce resistance in a circuit. You can test the resistance of the charge through the resistor with your multimeter. To test resistance, choose the ohm scale that is closest to the rating of the resistor. For example, I tested a resistor that I believed about 200 ohms, but since I was not sure, I chose the 2K setting.

Next, plug the black test lead into the COM port and the red test lead into the port labeled **Hz VΩ**. Now, touch a probe to one side of a resistor and the other probe to the other side. It doesn't matter which side you choose—a resistor works in both directions. Notice the readout. The meter reads one of three things: 0.00, 1, or the actual resistor value.

In this case, the meter reads 0.219, meaning this resistor has a value of 220Ω. Recall, I used the 2K scale, which means a resistor of 1K would read 1.0. Since the value is a decimal, I can move the decimal point to the left to get a whole number.

If the multimeter displays another value, such as 0 or 1, it indicates the scale is wrong and you should try a higher scale. This isn't a problem. It just means you need to choose a larger scale. On the other hand, if the display shows 0 or a really small number, you need to choose a lower scale. I like to go one tick of the knob either way when I am testing resistance in an unknown component or circuit.

Figure 8-11 shows an example of measuring resistance for a resistor. The display reads, 219. I am testing a resistor rated at 220 ohms. The reason it is 219 instead of 220 is because the resistor I am using is rated at 220 +/- 5%. Thus, the acceptable range for this resistor is 209 ohms to 231 ohms.

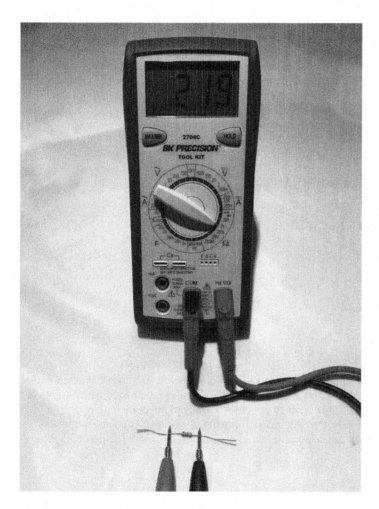

Figure 8-11. *Measuring resistance of a resistor*

Now you know how to test a resistor to discover its rating. As you will see, those rings around the body of the resistor are the primary way you know its rating but you can always test it if you're unsure, someone has painted over it (hey, it happens), or you're too lazy to look it up.

Electronic Components

Aside from learning how to use a multimeter and possibly learning to solder, you also need to know something about the electronic components available to build your projects. In this section, I describe some common components—listed in alphabetical order by name—that you encounter when building IoT solutions. I also cover breakout boards and logic circuits, which are small circuits built with a set of components that provide a feature or solve a problem. For example, you can get breakout boards for USB host connections, Ethernet modules, logic shifters, real time clocks, and more.

Button

A *button* (sometimes called a *momentary button*) is a mechanism that makes a connection when pressed. More specifically, a button connects two or more poles together while it is pressed. A common (and perhaps over used) example of a button is a home doorbell. When pressed, it completes a circuit that triggers a chime, bell, tone, or music to play. Some older doorbells continue to sound while the button is pressed.

In IoT projects, you use buttons to trigger events, start and stop actions, and similar operations. A button is a simple form of a switch but unlike a switch, you must continue to press the button to make the electrical connections. Most buttons have at least two legs (or pins) that are connected when the button is pressed. Some have more than two legs connected in pairs and some of those can permit multiple connections. Figure 8-12 shows a number of buttons.

Figure 8-12. *Momentary buttons*

There is a special variant of a momentary button called a *latching momentary button*. This version uses a notch or detent to keep the poles connected until it is pushed again. If you've seen a button on a stereo or in your car that remains depressed until pressed again, it is likely a latching momentary button.

There are all kinds of buttons from those that can be used with breadboards (the spacing of the pins allow it to be plugged into a breadboard), can be mounted in a panel, or those made for soldering to printed circuit boards.

Capacitor

A *capacitor* is designed to store charges. As current flows through the capacitor, it accumulates charge and can discharge after the current is disconnected. In this way, it is like a battery but unlike a battery, a capacitor charges and discharges very fast. You use capacitors for all manner of current storage from blocking current, reducing noise in power supplies, in audio circuits, and more. Figure 8-13 shows a number of capacitors.

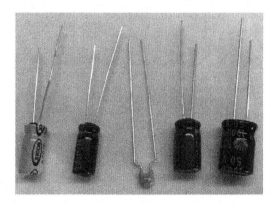

Figure 8-13. *Capacitors*

There are several types of capacitors but you most often encounter capacitors when building power supplies for IoT projects. Most capacitors have two legs (pins) that are polarized. That is, one is positive and the other negative. Be sure to connect the capacitor with the correct polarity in your circuit.

Diode

A *diode* is designed to allow current to flow in only one direction. Most are marked with an arrow pointing to a line, which indicates the direction of flow. A diode is often used as rectifiers in AC-to-DC converters (devices that convert AC to DC voltage), used in conjunction with other components to suppress voltage spikes, or protect components from reversed voltage. Often used to protect against current flowing into a device.

Most diodes are shaped like a small cylinder, are usually black with silver writing, and have two legs. They look a little like resistors. You use a special variant called a *Zener diode* in power supplies to help regulate voltages. Figure 8-14 shows a number of Zener diodes.

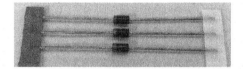

Figure 8-14. *Diodes*

Fuse

A fuse is designed to protect a device (actually the entire circuit) from current greater than what the components can safely operate. Fuses are placed inline on the positive pole. When too much current flows through the fuse, the internal parts trigger a break in the flow of current.

Some fuses use a special wire inside that melts or breaks (thereby rendering it useless but protecting your equipment) while other fuses use a mechanism that operates like a switch (many of these are resettable). When this happens, you say the fuse has "blown" or "tripped." Fuses are rated at a certain current in amperage, indicating the maximum amps that the fuse permits to flow without tripping.

Fuses come in many shapes and varieties. They work with either AC or DC voltage. The fuses that you use are of the disposable variety. Figure 8-15 shows an example of two fuses: an automotive-style blade fuse on the left and a glass cartridge fuse on the right.

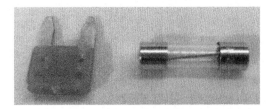

Figure 8-15. *Fuses*

If you are familiar with your home's electrical panel that contains the circuit breakers, they are resettable fuses. So the next time one of them goes "click" and the lights go out, you can say, "Hey, a fuse has tripped!" Better still, now you know why—you have exceeded the maximum rating of the circuit breaker.

Which is probably fine in situations where you accidentally left that infrared heater on when you dropped the toast and started the microwave (it happens), but if you are tripping breakers frequently without any load, you should call an electrician to have the circuit checked.

Light Emitting Diode (LED)

Recall from Chapter 3 that an LED is a special diode that produces light when powered.

As you learned in Chapter 3, an LED has two legs: the longer leg is positive and the shorter is negative. LEDs also have a flat edge that also indicates the negative leg. They come in a variety of sizes ranging from as small as 3mm to 10mm. Figure 8-16 shows an example of some smaller LEDs.

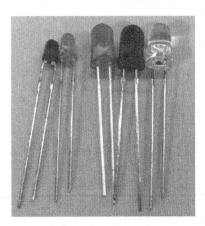

Figure 8-16. *Light emitting diodes*

Recall you also need to use a resistor with an LED. You need this to help reduce the flow of the circuit to lower the current flowing through the LED. LEDs can be used with lower current (they burn a bit dimmer than normal) but should not be used with higher current.

To determine what size resistor you need, you need to know several things about the LED. This data is available from the manufacturer who provides the data in the form of a data sheet or in the case of commercially packaged products, lists the data on the package. The data you need includes the maximum voltage, the supply voltage (how many volts are coming to the LED), and the current rating of the LED.

For example, if I have an LED like the one you used in the last chapter, in this case a 5mm red LED, you find on Adafruit's web site (www.adafruit.com/products/297) that the LED operates at 1.8V–2.2V and 20mA of current. Let's say you want to use this with a 5V supply voltage. You can then take these values and plug them into this formula:[4]

```
R = (Vcc-Vf)/I
```

Using more descriptive names for the variable, you get the following.

```
Resistor = (Volts_supply - Volts_forward) / Desired_current
```

Plugging in the data, you get this result. Note that you have mA so you must use the correct decimal value (divide by 1000). In this case, 0.020 and you pick a voltage in the middle.

```
Resistor = (5 - 2.0) / 0.020
         = 3.0 / 0.020
         = 150
```

Thus, you need a resistor of 150 ohms. Cool. Sometimes the formula produces a value that does not match any existing resistors. In that case, choose one closest to the value but a bit higher. Remember, you want to limit current and thus err on the side of more restrictive than less restrictive. For example, if you found you need a resistor of 95 ohms, you can use one rated at 100 ohms, which is safer than using one rated at 90 ohms.

■ **Tip** Always err on the side of the more restrictive resistor when the formula produces a value for which there is no resistor available.

Also, if you use LEDs in serial or parallel, the formula is a little different. See https://learn.adafruit.com/all-about-leds for more information about using LEDs in your projects and calculating the size of resistors to use with LEDs.

Relay

A relay is an interesting component that helps you control higher voltages with lower voltage circuits. For example, suppose you wanted to control a device that is powered by 12V from your Raspberry Pi, which only produces a maximum of 5V. A relay can be used with a 5V circuit to turn on (or relay) power from that higher source. In this example, you would use the Raspberry Pi's output to trigger the relay to switch on the 12V power. Thus, relays are a form of switch. Figure 8-17 shows a typical relay and how the pins are arranged.

Figure 8-17. *Relay*

---

[4]A variant of Ohm's law (https://en.wikipedia.org/wiki/Ohm's_law).

Relays can take a lot of different forms and typically have slightly different wiring options, such as where the supply voltage is attached and where the trigger voltage attaches as well as whether the initial state is open (no flow) or close (flow) and thus the behavior of how it controls voltage. Some relays come mounted on a PCB with clearly marked terminals that show where to change the switching feature and where everything plugs in. If you want to use relays in your projects, always check the datasheet to make sure that you are wiring it correctly, based on its configuration.

You can also use relays to allow your DC circuit to turn AC appliances on and off like PowerSwitch Tail from Adafruit (www.adafruit.com/products/268).

Resistor

A resistor is one of the standard building blocks of electronics. Its job is to impede current and impose a reduction in voltage (which is converted to heat). Its effect, known as resistance, is measured in ohms. A resistor can be used to reduce voltage to other components, limiting frequency response, or protect sensitive components from over voltage. Figure 8-18 shows a number of resistors.

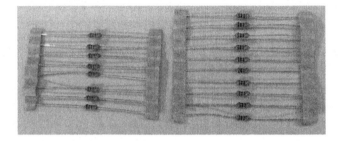

Figure 8-18. Resistors

When a resistor is used to pull up voltage (by attaching one end to positive voltage) or pull down voltage (by attaching one end to ground) (resistors are bidirectional), it eliminates the possibility of the voltage floating in an indeterminate state. Thus a pull-up resistor ensures that the stable state is positive voltage, and a pull-down resistor ensures that the stable state is zero voltage (ground).

Switch

A switch is designed to control the flow of current between two or more pins. Switches come in all manner of shapes, sizes, and packaging. Some are designed as a simple on/off while others can be used to change current from one set of pins to another. Like buttons, switches come in a variety of mounting options from PCB (also called a *through hole*) to panel mount for mounting in enclosures. Figure 8-19 shows a variety of switches.

Figure 8-19. *Various switches*

Switches that have only one pole (leg or side) are called *single-pole switches*. Switches that can divert current from one set of poles to another set are called *two-pole switches*. Switches where there is only one secondary connection per pole are called *single-throw switches*. Switches that disconnect from one set of poles and connect to another while maintaining a common input are called *double-throw switches*. These are often combined together and form the switch type (or kind) as follows.

- *SPST*: Single pole, single throw

- *DPST*: Double pole, single throw

- *SPDT*: Single pole, double throw

- *DPDT*: Double pole, double throw

- *3PDT*: Three pole, double throw

There may be other variants that you could encounter. I like to keep it simple: if I have just an on/off situation, I want a single throw switch. How many poles depends on how many wires or circuits I want to turn on or off at the same time. I use these for double throw switches when I have an "A" condition and a "B" condition in which I want A on when B is off, and vice-versa. I sometimes use multiple throw switches when I want A, B, and off situations. I use the center position (throw) as off. You can be very creative with switches!

Transistor

A transistor (a bipolar transistor) is designed to switch current on/off in a cycle or amplify fluctuations in current. Interestingly, transistors used to amplify current replaced vacuum tubes. If you are an audiophile, you likely know a great deal about vacuum tubes. When a resistor operates in switching mode, it behaves similar to a relay but its "off" position still allows a small amount of current to flow. Transistors are used in audio equipment, signal processing, and switching power supplies. Figure 8-20 shows two varieties of transistors.

Figure 8-20. *Transistors*

Transistors come in all manner of varieties, packaging, and ratings that make it suitable for one solution or another.

Voltage Regulator

A voltage regulator (linear voltage regulator) is designed to keep the flow of current constant. Voltage regulators often appear in electronics when you need to condition or lower current from a source. For example, you want to supply 5V to a circuit but only have a 9V power supply. Voltage regulators accomplish this (roughly) by taking current in and dissipating the excess current through a heat sink. Thus, voltage regulators have three legs: positive current in, negative, and positive current out. They are typically shaped like those shown in figure 8-21 but other varieties exist.

Figure 8-21. *Voltage regulators*

The small hole in the plate that extends out of the voltage regulator is where the heat sink is mounted. Voltage regulators are often numbered to match their rating; for example, a LM7805 produces 5V, whereas a LM7833 produces 3.3V.

An example of using a voltage regulator to supply power to a 3.3V circuit on a breadboard is shown in Figure 8-22. This circuit was designed with capacitors to help smooth or condition the power. Notice that the capacitors are rated by uF, which means *microfarad*.

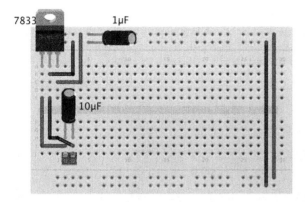

Figure 8-22. *Power supply circuit on a breadboard with voltage regulator*

Breakout Boards and Circuits

Breakout boards are your modular building blocks for IoT solutions. They typically combine several components together to form a function, such as measuring temperature, enabling reading GPS data, communicating via cellular services, and more. Figure 8-23 shows two breakout boards. On the left is an Adafruit AC/DC converter (www.adafruit.com/products/1083) and on the right is an Adafruit barometric pressure sensor breakout board (www.adafruit.com/products/391).

Figure 8-23. *Breakout boards*

Whenever you design a circuit or IoT solution, you should consider using breakout boards as much as possible because they simplify the use of the components. Take the Barometric Pressure Sensor for example, Adafruit has designed this board so that all you need to do to use it is attach power and connect it to your IoT device on its I2C bus. An I2C bus is a fast digital protocol that uses two wires (plus power and ground) to read data from circuits (or devices).

Thus, there is no need to worry about how to connect the sensor to other components to use it— just connect it like any I2C device and start reading data! You use several breakout boards in the projects later in this book.

Using a Breadboard to Build Circuits

If you have been following along with the projects thus far in the book, you have already encountered a breadboard to make a very simple circuit. Recall from Chapter 3 that a breadboard is a tool you use to plug components into to form circuits. Technically, you're using a solderless breadboard. A solder breadboard has the same layout only it has only through-hole solder points on a PCB.

A breadboard allows you to create prototypes for your circuits or simply temporary circuits without having to spend the time (and cost) to make the printed circuit board. Prototyping is the process of experimenting with a circuit by building and testing your ideas. In fact, once you've got your circuit to work correctly, you can use the breadboard layout to help you design a PCB. Figure 8-24 shows a number of breadboards.

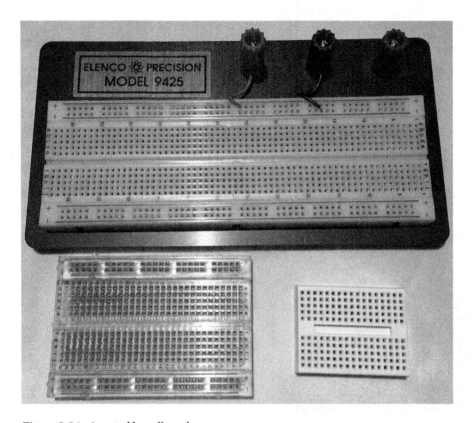

Figure 8-24. *Assorted breadboards*

WHY ARE THEY CALLED BREADBOARDS?

In the grand old days of microelectronics and discrete components became widely available for experimentation, when you wanted to prototype a circuit, some would use a piece of wood with nails driven into it (sometimes in a grid pattern) where connections were made (called *runs*) by wrapping wire around the nails. Some actually used a breadboard from the kitchen to build their wire wrap prototypes. The name has stuck ever since.

Most breadboards (there are several varieties) have a center groove (called a *ravine*) or a printed line down the center of the board. This signifies the terminal strips that run perpendicular to the channel are not connected. That is, the terminal strip on one side is not connected to the other side. This allows you to plug integrated circuits (IC) or chip that are packaged as two rows of pins. Thus, you can plug the IC into the breadboard with one set of pins on each side of the breadboard. You see this in the following example.

Most breadboards also have one or more sets of power rails that are connected together parallel to the ravine. If there are two sets, the sets are not connected together. The power rails may have a colored reference line but this is only for reference; you can make either one positive with the other negative. Finally, some breadboards number the terminal strip rows. These are for reference only and have no other meaning. However, they can be handy for making notes in your engineering notebook. Figure 8-25 shows the nomenclature of a breadboard and how the terminal strips and power rails are connected together.

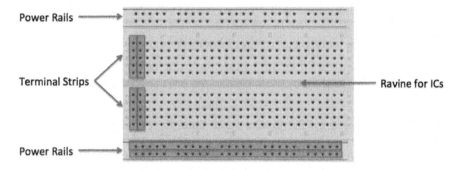

Figure 8-25. *Breadboard layout*

■ **Note**   The sets of power rails are not connected together. If you want to have power on both sides of the breadboard, you must use jumpers to connect them.

It is sometimes desirable to test a circuit out separately from code. For example, if you want to make sure that all of your devices are connected together properly, you can use a breadboard power supply to power the circuit. This way, if something goes horribly wrong, you don't risk damaging your IoT device. Most breadboard power supplies are built on a small PCB with a barrel jack for a wall wart power supply, two sets of pins to plug into the power rails on the breadboard, an off switch (very handy), and some can generate different voltages. Figure 8-26 shows one of my favorite breadboard power supplies from SparkFun (www.sparkfun.com/products/13157).

Figure 8-26. *Breadboard power supply*

Should your circuits require more room than what is available on a single breadboard, you can use multiple breadboards by simply jumping the power rails and continuing the circuit. To facilitate this, some breadboards can be connected together using small nubs and slots on the side. Finally, most breadboards also come with an adhesive backing that you can use to mount on a plate or inside an enclosure or similar workspace. If you decide to use the adhesive backing, be forewarned that they cannot be unstuck easily—they stay put quite nicely.

| **FRITZING: A BREADBOARDING SOFTWARE APPLICATION** |

The drawings of breadboards in this book were made with a program named Fritzing (`http://fritzing.org/home/`). This open source application allows you to create a digital representation of a circuit on a breadboard. It is really quite handy to use. If you find yourself wanting to design a prototype circuit, using Fritzing can help save you a lot of trial and error. As a bonus, Fritzing allows you to see the same circuit in an electronic schematic or PCB layout view. I recommend downloading and trying this application out.

Now that you know more about how breadboards work, let's discuss the component your IoT solutions employ to collect data: sensors.

What Are Sensors?

A *sensor* is a device that measures phenomena of the physical world. These phenomena can be things you see, like light, gases, water vapor, and so on. They can also be things you feel, like temperature, electricity,[5] water, wind, and so on. Humans have senses that act like sensors, allowing you to experience the world around you. However, there are some things your body can't see or feel, such as radiation, radio waves, voltage, and amperage. Upon measuring these phenomena, it's the sensors' job to convey a measurement in the form of either a voltage representation or a number.

There are many forms of sensors. They're typically low-cost devices designed for a single purpose and with a limited capability for processing. Most simple sensors are discrete components; even those that have more sophisticated parts can be treated as separate components. Sensors are either analog or digital and are typically designed to measure only one thing. But an increasing number of sensor modules are designed to measure a set of related phenomena, such as the USB weather board from SparkFun Electronics (`www.sparkfun.com/products/10586`).

The following sections examine how sensors measure data, how to store that data, and examples of some common sensors.

How Sensors Measure

Sensors are electronic devices that generate a voltage based on the unique properties of their chemical and mechanical construction. They don't actually manipulate the phenomena they're designed to measure. Rather, sensors sample some physical variable and turn it into a proportional electric signal (voltage, current, digital, and so on).

For example, a humidity sensor measures the concentration of water (moisture) in the air. Humidity sensors react to these phenomena and generate a voltage that the microcontroller or similar device can then read and use to calculate a value on a scale. A basic, low-cost humidity sensor is the DHT-22 available from most electronic stores.

The DHT-22 is designed to measure temperature as well as humidity. It generates a digital signal on the output (data pin). Although simple to use, it's a bit slow and should be used to track data at a reasonably slow rate (no more frequently than about once every 3 or 4 seconds).

When this sensor generates data, that data is transmitted as a series of high (interpreted as a 1) and low (interpreted as a 0) voltages that the microcontroller can read and use to form a value. In this case, the microcontroller reads a value that is 40 bits in length (40 pulses of high or low voltage)—that is, 5 bytes— from the sensor and places it in a program variable. The first two bytes are the value for humidity, the second two are for temperature, and the fifth byte is the checksum value to ensure an accurate read. Fortunately, all of this hard work is done for you in the form of a special library designed for the DHT-22 and similar sensors.

---

[5]Shocking, isn't it?

The DHT-22 produces a digital value. Not all sensors do this; some generate a voltage range instead. These are called *analog sensors*. Let's take a moment to understand the differences. This becomes essential information as you plan and build your sensor nodes.

Analog Sensors

Analog sensors are devices that generate a voltage range, typically between 0 and 5 volts. An analog-to-digital circuit is needed to convert the voltage to a number. But it isn't that simple (is it ever?). Analog sensors work like resistors and, when connected to GPIO pins, often require another resistor to "pull up" or "pull down" the voltage to avoid spurious changes in voltage known as *floating*. This is because voltage flowing through resistors is continuous in both time and amplitude.

Thus, even when the sensor isn't generating a value or measurement, there is still a flow of voltage through the sensor that can cause spurious readings. Your projects require a clear distinction between OFF (zero voltage) or ON (positive voltage). Pull-up and pull-down resistors ensure that you have one of these two states. It's the responsibility of the A/D converter to take the voltage read from the sensor and converts it to a value that can be interpreted as data.

When sampled (when a value is read from a sensor), the voltage read must be interpreted as a value in the range specified for the given sensor. Remember that a value of, say, 2 volts from one analog sensor may not mean the same thing as 2 volts from another analog sensor. Each sensor's data sheet shows you how to interpret these values.

As you can see, working with analog sensors is a lot more complicated than using the DHT-22 digital sensor. With a little practice, you find that most analog sensors aren't difficult to use once you understand how to attach them to a microcontroller and how to interpret their voltage on the scale in which the sensor is calibrated to work.

Digital Sensors

Digital sensors like the DHT-22 are designed to produce a string of bits using serial transmission (one bit at a time). However, some digital sensors produce data via parallel transmission (one or more bytes[6] at a time). As described previously, the bits are represented as voltage, where high voltage (say, 5 volts) or ON is 1 and low voltage (0 or even -5 volts) or OFF is 0. These sequences of ON and OFF values are called *discrete values* because the sensor is producing one or the other in pulses—it's either ON or OFF.

Digital sensors can be sampled more frequently than analog signals because they generate the data more quickly and because no additional circuitry is needed to read the values (such as A/D converters and logic or software to convert the values to a scale). As a result, digital sensors are generally more accurate and reliable than analog sensors. But the accuracy of a digital sensor is directly proportional to the number of bits it uses for sampling data.

The most common form of digital sensor is the pushbutton or switch. What, a button is a sensor? Why, yes, it's a sensor. Consider for a moment the sensor attached to a window in a home security system. It's a simple switch that is closed when the window is closed and open when the window is open. When the switch is wired into a circuit, the flow of current is constant and unbroken (measuring positive volts using a pull-up resistor) when the window is closed and the switch is closed, but the current is broken (measuring zero volts) when the window and switch is open. This is the most basic of ON and OFF sensors.

Most digital sensors are actually small circuits of several components designed to generate digital data. Unlike analog sensors, reading their data is easy because the values can be used directly without conversion (except to other scales or units of measure). Some may suggest this is more difficult than using analog sensors, but that depends on your point of view. An electronics enthusiast would see working with analog sensors as easier, whereas a programmer would think digital sensors are simpler to use.

---

[6]This depends on the width of the parallel buffer. An 8-bit buffer can communicate 1 byte at a time; a 16-bit buffer can communicate 2 bytes at a time, and so on.

Now let's take a look at some of the sensors available and the types of phenomena they measure.

Examples of Sensors

An IoT solution that observes something may use at least one sensor and a means to read and interpret the data. You may be thinking of all manner of useful things you can measure in your home or office, or even in your yard or surroundings. You may want to measure the temperature changes in your new sun room, detect when the mail carrier has tossed the latest circular in your mailbox, or perhaps keep a log of how many times your dog uses his doggy door. I hope that by now you can see these are just the tip of the iceberg when it comes to imagining what you can measure.

What types of sensors are available? The following list describes some of the more popular sensors and what they measure. This is just a sampling of what is available. Perusing the catalogs of online electronics vendors like Mouser Electronics (`www.mouser.com`), SparkFun Electronics (`www.sparkfun.com`) and Adafruit Industries (`www.adafruit.com`) reveal many more examples.

- *Accelerometers*: These sensors measure motion or movement of the sensor or whatever it's attached to. They're designed to sense motion (velocity, inclination, vibration, and so on) on several axes. Some include gyroscopic features. Most are digital sensors. A Wii Nunchuck (or WiiChuck) contains a sophisticated accelerometer for tracking movement. Aha: now you know the secret of those funny little thingamabobs that came with your Wii.

- *Audio sensors*: Perhaps this is obvious, but microphones are used to measure sound. Most are analog, but some of the better security and surveillance sensors have digital variants for higher compression of transmitted data.

- *Barcode readers*: These sensors are designed to read barcodes. Most often, barcode readers generate digital data representing the numeric equivalent of a barcode. Such sensors are often used in inventory-tracking systems to track equipment through a plant or during transport. They're plentiful, and many are economically priced, enabling you to incorporate them into your own projects.

- *RFID sensors*: Radio frequency identification uses a passive device (sometimes called an *RFID tag*) to communicate data using radio frequencies through electromagnetic induction. For example, an RFID tag can be a credit-card-sized plastic card, a label, or something similar that contains a special antenna, typically in the form of a coil, thin wire, or foil layer that is tuned to a specific frequency. When the tag is placed in close proximity to the reader, the reader emits a radio signal; the tag can use the electromagnet energy to transmit a nonvolatile message embedded in the antenna, in the form of radio signals, which is then converted to an alphanumeric string.[7]

- *Biometric sensors*: A sensor that reads fingerprints, irises, or palm prints contains a special sensor designed to recognize patterns. Given the uniqueness inherit in patterns, such as fingerprints and palm prints, they make excellent components for a secure access system. Most biometric sensors produce a block of digital data that represents the fingerprint or palm print.

---

[7]http://en.wikipedia.org/wiki/Radio-frequency_identification

- *Capacitive sensors*: A special application of capacitive sensors, pulse sensors are designed to measure your pulse rate and typically use a fingertip for the sensing site. Special devices known as pulse oximeters (called *pulse-ox* by some medical professionals) measure pulse rate with a capacitive sensor and determine the oxygen content of blood with a light sensor. If you own modern electronic devices, you may have encountered touch-sensitive buttons that use special capacitive sensors to detect touch and pressure.

- *Coin sensors*: This is one of the most unusual types of sensors.[8] These devices are like the coin slots on a typical vending machine. Like their commercial equivalent, they can be calibrated to sense when a certain size of coin is inserted. Although not as sophisticated as commercial units that can distinguish fake coins from real ones, coin sensors can be used to add a new dimension to your projects. Imagine a coin-operated Wi-Fi station. Now, that should keep the kids from spending too much time on the Internet!

- *Current sensors*: These are designed to measure voltage and amperage. Some are designed to measure change, whereas others measure load.

- *Flex/Force sensors:* Resistance sensors measure flexes in a piece of material or the force or impact of pressure on the sensor. Flex sensors may be useful for measuring torsional effects or as a means to measure finger movements (like in a Nintendo Power Glove). Flex-sensor resistance increases when the sensor is flexed.

- *Gas sensors*: There are a great many types of gas sensors. Some measure potentially harmful gases, such as LPG and methane, and other gases, such as hydrogen, oxygen, and so on. Other gas sensors are combined with light sensors to sense smoke or pollutants in the air. The next time you hear that telltale and often annoying low-battery warning beep[9] from your smoke detector, think about what that device contains. Why, it's a sensor node!

- *Light sensors*: Sensors that measure the intensity or lack of light are special types of resistors: light-dependent resistors (LDRs), sometimes called *photo resistors* or *photocells*. Thus, they're analog by nature. If you own a Mac laptop, chances are you've seen a photo resistor in action when your illuminated keyboard turns itself on in low light. Special forms of light sensors can detect other light spectrums, such as infrared (as in older TV remotes).

- *Liquid-flow sensors*: These sensors resemble valves and are placed in-line in plumbing systems. They measure the flow of liquid as it passes through. Basic flow sensors use a spinning wheel and a magnet to generate a Hall effect (rapid ON/OFF sequences whose frequency equates to how much water has passed).

- *Liquid-level sensors*: A special resistive solid-state device can be used to measure the relative height of a body of water. One example generates low resistance when the water level is high and higher resistance when the level is low.

- *Location sensors*: Modern smartphones have GPS sensors for sensing location, and of course GPS devices use the GPS technology to help you navigate. Fortunately, GPS sensors are available in low-cost forms, enabling you to add location sensing to your project. GPS sensors generate digital data in the form of longitude and latitude, but some can also sense altitude.

---

[8]www.sparkfun.com/products/11719
[9]I for one can never tell which detector is beeping, so I replace the batteries in all of them.

- *Magnetic-stripe readers*: These sensors read data from magnetic stripes (like that on a credit card) and return the digital form of the alphanumeric data (the actual strings).

- *Magnetometers*: These sensors measure orientation via the strength of magnetic fields. A compass is a sensor for finding magnetic north. Some magnetometers offer multiple axes to allow even finer detection of magnetic fields.

- *Proximity sensors*: Often thought of as distance sensors, proximity sensors use infrared or sound waves to detect distance or the range to/from an object. Made popular by low-cost robotics kits, the Parallax Ultrasonic Sensor uses sound waves to measure distance by sensing the amount of time between pulse sent and pulse received (the echo). For approximate distance measuring,[10] it's a simple math problem to convert the time to distance. How cool is that?

- *Radiation sensors*: Among the more serious sensors are those that detect radiation. This can also be electromagnetic radiation (there are sensors for that too), but a Geiger counter uses radiation sensors to detect harmful ionizing. In fact, it's possible to build your very own Geiger counter using a sensor and an Arduino (and a few electronic components).

- *Speed sensors*: Similar to flow sensors, simple speed sensors like those found on many bicycles use a magnet and a reed switch to generate a Hall effect. The frequency combined with the circumference of the wheel can be used to calculate speed and, over time, distance traveled. The speed sensor on the wheel and fork provides the data for the monitor on your handlebars.

- *Switches and pushbuttons*: These are the most basic of digital sensors used to detect if something is set (ON) or reset (OFF).

- *Tilt switches*: These sensors can detect when a device is tilted one way or another. Although very simple, they can be useful for low-cost motion-detection sensors. They are digital and are essentially switches.

- *Touch sensors*: The touch-sensitive membranes formed into keypads, keyboards, pointing devices, and the like are an interesting form of sensor. You can use touch-sensitive devices like these for collecting data from humans.

- *Video sensors*: As mentioned previously, it's possible to obtain very small video sensors that use cameras and circuitry to capture images and transmit them as digital data.

- *Weather sensors*: Sensors for temperature, barometric pressure, rain fall, humidity, wind speed, and so on are all classified as weather sensors. Most generate digital data and can be combined to create comprehensive environmental solutions. Yes, it's possible to build your own weather station from about a dozen inexpensive sensors, an Arduino (or a Raspberry Pi), and a bit of programming to interpret and combine the data.

---

[10]Accuracy may depend on environmental variables, such as elevation, temperature, and so on.

Summary

Learning how to work with electronics as a hobby or as a means to create an IoT solution does not require a lifetime of study or a change of vocation. Indeed, learning how to work with electronics is all part of the fun of experimenting with the IoT!

As you've seen in this chapter, knowing about the types of components available, the types of sensors, and a bit of key knowledge of how to use a multimeter goes a long ways toward becoming proficient with electronics. You also learned about one of the key components of an IoT solution—sensors. You discovered two ways they communicate and a bit of what types of sensors are available.

In the next chapter, you take a look at a couple of special kits for experimenting with IoT solutions. As you will see, the kits provide a set of electronic components as well as several tools and accessories, such as a breadboard and a power supply.

■ ■ ■

## CHAPTER 9

The Adafruit Microsoft IoT Pack for Raspberry Pi

When working with electronics projects like those in this book, it is often the case that you have to acquire a host of components and tools in order to get started. The projects so far in this book have minimized the components needed and I have listed what you need for each project. However, if you have little experience with electronic components, you may not know what to buy. Fortunately, some vendors such as Adafruit are packaging electronic components, accessories, and even some tools together in a kit making it simple to get started—you just buy the kit!

Indeed, some kits package together a number of common components, such as resistors, LEDs, and a breadboard and jumpers—all the things that you need to get started if you already have the basic parts (low-cost computing board, power supply, etc.). At least one kit goes a bit further providing a more complete set of components and accessories including a development board (e.g., a Raspberry Pi), power supply, and more—everything you need to build an IoT solution using the Windows 10 IoT Core.

■ **Tip** For a complete list of the components used in this book and sources for purchasing them and related tools, see the Appendix.

This chapter explores the Adafruit Microsoft IoT Pack for Raspberry Pi 3 and demonstrates a small project that uses the components in the kit (well, mostly) to read data from a simple sensor. Let's begin with a look at what is in the kit.

Overview

The Microsoft IoT Pack for Raspberry Pi 3 (hence kit) is the result of collaboration between Microsoft and Adafruit to provide a one-stop shopping solution for those who want to explore IoT solutions using Windows 10. In fact, the kit comes with everything you need to run Windows 10 IoT Core including the Raspberry Pi 3, power supply, and a micro-SD card with the Windows 10 IoT Core image installed. For most, the kit is the best way to get started using Windows 10 and your Raspberry Pi.

The kit comes in two varieties: one with the Raspberry Pi (www.adafruit.com/products/2733) and one without the Raspberry Pi (www.adafruit.com/products/2702) for those who already own a Raspberry Pi 2 or 3.[1]

---

[1]While the Raspberry Pi 3 is the latest board, the Raspberry Pi 2 is more than capable for implementing any IOT project.

© Charles Bell 2016
C. Bell, *Windows 10 for the Internet of Things*, DOI 10.1007/978-1-4842-2108-2_9

Adafruit has made several updates to the kit including keeping up with the latest releases of the Windows 10 IoT Core. The kit has been a huge success and sells out regularly. Fortunately, you will not have to wait more than a few weeks for them to restock.

The kit with the Raspberry Pi costs about $114.95 and the kit without the Raspberry Pi costs about $75.00. Clearly, if you already have a Raspberry Pi, you can save some money there. In fact, for those who want to use a different low-cost computer board, you can buy the kit without the Raspberry Pi—except for the micro-SD card with Windows 10 and possibly the power supply, all the other components work with other boards.

■ **Tip** You can also use the kit with Raspbian Linux and Python. Adafruit has a long list of tutorials to explore.

The kit comes with a number of handy components including prototyping tools and a few sensors. It is even certified for use with Microsoft Azure (Microsoft's cloud computing platform)! Figure 9-1 shows a photo of all the components included in the kit. I discuss all the components in the next section.

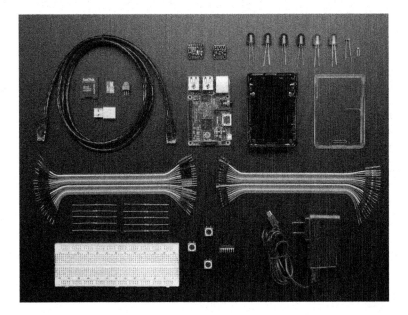

Figure 9-1. *The Adafruit Windows IoT Pack for Raspberry Pi 3 (courtesy of adafruit.com)*

Components Included

As you can see, there are a lot of pieces in the kit. There are three categories of components: electronic components, accessories for the Raspberry Pi, and sensors.

The electronic components provided in the kit include the following.

- (2) breadboard trim potentiometer

- (5) 10K 5% 1/4W resistor

- (5) 560 ohm 5% 1/4W resistor

- (2) diffused 10mm blue LED

- (2) diffused 10mm red LED

- (2) diffused 10mm green LED

- (1) electrolytic capacitor: 1.0uF

- (3) 12mm tactile switches[2]

The list of accessories in the kit is long. The following includes all the accessories included in the kit. I describe some of these in more detail.

- *Adafruit Raspberry Pi B+ case, smoke base/clear top*: An excellent case to protect your Pi from accidents.

- *Full-size breadboard*: Plenty of space to spread out your circuits.

- *Premium male-to-male jumper wires, 20 × 6 inches (150mm)*: Used to jump from one port to another on the breadboard. They're extra long and come molded in a ribbon so you can peel off only those you need.

- *Premium female-to-male extension jumper wires, 20 × 6 inches*: Used to jumper from male GPIO pins to the breadboard ports. They also come molded in a ribbon.

- *Miniature Wi-Fi module*: A Raspberry Pi–approved Wi-Fi dongle (not needed for the Raspberry Pi 3).

- *5V 2A switching power supply with a 6-foot micro-USB cable*: Meets the Raspberry Pi requirements for power.

- *MCP3008 – 8-channel 10-bit ADC with SPI interface*: A breakout board that you can use to expand the number of SPI interface channels for larger IoT projects.

- *Ethernet cable, 5-foot*: A nice touch considering the kit has a Wi-Fi dongle—good to have a backup plan!

- *8GB class 10 SD/micro-SD memory card*: Windows 10 IoT core preloaded!

The sensors included with the kit are an unexpected surprise. They provide what you need to create some interesting IoT solutions. Best of all, they are packaged as breakout boards making them easy to wire into our circuits. The following lists the sensors included in the kit.

- *1 photocell*: A simple component used to measure light.

- *Assembled Adafruit BME280 temperature, pressure, and humidity sensor*: Measures temperature, barometric pressure, and humidity.

- *Assembled TCS34725 RGB color sensor*: Measures color. Comes with an infrared filter and white LED.

Some of the parts in this kit require a bit more explanation. The following sections describe some of the more interesting parts in more detail.

Environmental Sensor: BME280

This sensor is great for all sorts of environmental sensor projects. It features both I2C[3] and SPI[4] interfaces making this a very versatile breakout board. Figure 9-2 shows an image of the BME280 sensor breakout board.

---

[2]Technically, switches are the simplest of all sensors.
[3]https://en.wikipedia.org/wiki/I%C2%B2C
[4]https://en.wikipedia.org/wiki/Serial_Peripheral_Interface_Bus

Figure 9-2. *Environmental sensor: BME280 (courtesy of adafruit.com)*

■ **Note**   The photos show the breakout boards without the headers (row of pins) soldered. The components included in the kit come with the headers soldered in place.

The sensor on the board has a ±3% accuracy for measuring humidity, barometric pressure with ±1 hPa absolute accuracy, and temperature within ±1.0°C accuracy. In addition, the breakout board includes a 3.3V regulator and level shifting so you can use it with a 3V or 5V connections. For more information about how to use the sensor, see the tutorial at `https://learn.adafruit.com/adafruit-bme280-humidity-barometric-pressure-temperature-sensor-breakout`.

Color Sensor: TCS34725

If you want to measure light beyond the basic intensity that the photocell sensor provides, such as determining the color of light, you can use this sensor to add that capability to your IoT projects. Adafruit combines a highly accurate color sensor, the TCS34725, and bundles it with other components to make a sensor capable of "seeing" infrared and more. Figure 9-3 shows a photo of the color sensor.

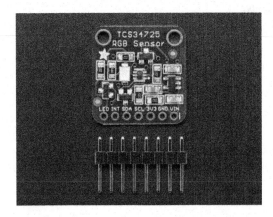

Figure 9-3. *Color sensor: TCS34725 (courtesy of adafruit.com)*

Like the environmental sensor, it has a 3.3V regulator with level shifting for the I2C pins so they can be used with 3.3V or 5V. You can find out more about the color sensor at http://learn.adafruit.com/adafruit-color-sensors/overview.

8-Channel 10-Bit ADC with SPI Interface: MCP3008

If you need more analog inputs for your IoT project, Adafruit has provided a nifty integrated circuit in the form of the MCP3008 that you can use to add additional inputs. It uses an SPI interface so you only need to use 4 pins to connect to the chip. Figure 9-4 shows a photo of the IC.

Figure 9-4. *8 Channel 10-bit ADC with SPI interface (courtesy of adafruit.com)*

If you would like to see how to use this IC, see http://learn.adafruit.com/reading-a-analog-in-and-controlling-audio-volume-with-the-raspberry-pi.

Are There Alternatives?

If you are planning to use a board other than the Raspberry Pi, are on a more limited hobby budget, or want only the bare essentials, there are alternatives to the Microsoft IoT Pack from Adafruit. In fact, Adafruit sells another kit that includes almost everything you need for the projects in this book. It doesn't come with sensors, but all the basic bits and bobs are in there, and you can always buy the sensors separately.

The Adafruit Parts Pal comes packaged in a small plastic case with a host of electronic components (www.adafruit.com/products/2975). Figure 9-5 shows the Parts Pal kit.

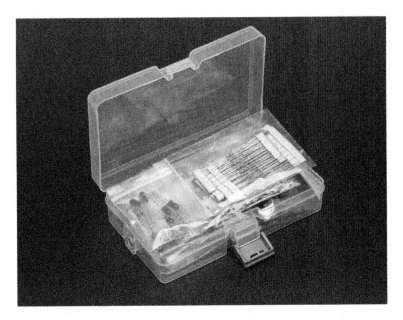

Figure 9-5. *Adafruit Parts Pal (courtesy of adafruit.com)*

The kit includes the following components: prototyping tools, LEDs, capacitors, resistors, some basic sensors, and more. In fact, there are more components in this kit than the Windows IoT Pack for the Raspberry Pi 3. Better still, the kit costs only $19.95 making it a good deal (and the case is a great bonus).

- (1) storage box with latch
- (1) half-size breadboard
- (20) jumper wires: male-to-male, 3 inches (75mm)
- (10) jumper wires: male-to-male, 6 inches (150mm)
- (5) 5mm diffused green LEDs
- (5) 5mm diffused red LEDs
- (1) 10mm diffused common-anode RGB LED
- (10) 1.0uF ceramic capacitors
- (10) 0.1uF ceramic capacitors
- (10) 0.01uF ceramic capacitors
- (5) 10uF 50V electrolytic capacitors
- (5) 100uF 16V electrolytic capacitors
- (10) 560 ohm 5% axial resistors
- (10) 1K ohm 5% axial resistors
- (10) 10K ohm 5% axial resistors
- (10) 47K ohm 5% axial resistors

- (5) 1N4001 diodes

- (5) 1N4148 signal diodes

- (5) NPN transistor PN2222 TO-92

- (5) PNP transistor PN2907 TO-92

- (2) 5V 1.5A linear voltage regulator, 7805 TO-220

- (1) 3.3V 800mA linear voltage regulator, LD1117-3.3 TO-220

- (1) TLC555 wide-voltage range, low-power 555 timer

- (1) photocell

- (1) thermistor (breadboard version)

- (1) vibration sensor switch

- (1) 10K breadboard trim potentiometer

- (1) 1K breadboard trim potentiometer

- (1) Piezo buzzer

- (5) 6mm tactile switches

- (3) SPDT slide switches

- (1) 40-pin break-away male header strip

- (1) 40-pin female header strip

The only thing that I feel is missing are the male/female jumpers, but you can buy them separately (www.adafruit.com/product/1954). For only $1.95 more, they're worth adding to your order!

■ **Tip** If you want to save some money and don't need the accessories in the Windows IoT Pack for Raspberry Pi 3, you should consider buying the Adafruit Parts Pal and male/female jumpers. With a cost of about $22, they're a great bargain.

Now, let's put our new kit to work with a simple project that uses a very simple sensor.

Example Project: A Simple Sensor

The projects thus far in the book have not used a sensor or read any input other than interacting with the user (which is still a form of sensing). In this project, you see how to write an IoT solution that uses a simple sensor (a pushbutton) and models a real-life solution that uses sensors. No matter which kit you decide to buy, each contains a pushbutton.

While the pushbutton is easy to use, the code is a bit more complicated than the examples from previous chapters. This is not due to the complexity of the problem, rather, due to a new concept that you must consider when writing applications without graphical interfaces that use facilities from the UWP libraries. That is, you have a problem dealing with scope of the threads. You'll see this in the following code section.

The solution you're modeling is a subset of a typical traffic light in an urban setting - a pedestrian crosswalk pushbutton. More specifically, you implement a single traffic light for a one-way street with only a single crosswalk button so that you can keep the circuit simple. You can extend the circuit to include two buttons if you would like and I encourage you to do so once you've mastered the project as written.

So how does this pedestrian crosswalk button work? When a pedestrian presses the crosswalk request button, the traffic light cycles from green to yellow to red, and then the walk signal cycles from DON'T WALK to WALK. A yellow LED is used for WALK and a red LED is used for DON'T WALK. After some time, the walk light flashes, warning the pedestrian that the traffic cycles back to green soon.

Thus, if you watch how the traffic lights work when you signal that you want to cross the street (at least in some US cities), notice that there are several states that the lights go through. I have simplified the states a bit as follows. I use *cycle* to indicate one light is turned off and another is turned on.

1. In the default state, the traffic light is green and the walk light is red.

2. When a pedestrian presses the walk button, the traffic light waits a few seconds and then cycles to yellow.

3. After a few seconds, the traffic light cycles to red.

4. After a few seconds, the walk light cycles to yellow.

5. After a few seconds, the yellow walk light begins to blink.

6. After a few seconds, the walk light cycles to red.

7. After a few seconds, the traffic light cycles to green—returns to state (1).

You learn how to write code to execute these states in a later section, but first, let's look at the hardware that you need to build a circuit for the project.

Required Components

The following lists the components you need. You can find these components in either of the kits mentioned previously or you can purchase the components separately from Adafruit (www.adafruit.com), SparkFun (www.sparkfun.com), or any electronics store that carries electronic components.

- (1) pushbutton (breadboard pin spacing)
- (2) red LEDs
- (2) yellow LEDs (or blue is OK)
- (1) green LED
- (5) 150 ohm resistors (see upcoming notes)
- jumper wires
- breadboard (full size recommended but half size is OK)
- Raspberry Pi 2 or 3
- power supply

Set up the Hardware

This project has many more connections than the projects thus far in the book. In order to set up the hardware correctly and make all the connections that you need, it is always best to make a plan for how things should connect.

For example, since you are using five LEDs and a pushbutton, as well as at least one connection to ground, you make seven connections to the GPIO header. Keeping all of those connections straight and planning which GPIO pins to use could be tricky if you didn't have a plan. I like to call my wiring plans "maps" because they map the connections from the GPIO header to the breadboard. Table 9-1 shows the map I designed for this project. Notice I leave a space for you to make any notes as you learn more about the connections and code.

Table 9-1. *Connection Map for Pedestrian Crossing Project*

GPIO	Connection	Function	Notes
4	Resistor for LED	Red traffic light	
5	Resistor for LED	Yellow traffic light	
6	Resistor for LED	Green traffic light	
16	Button	Walk request	
20	Resistor for LED	Red walk light	
21	Resistor for LED	Green walk light	
GND	Breadboard power rail	Ground	

Recall that you must use a resistor when connecting an LED directly to power (in this case a GPIO pin set to HIGH) because the LED does not operate at 5V. Furthermore, LEDs are not all rated the same. Their power requirements can vary from one manufacturer to another as well as one color to another. That is, a green LED may have different power requirements than a blue LED. Thus, you must check the manufacturer (or vendor) to get the data sheet for the LED and write down the power requirements. Table 9-2 shows a number of different LEDs from Adafruit and SparkFun. I used Ohm's law (see Chapter 7) to figure out the right size LED.

Table 9-2. *Various LEDs and Resistors for 5V*

Source	LED	Power Requirements	Resistor Needed
Adafruit Microsoft IoT Pack	10mm red	1.85–2.5V, 20mA	150 ohms
	10mm blue	3.0–3.4V, 20mA	100 ohms
	10mm green	2.2–2.5V, 20mA	150 ohms
SparkFun	3mm red	1.9–2.3V, 20mA	180 ohms
	3mm yellow	2.0–2.4V, 20mA	150 ohms
	3mm green	2.0–2.5V, 20mA	150 ohms

If you do not have the resistor listed, you can use the next higher resistor value or the one closest to but greater than the value listed. You can use the next higher value resistor safely because the higher the value of the resistor, the less current is fed to the LED. You should not use a smaller resistor value because too much current damages the LED.

■ **Tip**   There are several online LED resistor calculators. I used the one at `http://led.linear1.org/1led.wiz` for this data.

COOL GADGET: GPIO REFERENCE CARD

There is a very cool gadget that helps you sort out the connections with ease. It is called the GPIO Reference Card for Raspberry Pi 2 or 3 and is available from Adafruit (www.adafruit.com/products/2263). The following shows how handy it is to use when making connections to the GPIO.

To use this gadget, place it over the GPIO pins on your Raspberry Pi. Now, when you make the connections, you can clearly see which pin to use! How cool is that?

Since you are going to model a traffic light and a walk light, it would be best if you arrange the components so that the three LEDs for the traffic light are grouped together and likewise the walk light are grouped together. I arranged the components on my breadboard in this way, as shown in Figure 9-6.

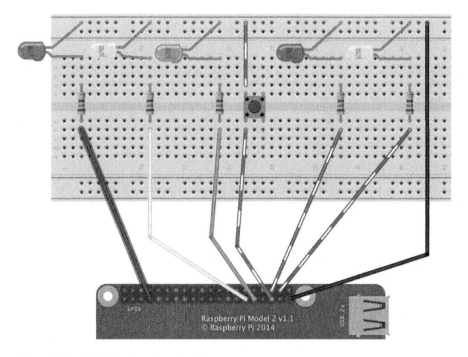

Figure 9-6. *Pedestrian crossing circuit*

Notice how I arranged the LEDs. More specifically, notice that I plugged the negative leg of each LED into the negative side of the power rail on the breadboard. This allows me to make one connection from one of the ground pins on the GPIO to the breadboard rail, which I can then use to plug in the ground side of the components. For example, each positive leg of each LED is plugged into the breadboard so that you can plug the corresponding resistor across the DIP trough and connect those to the appropriate GPIO pin. Recall there are no connections from one side of the trough to the other.

Also notice that I placed the button in the center spanning the DIP trough. One side is connected to ground and the other to the appropriate GPIO pin. If you unsure which way to orient the pushbutton, you can use a multimeter to test continuity among the pins. Use the pins where continuity is found (the multimeter beeps) when the pushbutton is pressed but no connection when released.

■ **Note**   If you do not have a yellow LED, the blue LED in the kit can be used in its place. Just be sure to use the right resistors as described earlier.

If you are following along with this chapter working on the project, go ahead and make the hardware connections now. Don't power on the board yet, but do double- and triple-check the connections.

Write the Code

The code for this project is a bit more involved than the previous projects. This is partly because of the extra LEDs, but more so because you want to use a button and need to write code to determine when the sensor indicates the event (the button is pressed). You also need to use the `DispatcherTimer` class to control the light sequence like you did in the last project.[5] More importantly, you will not use a user interface, which simplifies the code a bit (but you can add a user interface if you want).

Just like you planned the GPIO connections, you also need to plan the code so that all goes as well as possible on the first attempt. It is a rare case that your code works on the first implementation (unless you're following an example like this one). The following lists the major decisions and features/areas of the code. I explain each in upcoming sections.

- Which project template do you want to use?

- How should the lights work?

- How do you read the button events?

Perhaps the more important question is which language will you use? I've decided to use C#, but you could implement this project in C++. If you are a big C++ fan, I encourage you to do just that using the following as a pattern!

New Project

You may be wondering what project template to use. Since you will not have a user interface, you may be tempted to use a headless, background application. However, you cannot use such a template because you want to use the dispatcher timer, which is only available in the headed project types. You can still use a blank template and can run the application without a user interface, but to get the support for the timer, you need to use a headed application template.

Thus, open a new project and choose the **Blank App (Universal Windows)** project choice from the **C# ➤ Windows ➤ Universal** selection in the tree on the left in the **New Project** dialog. Use the name **PedestrianCrossing** for the project name. You can save the project wherever you like or use the default. Figure 9-7 shows an example of the project template you should choose.

---

[5]There are other ways to do this, but this uses a technique you've seen previously.

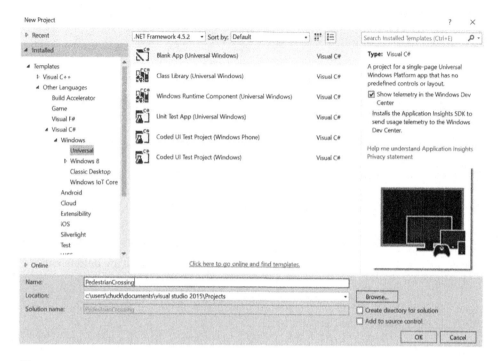

Figure 9-7. *New Project Template selection*

Once the project opens, double-click the `MainPage.xaml.cs` file. This is where you put all the code for the project. There are two namespaces that you need to include. Go ahead and add those now, as shown here.

```
using Windows.Devices.Gpio;    // GPIO header
using Windows.UI.Core;         // DispatcherTimer
```

Next, you need to add some variables and constants. Let's start with constants for the lights. You use this constant with an array for each of the traffic and walk lights. By using constants instead of integers, you make the code easier to read.[6]

```
// Light constants
private const int RED = 0;
private const int YELLOW = 1;
private const int GREEN = 2;
```

Next, let's define the pins for each of our components: the traffic light, button, and the walk lights.

```
// Traffic light pins
private int[] TRAFFIC_PINS = { 4, 5, 6 };
```

---

[6]Which is always a good choice and worth the effort.

```
// Button pin
private const int BUTTON_PIN = 19;

// Walk light pins
private int[] WALK_PINS = { 20, 21 };
```

Next, you add constants to describe the states or stages that the lights cycle through. In this case, the values (integers) represent the time sequence in seconds when the lights change. Once again, using constants makes the code easier to read, as you shall see later.

```
// State constants
private const int GREEN_TO_YELLOW = 4;
private const int YELLOW_TO_RED = 8;
private const int WALK_ON = 12;
private const int WALK_WARNING = 22;
private const int WALK_OFF = 30;
```

Now you can create the variables to hold the GPIO pin instances for the traffic light, button, and walk light.

```
// Traffic light pin variables
private GpioPin[] Traffic_light = new GpioPin[3];

// Walk light pin variables
private GpioPin[] Walk_light = new GpioPin[2];

// Button pin variable
private GpioPin Button;
```

Since you decide to use the DispatcherTimer class, you can create the variable for that too.

```
// Add a Dispatcher Timer
private DispatcherTimer walkTimer;
```

Next, you need a variable to use for counting the seconds that have elapsed since the light sequence was started. This allows us to keep certain LEDs on for a specific time period.

```
// Variable for counting seconds elapsed
private int secondsElapsed = 0;
```

Finally, recall from the last project that you need an InitGPIO() method, as well as code, to set up the GPIO. As you discovered in the last project, you use this method to set up the GPIO pins for all the LEDs and buttons. Listing 9-1 shows the completed code for this method.

Listing 9-1. GPIO Initialization Code

```
// Setup the GPIO initial states
private void InitGPIO()
{
    var gpio = GpioController.GetDefault();

    // Do nothing if there is no GPIO controller
```

```
if (gpio == null)
{
    return;
}
// Initialize the GPIO pins
for (int i = 0; i < 3; i++)
{
    this.Traffic_light[i] = gpio.OpenPin(TRAFFIC_PINS[i]);
    this.Traffic_light[i].SetDriveMode(GpioPinDriveMode.Output);
}
this.Button = gpio.OpenPin(BUTTON_PIN);
for (int i = 0; i < 2; i++)
{
    this.Walk_light[i] = gpio.OpenPin(WALK_PINS[i]);
    this.Walk_light[i].SetDriveMode(GpioPinDriveMode.Output);
}
this.Traffic_light[RED].Write(GpioPinValue.Low);
this.Traffic_light[YELLOW].Write(GpioPinValue.Low);
this.Traffic_light[GREEN].Write(GpioPinValue.High);
this.Walk_light[RED].Write(GpioPinValue.High);
this.Walk_light[YELLOW].Write(GpioPinValue.Low);
}
```

Notice how I used the arrays to initialize the LEDs. Isn't the code easier to read with the constants? Take a moment to ensure that you understand how the LEDs are set up. That is, the green LED in the traffic light and the red LED in the walk light are on (set to HIGH) at the initial state. This mimics the scenario where the pedestrian approaches a busy street where traffic is flowing (hence the green traffic light).

Finally, I added code to return from the method if the GPIO library cannot be initialized. Recall you used a text label on the user interface to communicate an error. Here, the absence of any LEDs illuminating indicates that something is wrong.

Be sure to add the method call to the MainPage() method right after the InitializeComponent() call. Also, you set the initial value for secondsElapsed to 0.

```
public MainPage()
{
    this.InitializeComponent();
    InitGPIO();
    this.secondsElapsed = 0;
}
```

That may seem like a lot of initialization code and I suppose it is compared to the previous projects, but you use all of these in the rest of the code. Now that you have the basic project code, let's add the code for controlling the LEDs.

Light Sequences

Recall from the description of the project that you want the lights to cycle through several stages based on time. Thus, you use the DispatcherTimer class to start a timer that you then count as seconds expired to control the stages. This is done by making a new method to fire whenever the timer event occurs. You named the DispatcherTimer class walkTimer earlier, so you use WalkTimer_Tick() for the event method.

Inside this method, you use the secondsElapsed variable to count the seconds (tick events) and when the count reaches the number you assigned to the state changes, you change the light. For example, when the sescondsElapsed reaches 4 (GREEN_TO_YELLOW), you execute state (2) from earlier, where the green traffic light is turned off and the yellow traffic light is turned on. You do the same for the other states. Listing 9-2 shows the code you need to control the light sequences.

Listing 9-2. Light Sequence Code

```
// Here you do the lights state change if and only if elapsed_seconds > 0
private void WalkTimer_Tick(object sender, object e)
{
    // Change green to yellow
    if (this.secondsElapsed == GREEN_TO_YELLOW)
    {
        this.Traffic_light[GREEN].Write(GpioPinValue.Low);
        this.Traffic_light[YELLOW].Write(GpioPinValue.High);
    }

    else if (this.secondsElapsed == YELLOW_TO_RED)
    {
        this.Traffic_light[YELLOW].Write(GpioPinValue.Low);
        this.Traffic_light[RED].Write(GpioPinValue.High);
    }
    else if (this.secondsElapsed == WALK_ON)
    {
        this.Walk_light[RED].Write(GpioPinValue.Low);
        this.Walk_light[YELLOW].Write(GpioPinValue.High);
    }
    else if ((this.secondsElapsed >= WALK_WARNING) &&
            (this.secondsElapsed < WALK_OFF))
    {
        // Blink the walk warning light
        if ((secondsElapsed % 2) == 0)
        {
            this.Walk_light[YELLOW].Write(GpioPinValue.Low);
        }
        else
        {
            this.Walk_light[YELLOW].Write(GpioPinValue.High);
        }
    }
    else if (this.secondsElapsed == WALK_OFF)
    {
        this.Walk_light[YELLOW].Write(GpioPinValue.Low);
        this.Walk_light[RED].Write(GpioPinValue.High);
        this.Traffic_light[RED].Write(GpioPinValue.Low);
        this.Traffic_light[GREEN].Write(GpioPinValue.High);
        this.secondsElapsed = 0;
        this.walkTimer.Stop();
```

```
        return;
    }
    // increment the counter
    this.secondsElapsed += 1;
}
```

Notice you used a return in the state where you turn the walk light off. This terminates the code so that the seconds elapsed variable is not increments. You also reset the secondsElapsed variable so that the next pedestrian can initiate the crossing. You also increment the variable each time the event fires otherwise.

Button

The button is a sensor that you read from the GPIO. In order to do so, you need to add some code to the InitGPIO() method to set up the button. This code is very similar to the LED startup code as shown here.

```
// Check if input pull-up resistors are supported
if (this.Button.IsDriveModeSupported(GpioPinDriveMode.InputPullUp))
    this.Button.SetDriveMode(GpioPinDriveMode.InputPullUp);
else
    this.Button.SetDriveMode(GpioPinDriveMode.Input);

// Set a debounce timeout to filter out switch bounce noise from a button press
this.Button.DebounceTimeout = TimeSpan.FromMilliseconds(50);

// Register for the ValueChanged event so our Button_ValueChanged
// function is called when the button is pressed
this.Button.ValueChanged += Button_ValueChanged;
```

Notice you first check to see if the GPIO header has pull up resistors (it "pulls" the voltage high)[7] and if it does, you use the GpioPinDriveMode.InputPullUp mode for the GPIO pin. If it does not have pull up resistors, you simply set the mode to GpioPinDriveMode.Input. Recall you used the SetDriveMode() method for the LEDs but set the mode to GpioPinDriveMode.Output.

You also need to create a method to fire when the button is pressed. You'll name that method Button ValueChanged(). You need to assign the ValueChanged attribute of the button to this method. You see that in the last line of code.

Now you can write the code to execute when the button is pressed. This may seem like a simple thing, but there is one aspect you must consider. When a button is pressed, it may not make a connection right away or it may be the case that there is some hesitation on the user's part. This creates a condition where the event could trigger prematurely. This is called bouncing and can be controlled with a bit more code. In this case, you check the edge attribute of the button to see if the state is on the trailing edge. In other words, that the button has been pressed for a period of time. The condition you use is (e.Edge == GpioPinEdge. FallingEdge).

There is one trickier bit. Once the button event fires, you cannot call back to the main code directly. This is because the button event is running simultaneously with the main code. More specifically, there are two threads involved. Thus, you need to make a call to the dispatcher to run a task. In this case, you want to simply turn on the timer. Listing 9-3 shows the method that you use to execute when the button is pressed.

---

[7]See https://learn.sparkfun.com/tutorials/pull-up-resistors.

Listing 9-3. Button Event Code

```
// Detect button press event
private void Button_ValueChanged(GpioPin sender, GpioPinValueChangedEventArgs e)
{
    // Pedestrian has pushed the button. Start timer for going red.
    if (e.Edge == GpioPinEdge.FallingEdge)
    {
        // Start the timer if and only if not in a cycle
        if (this.secondsElapsed == 0)
        {
            // need to invoke UI updates on the UI thread because this event
            // handler gets invoked on a separate thread.
            var task = Dispatcher.RunAsync(CoreDispatcherPriority.Normal, () =>
            {
                if (e.Edge == GpioPinEdge.FallingEdge)
                {
                    this.walkTimer.Start();
                }
            });
        }
    }
}
```

Notice the code to make the call to the asynchronous feature of the dispatcher. The code is formatted to read like normal code but it is actually a method inside the RunAsync() call. This is an advanced feature that you may have to use in some of your projects. For more detailed information about this feature, see the online documentation at https://msdn.microsoft.com/en-us/library/windows/apps/windows.ui.core. coredispatcher.runasync.

Notice one more thing. I use another conditional to only allow turning on the dispatcher if the secondsElapsed variable is 0. Thus, even if someone pressed the button once the light sequence starts, it is ignored until the variable returns to 0 at the end of the light sequence code. Cool, eh?

Completing the Code

OK, now you have all the pieces of code that you need, but you may not know where each piece goes. Rather than relist the entire code, I include the skeleton of the code in Listing 9-4 to help you put things in the right places. The ellipse is used to represent code omitted.

Listing 9-4. Pedestrian Crossing Code Layout

```
using System;
using System.Collections.Generic;
using System.IO;
using System.Linq;
using System.Runtime.InteropServices.WindowsRuntime;
using Windows.Foundation;
using Windows.Foundation.Collections;
using Windows.UI.Xaml;
using Windows.UI.Xaml.Controls;
using Windows.UI.Xaml.Controls.Primitives;
```

```
using Windows.UI.Xaml.Data;
using Windows.UI.Xaml.Input;
using Windows.UI.Xaml.Media;
using Windows.UI.Xaml.Navigation;
using Windows.Devices.Gpio;     // GPIO header
using Windows.UI.Core;          // DispatcherTimer

namespace PedestrianCrossing
{
    public sealed partial class MainPage : Page
    {
        // Light constants
...

        public MainPage()
        {
...
        }

        // Setup the GPIO initial states
        private void InitGPIO()
        {
...
        }

        // Detect button press event
        private void Button_ValueChanged(GpioPin sender,
                                    GpioPinValueChangedEventArgs e)
        {
...
        }

        // Here you do the lights state change iff elapsed_seconds > 0
        private void WalkTimer_Tick(object sender, object e)
        {
...
        }
    }
}
```

Wow, that is a lot of code! Clearly, this project is larger than the examples so far. If you have not written a project of this size, be sure to check the listing above to ensure that you have all the code in the right place. Once you have entered all the code, you should now attempt to compile the code. Correct any errors you find until the code compiles without errors or warnings.

Deploy and Execute

Now it is time to deploy our application! Be sure to fix any compilation errors first. Like you have with other applications, you want to compile the application in debug first (but you can compile in release mode if you'd prefer) and you must turn on the debugger on our board. You do this with the device portal.

Go ahead and power on your board, and then connect to the board. Recall you can simply navigate to the IP address (or name) of the board and append 80 to the address like this: 10.0.1.89/8080. Figure 9-8 shows the debugger panel for the device portal. Click the **Start** button to turn on the debugger.

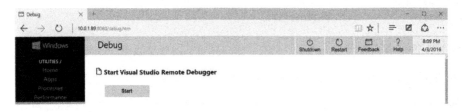

Figure 9-8. *Device Portal: debugger*

Next, you must set the project properties to target the device and run with the remote debugger. In this case, you must modify two settings: the **Remote machine** name and the **Authentication Mode**. Set the **Remote machine** name to the IP address of your device with the port specified by the remote debugger when you started it from the device portal. For example, my device gave me the port number 8116 so I use 10.0.1.89:8116. For the **Authentication Mode**, set it to **None**. Figure 9-9 shows the location of these options.

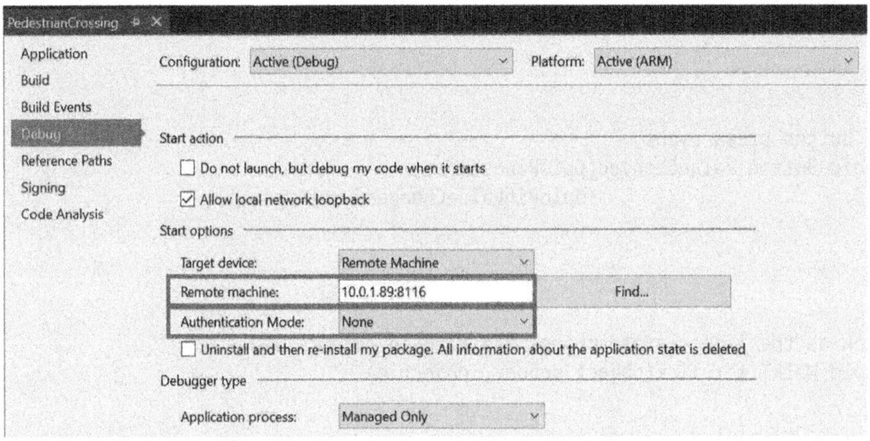

Figure 9-9. *Project properties debug*

Now you can deploy your application. Go ahead and do that now. Recall you can run the deployment from the Debug menu. Once the application deploys, you see the application appear in the list on the Apps pane in the device portal, as shown in Figure 9-10.

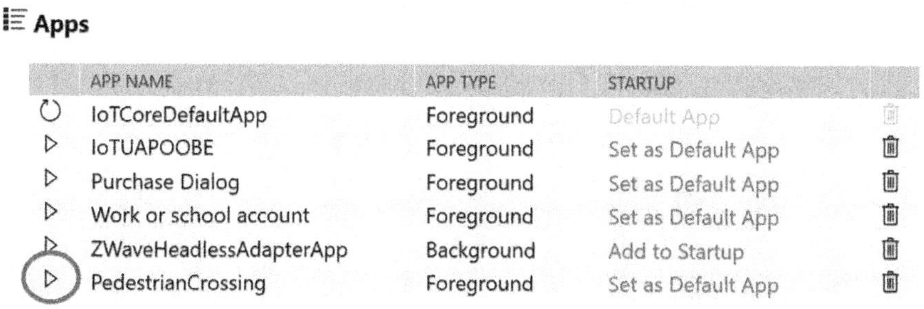

☰ Apps

	APP NAME	APP TYPE	STARTUP	
○	IoTCoreDefaultApp	Foreground	Default App	🗑
▷	IoTUAPOOBE	Foreground	Set as Default App	🗑
▷	Purchase Dialog	Foreground	Set as Default App	🗑
▷	Work or school account	Foreground	Set as Default App	🗑
▷	ZWaveHeadlessAdapterApp	Background	Add to Startup	🗑
▷	PedestrianCrossing	Foreground	Set as Default App	🗑

Figure 9-10. Device Portal: Apps pane

Notice there is a small triangle or arrow next to PedestrianCrossing in the list. This indicates you can click it to start the application. When the application is running, the icon changes to a square box, which, when clicked, stops the application. Finally, you can also set the app as the default application to start when the device boots and you can delete the application using the trashcan icon.

■ **Note** If you are seeing a slightly different list, it may be because your image is older than the version that I am using. Older versions had a more primitive set of controls for the Apps pane.

If everything worked correctly, you should see the lights change when the button is pressed. If the lights don't change, be sure you have the pushbutton oriented correctly. Once it is working, try it out a few times to ensure that you see the lights sequence several times. If you like this project, I encourage you to experiment with it further by either adding another button or adding a second traffic light for an intersection rather than a one-way street.

Summary

Learning how to build IoT projects requires a number of electronic components. Fortunately, vendors such as Adafruit sell electronic component kits that have a wide array of common electronic components, such as LEDs, resistors, jumper wires, breadboards, and more. If you are planning to explore Windows IoT projects and especially if you want to work on the more advanced projects in the next few chapters, you should consider buying one (or both) of the kits described in this chapter or a similar kit from another vendor. Just be sure you get the data sheets for all the components in the kits.

You also explored a project that introduces how to read sensors. If you have been implementing all the projects in this book so far or have been reading through them in some detail, you now have the skills (or at least examples of how) to read and write values on GPIO pins. Indeed, you have now seen the basics for writing any form of IoT project.

The next chapter begins a series of example IoT projects that you can use to learn more about building IoT solutions. I recommend working through as many of them as you can. Some require additional hardware that you may or may not want to acquire. Should you decide to not implement some of the projects, you can read through the projects to learn how to use the components and get ideas for your own designs.

▓ ▓ ▓

## CHAPTER 10

Project 1: Building an LED Power Meter

You've seen a lot of examples of powering LEDs—turning them on and off from our IoT device.[1] There are a lot of interesting things you can do with LEDs. For example, have you wondered how a power meter works? If you have used high end audio equipment, such as a studio sound board, you may have seen a power meter that has several segments ranging from green to yellow to red where green means the level is low, yellow is medium, and red is high. You can duplicate this behavior with a set of LEDs and your IoT device. Yes, you're going to build a fancy LED power meter!

While this seems very simple and in concept it is, this project helps you learn quite a lot. You will see how to use a potentiometer as a variable input device read from an analog to digital converter (ADC), learn how to set up and use a serial peripheral interface (SPI), discover a powerful debugging technique, and learn how to create a class to encapsulate functionality. Clearly, there is a lot to discover, so let's get started.

▓ **Note**   Although this project is written in C#, other than the syntax and mechanics of building, the concepts of using a class are the same in C++.

Overview

You will design and implement an LED power meter that allows you to simulate controlling power and displaying the result as a percentage of a set of LEDs where no LEDs on means minimum and all LEDs on means maximum. More specifically, you use a component called a *potentiometer* to read and interpret its value as a percentage of its maximum range. You will use the potentiometer through the ADC to control the LEDs.

The potentiometer is a special rotary component that can vary resistance as you turn it. Thus, potentiometers are rated at their maximum resistance. There are a variety of potentiometers packaged in a variety of ways, but you will use a simple 10K ohm potentiometer that you can plug into a breadboard.

Although you will use a series of LEDs for the power meter, you can buy components where LEDs are arranged in a bar. In fact, they're often called *bar graph LEDs*. If you want to take this project a bit further, you can find a number of different bar graph LEDs like those from Adafruit (www.adafruit.com/categories/279).

---

[1]And here's another one!

© Charles Bell 2016
C. Bell, *Windows 10 for the Internet of Things*, DOI 10.1007/978-1-4842-2108-2_10

You need to use the ADC because you want to read analog values. The Raspberry Pi (and many other boards) does not have analog-to-digital logic. That is, the GPIO pins are digital only. The ADC acts like a "bridge" between digital and analog devices. The ADC you will use is the MCP3008 from the Adafruit Microsoft IoT Pack for Raspberry Pi (see Chapter 9). The ADC is an integrated circuit (or chip) that you will plug into a breadboard and wire it to your Raspberry Pi. As a side benefit, the MCP3008 has eight analog input pins but you only need four pins on the GPIO to access it so you're gaining four more pins. This may not be important for this simple project, but for project with many sensors or devices, saving four pins may enable you to fully implement your ideas.

The MCP3008 uses the Serial Peripheral Interface (SPI) bus to communicate with the Raspberry Pi. The SPI bus is an interface developed by Motorola as a synchronous serial clocked, full-duplex master/slave protocol.[2] In other words, data is transmitted in a synchronized manner timed to a clock signal. The protocol supports full duplex (which means both transmit and receive at the same time). The four wires therefore are one for the clock, one for transmit, one for receive, and one additional wire used to select the chip (the Raspberry Pi can support two SPI buses but only in master mode).

One interesting thing about the SPI is that in order to receive data (say a byte), you must first send data (a byte), which sounds really weird but it turns out the first transmission can be thought of as a command to read data and the response is the return of the command. You'll see how this works in the code.

Let's look at the components that you need, and then look at how to wire everything together.

Required Components

The following lists the components that you need. You can find these components in either of the kits mentioned in Chapter 9 or you can purchase the components separately from Adafruit (www.adafruit.com), SparkFun (www.sparkfun.com), or any electronics store that carries electronic components.

- (1) 10K ohm potentiometer (breadboard pin spacing)
- (2) red LEDs
- (2) yellow LEDs (or blue is OK)
- (1) green LED
- (5) 150 ohm resistors (or appropriate for your LEDs)
- MCP3008 ADC chip
- Jumper wires: (8) male-to-male, (11) male-to-female
- Breadboard (full size recommended but half size is OK)
- Raspberry Pi 2 or 3
- Power supply

Set up the Hardware

This project like the project in Chapter 9 has a lot of connections. Thus, you will make a plan for how things should connect. To connect the components to the Raspberry Pi, you need four pins for the ADC, five for the LEDs, and one each for power and ground. You will also need to make a number of connections on the breadboard to configure the ADC chip and connect the potentiometer to the ADC. Table 10-1 shows the map I designed for this project. I list the physical pin numbers in parenthesis for the named pins. You will use male-to-female jumper wires to make these connections.

---

[2]For more information about SPI, see https://en.wikipedia.org/wiki/Serial_Peripheral_Interface_Bus.

Table 10-1. *Connection Map for Power Meter Project*

GPIO	Connection	Function	Notes
3.3V (1)	Breadboard power rail	Power	
GND (6)	Breadboard ground rail	GND	
MOSI (19)	SPI	MCP3008 pin 4	
MISO (21)	SPI	MCP3008 pin 4	
SCLK (23)	SPI	MCP3008 pin 3	
CC0 (24)	SPI	MCP3008 pin 5	
17	Red LED #1	Meter 81-100%	
18	RED LED #2	Meter 61-80%	
19	Yellow LED #1	Meter 41-60%	
20	Yellow LED #2	Meter 21-40%	
21	Green LED	Meter 0-21%	

■ **Tip**   Refer to Chapter 9 for the size of resistors needed for your LEDs.

Next, you need to make a number of connections on the breadboard. For these, you use male-to-male jumpers. Table 10-2 shows the connections needed on the breadboard.

Table 10-2. *Connections on the Breadboard*

From	To	Notes
Breadboard power	Potentiometer pin #1	
Potentiometer pin #2	ADC Channel 0 (pin 1)	
Breadboard GND	Potentiometer pin #3	
Breadboard power	ADC VDD (pin 16)	
Breadboard power	ADC VREF (pin 15)	
Breadboard GND	ADC AGND (pin 14)	
Breadboard GND	ADC GND (pin 9)	
Breadboard GND rail	Breadboard GND rail jump	

Clearly, that's a lot of wires! Don't worry too much about neatness when you build this project. Rather, concentrate on making sure everything is connected correctly. I recommend spending some time to carefully check your connections. There's so many that it is easy to get some plugged in the wrong place. Figure 10-1 shows what my project looked like.

Figure 10-1. *Example power meter connections*

As you can see in the photo, I'm cheating a bit by using the Adafruit GPIO Reference Card for Raspberry Pi 2 or 3 (www.adafruit.com/products/2263), which makes locating the SPI pins much easier than counting pin numbers.

The MCP3008 chip is in the center of the breadboard. You cannot see it in this photo, but chips have a small semicircular notch on one end. This indicates the side that contains pin 1 so that you can orient the chip correctly. In this case, you orient the chip on the breadboard with pins 1–8 on the far side of the breadboard (away from the Raspberry Pi as shown in the photo). This is because the SPI interface pins are located on the nearer side (pins 9–16). Orienting the pins closest to the Raspberry Pi makes the connections a bit easier. Figure 10-2 shows the pin layout of the MCP3008. Notice how the pins are numbered. Most are self-explanatory like the channel pins, but all are documented on the data sheet from Adafruit (https://cdn-shop.adafruit.com/datasheets/MCP3008.pdf).

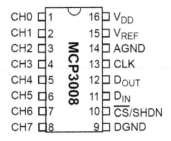

Figure 10-2. *MCP3008 pinout*

Figure 10-3 shows all the connections needed.

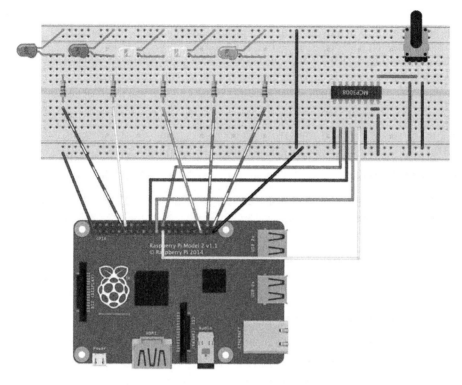

Figure 10-3. *Connections for the power meter project*

Notice how I arranged the LEDs. I placed the red LEDs to the left, the yellow in the middle, and the green to the right. More specifically, I plugged the negative leg of each LED into the GND rail on the breadboard. This allows me to make one connection from one of the ground pins on the GPIO to the breadboard rail, which I can then use to plug in the GND leg of the components. For example, each positive leg of each LED is plugged into the breadboard so that you can plug the corresponding resistor across the DIP trough and connect those to the appropriate GPIO pin. Also, I placed the MCP3008 in the center spanning the DIP trough. Make sure that you plug it in so that pins 1–8 are on one side of the breadboard and that pins 9–16 are on the other side.

If you are following along with this chapter working on the project, make the hardware connections now. Don't power on the board yet; but do double- and triple-check the connections.

Write the Code

Now it's time to write the code for our example. Since you are working with several new components, I introduce the code for each in turn. The code isn't overly complicated, but may not be as clear as some of the code from previous projects. I've decided to use C#, but you could implement this project in C++. If you are a big C++ fan, I encourage you to do just that using the following as a pattern!

■ **Note** Since you have learned all the basics of creating projects in Visual Studio, including how to build and deploy applications, I omit the details of the common operations for brevity.

The project uses a potentiometer. You read its value (via the ADC) and convert that to a scale that you can use to decide how many LEDs to turn on (and consequently those that need to be turned off). You also use a new technique for debugging the code. Thus, you need to start a new project, set up the SPI interface, write code to communicate with the ADC, and code to turn the LEDs on and off. You will use the DispatcherTimer class to periodically check the potentiometer and control the light sequence like you did in the last project.[3] Let's talk about the debugging feature first, and then we will walk through building the project.

Debug Output

Visual Studio supports an interesting and very powerful feature that permits you to insert print statements (and more!) in the code to help you debug your code. It is similar in some ways to the old school print statement trace (a log of statements written to a file), which are written as the code runs. However, in this case, the statements appear in the Output window of Visual Studio. To use the feature, add the following namespace to your code.

```
using System.Diagnostics;          // add this for debugging
```

You use the Debug class from this namespace to write out values so you can see them in the output window as you run the code in debug. The following code shows some examples of the methods you use to report data. Some are informational (the proverbial, "I'm here!"), while others show how you can print out the values of variables.

```
Debug.WriteLine("Sorry, the GPIO cannot be initialized. Drat.");
Debug.Write("Val read = ");
Debug.WriteLine(valRead);
```

Here I use two methods: Write(), which writes a string without a carriage return/line feed (CRLF) symbol (sometimes called a *newline*), and Writeline(), which writes the CRLF.

The output of this code is shown in the output window. The following shows an excerpt of the output from this project. It is to see the progress of the code. There is much more to this class, but this gives you a taste for what is possible with this alternative debugging technique.

```
...
GPIO ready.
Setting up GPIO pin 21.
Setting up GPIO pin 20.
Setting up GPIO pin 19.
Setting up GPIO pin 18.
Setting up GPIO pin 17.
Setting up the MCP3008.
...
```

■ **Tip** For more information about the system diagnostics debug class, see https://msdn.microsoft.com/en-us/library/system.diagnostics.debug%28v=vs.110%29.aspx.

---

[3]There are other ways to do this, but this uses a technique we've seen previously.

You must compile your code in debug mode and run the debugger to see the output. If you compile in release mode, the debug code is ignored automatically and not included in the executable code (binary file).

New Project

You will use the same project template as the last project—the Blank App (Universal Windows) template. Use the name **PowerMeter** for the project name. You can save the project wherever you like or use the default location. Once the project opens, double-click the MainPage.xaml.cs file. This is where you put most of the code for the project. There are three namespaces you need to include. Go ahead and add those now, as shown next.

```
using Windows.Devices.Gpio;     // add this for GPIO pins
using System.Diagnostics;       // add this for debugging
using Windows.UI.Core;          // DispatcherTimer
```

■ **Tip**   You may have noticed some of the namespaces are greyed out. This indicates the namespace may not be needed. If you right-click on one, you can remove any unnecessary namespaces.

Next, you need to add some variables and constants. You add constants for the number of LEDs and maximum potentiometer value (1023). You also add variables for the meter pin numbers and the GPIO pin variables. Finally, you add a variable for the timer.

```
 // Power meter pins, variables
private const int numLEDs = 5;
private const float maxPotVal = 1023;
private int[] METER_PINS = { 21, 20, 19, 18, 17 };
private GpioPin[] Meter = new GpioPin[numLEDs];

// Add a Dispatch Timer
private DispatcherTimer meterTimer;
```

Next, you'll add a variable for the MCP3008 and instantiate an instance for the new class. In this case, you create a new class to encapsulate the code for the ADC. You'll see this in a later section.

```
// Instantiate the new ADC Chip class
ADC_MCP3008 adc = new ADC_MCP3008();
```

Last, you'll add a constant for the channel on the ADC that you will use. In this case, you use channel zero.

```
// Channel to read the potentiometer
private const int POT_CHANNEL = 0;
```

The code in the MainPage() function initializes the components, the GPIO, and the new ADC class. You also set up the timer. In this case, you use a value of 500, which is one-half a second. If you adjust this lower, the solution is a bit more responsive. You can try this once you've got it to work and tested. Listing 10-1 shows the complete MainPage() method.

Listing 10-1. MainPage Method

```
public MainPage()
{
    this.InitializeComponent();
    InitGPIO();                    // Initialize GPIO
    adc.Initialize();              // Call initialize() method for ADC

    // Setup the timer
    this.meterTimer = new DispatcherTimer();
    this.meterTimer.Interval = TimeSpan.FromMilliseconds(500);
    this.meterTimer.Tick += meterTimer_Tick;
    this.meterTimer.Start();
}
```

Finally, you must add the reference to the Windows 10 IoT Extensions from the project property page. You do this by right clicking the **References** item on the project item in the Solution Explorer.

Initialize GPIO

As usual, you must provide the InitGPIO() method you called in the MainPage() method.[4] Recall, you want to set up the GPIO controller and initialize the pins you are going to use. In this case, you must initialize the five LED pins and since you have them in an array, you can use a for loop to accomplish the task.

While you are at it, you'll also use a few diagnostic statements to record whether the GPIO controller succeeds or not and the state of the pins. You can see these calls are made using the Debug.Write(), which does not write an end of line and Debug.Writeline(), which does write the end of line character. Listing 10-2 shows the completed InitGPIO() method.

Listing 10-2. Initialize GPIO Method

```
// Setup the GPIO initial states
private void InitGPIO()
{
    var gpio = GpioController.GetDefault();

    // Show an error if there is no GPIO controller
    if (gpio == null)
    {
        Debug.WriteLine("Sorry, the GPIO cannot be initialized. Drat.");
        return;
    }
    Debug.WriteLine("GPIO ready.");
    // Initialize the GPIO pins
    for (int i = 0; i < numLEDs; i++)
    {
        this.Meter[i] = gpio.OpenPin(METER_PINS[i]);
        this.Meter[i].SetDriveMode(GpioPinDriveMode.Output);
```

---

[4]You don't have to name it InitGPIO(), but that is the common theme. Actually, any reasonable name is fine; but if you are going to share your code, InitGPIO() is a good name.

```
        this.Meter[i].Write(GpioPinValue.Low);
        Debug.Write("Setting up GPIO pin ");
        Debug.Write(METER_PINS[i]);
        Debug.WriteLine(".");
    }
}
```

Controlling the LEDs

Next, you need code to control the LEDs. You're using the timer class to fire an event every 500 milliseconds (one half of a second). In the `MainPage()` method, you assigned the name `meterTimer_Tick()` to the event. In this method, you must read the value from the potentiometer via the new ADC class (explanation is in the next section), and then calculate how many LEDs to turn on.

Since you are using five LEDs, you convert the value read from the potentiometer as a percentage, and then convert it to a scale from 0–5. You do this so you can use two loops: one to turn on those LEDs that should be on and another to turn the remainder off. Check the code to ensure that you understand how this works. You also see a copious amount of `Debug.Write*` methods to display the state of this method. Listing 10-3 shows the complete code for the `meterTimer_Tick()` method.

Listing 10-3. Controlling the LEDs

```
private void meterTimer_Tick(object sender, object e)
{
    float valRead = 0;
    int numLEDs_On = 0;

    Debug.WriteLine("Timer has fired the meterTimer_Tick() method.");

    // Read value from the ADC
    valRead = adc.getValue(POT_CHANNEL);
    Debug.Write("Val read = ");
    Debug.WriteLine(valRead);

    float percentCalc = (valRead / maxPotVal) * (float)10.0;
    numLEDs_On = (int)percentCalc / 2;
    Debug.Write("Number of LEDs to turn on = ");
    Debug.WriteLine(numLEDs_On);

    // Adjust power meter LEDs On or Off based on value read
    for (int i = 0; i < numLEDs_On; i++)
    {
        this.Meter[i].Write(GpioPinValue.High);
        Debug.Write("Setting pin ");
        Debug.Write(METER_PINS[i]);
        Debug.WriteLine(" HIGH.");
    }
    for (int i = numLEDs_On; i < numLEDs; i++)
    {
```

```
            this.Meter[i].Write(GpioPinValue.Low);
            Debug.Write("Setting pin ");
            Debug.Write(METER_PINS[i]);
            Debug.WriteLine(" LOW.");
        }
    }
}
```

Now that you've seen all the code for the main code file, let's look at how the methods are placed in the class. Listing 10-4 shows the skeleton of the MainPage class. Use this as a guide to place the code in the right places. I omit the details of the methods for brevity.

Listing 10-4. MainPage Code Layout

```
...

namespace PowerMeter
{
    public sealed partial class MainPage : Page
    {
...

        public MainPage()
        {
...

        }

        // Setup the GPIO initial states
        private void InitGPIO()
        {
...

        }

        private void meterTimer_Tick(object sender, object e)
        {
...

        }
    }
}
```

In the next section, you learn how to write the code to communicate with the ADC chip.

Code for the MCP3008

While you can write the code in this section in the project class MainPage, you'll discover an example of making your code easier to maintain using a class module to group the code for the ADC. You've already seen a couple of references to this in the previous code, including the constructor (ADC_MCP3008 adc = new ADC_MCP3008();), initialization (adc.Initialize();), and reading a value on a channel (valRead = adc. getValue(POT_CHANNEL);). Thus, you will need these functions, at the minimum.

This class is pretty simple in scope but for larger classes and the objects they model, you want to create a design for the class. That is, you should consider everything the class needs to do to model the object or concept.

To add a new class, right-click the project and choose **Add** ä **New Item…**. Name the class **ADC_ MCP3008**, as shown in Figure 10-4.

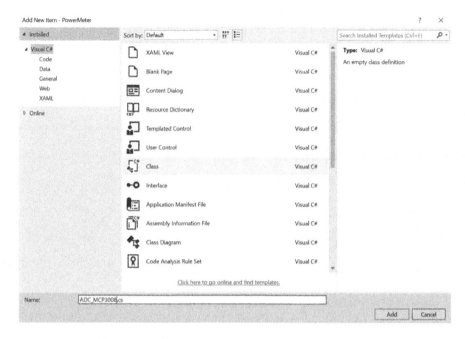

Figure 10-4. *Adding a new class*

When you create a class, you create a new file, as shown in Figure 10-4. Inside this new code file, you see a familiar skeleton. Of note is the same namespace as the main code and a list of namespaces that you will use. Open the file now and add the following namespaces. The need for each is shown in the comments.

```
using Windows.Devices.Spi;         // add this for SPI communication
using System.Diagnostics;          // add this for debugging
using Windows.Devices.Enumeration; // add this for DeviceInformation
```

Next, you need some variables and constants. You need a variable for the SPI device and constants for the MCP3008. In this case, you set constants for helping you read data as you shall see later in this section.

```
// SPI controller interface
private SpiDevice mcp_var;
const int SPI_CHIP_SELECT_LINE = 0;
const byte MCP3008_SingleEnded = 0x08;
const byte MCP3008_Differential = 0x00;
```

Next, you create the constructor method. You don't need to do anything here so you'll use a debug statement to indicate the class has been instantiated as follows.

```
public ADC_MCP3008()
{
    Debug.WriteLine("New class instance of ADC_MCP3008 created.");
}
```

Now that the preliminaries are set up, you can concentrate on the Initialize() and getValue() methods. The Initialize() method is used to set up the SPI interface for communicating with the ADC. I provide all the code that you need, but it is good to know what is going on. Listing 10-5 shows the Initialize() method.

Listing 10-5. Initializing the SPI Interface

```
// Setup the MCP3008 chip
public async void Initialize()
{
    Debug.WriteLine("Setting up the MCP3008.");
    try
    {
        // Settings for the SPI bus
        var SPI_settings = new SpiConnectionSettings(SPI_CHIP_SELECT_LINE);
        SPI_settings.ClockFrequency = 3600000;
        SPI_settings.Mode = SpiMode.Mode0;

        // Get the list of devices on the SPI bus and get a device instance
        string strDev = SpiDevice.GetDeviceSelector();
        var spidev = await DeviceInformation.FindAllAsync(strDev);

        // Create an SpiDevice with our bus controller and SPI settings
        mcp_var = await SpiDevice.FromIdAsync(spidev[0].Id, SPI_settings);

        if (mcp_var == null)
        {
            Debug.WriteLine("ERROR! SPI device {0} may be in used.", spidev[0].Id);
            return;
        }

    }
    catch (Exception e)
    {
        Debug.WriteLine("EXEPTION CAUGHT: " + e.Message + "\n" + e.StackTrace);
        throw;
    }
}
```

First, you set some variables of the SPI device class, including the line you will use (chip select = 0), the clock frequency (3.6 MHz) or 3.6 million times a second, and the mode (0), which is the default. Mode defines how the values are sampled.[5] Next, you get the device name (string) and then call a system method named await() that spawns a thread to wait for the chip's SPI interface to wake and set up. You do this because the SPI device is actually running on a separate thread.

You place all of this code in a try block in case the SPI bus fails (say if you misconnect one or more wires) using debug statements to track the progress and report success or failure. Don't worry too much if this code seems very strange. I should note that this may not be the only way to set up an SPI communication, but it is compact and captures a failed initialization.

---

[5]For mode explanations, see http://www.byteparadigm.com/applications/introduction-to-i2c-and-spi-protocols/.

Now you're ready to learn how to get data from the ADC. While the last method was a bit complex, reading a value is a bit more complex. Recall that you must write before you can read. In this case, you send an array of three bytes and you get back an array of three bytes, where the first byte sets the start bit, the second byte contains the command (shifted 4 bits to the left) and the channel select, and the third byte doesn't matter.

How did I discover this information? I read the data sheet! While it takes a bit of patience because many data sheets are written in a style known as "by engineers for engineers," which means that they can be a bit terse and not very easy to read. However, in this case you see a very good explanation of the data needed for reading and writing data. Figure 10-5 shows an excerpt from the data sheet.

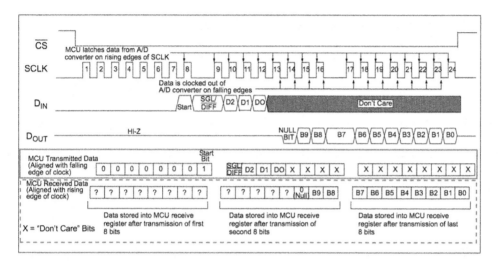

Figure 10-5. *Communicating with the MCP_3008*

I highlight the relevant portion for sending data to the chip in a solid box and the portion for receiving data in the dashed box. Here you see that the second byte for sending data holds the information for the single ended command and the command (0x8) shifted to the upper 4 bits (the order of bytes in memory is called *endianness*[6]).

Notice the data you get back from the chip. Here you see you get 10 bits followed by a 0 (null) that are returned in the third and part of the second byte. Thus, you need to write code to capture the 10 bits in the second and third byte, and then mask out the bits starting with the null bit. Since the bits in the second byte are the higher bits, you need to mask out the bits that you don't need, and then shift them so that you can interpret the two bytes as an integer (a common trick when working with byte streams). Listing 10-6 shows the completed getValue() method.

▪ **Tip**   Although this may seem confusing, the number and order of the bits is defined by the manufacturer and not a general case. You should always read the data sheet for each device to learn how to implement its protocol. Fortunately, when you are able to use custom libraries (as you will see in later chapters), the protocol is implemented for you.

---

[6]https://en.wikipedia.org/wiki/Endianness

Listing 10-6. Reading Data from the ADC

```
// Communicate with the chip via the SPI bus. You must encode the command
// as follows so that you can read the value returned (on byte boundaries).
public int getValue(byte whichChannel)
{
    byte command = whichChannel;
    command |= MCP3008_SingleEnded;
    command <<= 4;

    byte[] commandBuf = new byte[] { 0x01, command, 0x00 };
    byte[] readBuf = new byte[] { 0x00, 0x00, 0x00 };
    mcp_var.TransferFullDuplex(commandBuf, readBuf);

    int sample = readBuf[2] + ((readBuf[1] & 0x03) << 8);
    int s2 = sample & 0x3FF;
    return sample;
}
```

The first thing you may notice is the method takes the channel number as a parameter. This allows you to use the method to read other channels and indeed you do so in the next project.

Notice also how you perform the byte manipulations to set up the command, transmit it with the TransferFullDuplex() method, which returns three bytes to the read buffer. You then shift bytes and mask out the parts that you don't need, and then shift the remaining bytes so that you can read the value as an integer.

Once again, don't worry too much if this code seems overly complex. You can find out more about the chip from the data sheet at https://cdn-shop.adafruit.com/datasheets/MCP3008.pdf.

Now that you've seen the individual methods of this class, let's see how they are placed in context in the class. Listing 10-7 shows the skeleton for the ADC_MCP3008 class. I omit the details of the methods for brevity.

Listing 10-7. ADC_MCP3008 Class Layout

```
...

namespace PowerMeter
{
    class ADC_MCP3008
    {
...

        public ADC_MCP3008()
        {
            Debug.WriteLine("New class instance of ADC_MCP3008 created.");
        }

        public async void Initialize()
        {
...

        }

        public int getValue(byte whichChannel)
        {
...

        }
    }
}
```

■ **Tip** If you'd like to explore all the capabilities of the MCP3008, see the data sheet at `https://cdn-shop.` `adafruit.com/datasheets/MCP3008.pdf`.

OK, the code is ready for compilation! Be sure to check the earlier listings to ensure that you have all the code in the right place. Once you have entered all the code, you should now attempt to compile the code. Correct any errors that you find until the code compiles without errors or warnings.

Deploy and Execute

Now it is time to deploy the application! Be sure to fix and compilation errors first. Like you have with other applications, you want to compile the application in debug first (but you can compile in release mode if you'd prefer) and you must turn on the debugger on the board. You do this with the device portal.

Power on your board, and then connect to the board and turn on the debugger. Next, open the project properties to target the device and run with the remote debugger. Recall from the last chapter, you must modify two settings: the **Remote machine** name and the **Authentication Mode**. Set the **Remote machine** name to the IP address of your device with the port specified by the remote debugger when you started it from the device portal. For example, my device gave me the port number 8116 so I use 10.0.1.89:8116. Set the **Authentication Mode** to **None**.

Now you can deploy your application. Go ahead and do that now. You can run the deployment from the Debug menu, and when you do, you see the debug statements in the output window. Listing 10-8 shows an excerpt of the data you should see. When you start the debugger, the output window automatically opens and displays the debug output. If it does not, you can open the window manually and select debug from the drop-down menu.

Listing 10-8. Example Debug (Output Window)

```
...
New class instance of ADC_MCP3008 created.
...
GPIO ready.
Setting up GPIO pin 21.
Setting up GPIO pin 20.
Setting up GPIO pin 19.
Setting up GPIO pin 18.
Setting up GPIO pin 17.
Setting up the MCP3008.
...
Timer has fired the meterTimer_Tick() method.
Val read = 327
Number of LEDs to turn on = 1
Setting pin 21 HIGH.
Setting pin 20 LOW.
Setting pin 19 LOW.
Setting pin 18 LOW.
Setting pin 17 LOW.
...
Timer has fired the meterTimer_Tick() method.
Val read = 570
Number of LEDs to turn on = 2
```

```
Setting pin 21 HIGH.
Setting pin 20 HIGH.
Setting pin 19 LOW.
Setting pin 18 LOW.
Setting pin 17 LOW.
Timer has fired the meterTimer_Tick() method.
Val read = 571
```

Of course, you can run the application by starting it on your device using the Apps pane. Use the small triangle or arrow next to the application name to start the application, the square icon to stop it, and the trash can icon to delete it.

If everything worked correctly, you should see the lights change when the potentiometer is turned. If the lights don't change, be sure you have the potentiometer oriented correctly. Once it is working, try it out a few times to ensure that you see the lights sequence several times.

Summary

Learning how to build complex IoT projects requires learning how to use some components that may require a bit of effort to set up and use. You saw this with a very simple component—a potentiometer. In this case, since the Raspberry Pi does not have and ADC, you had to use the ADC to "bridge" the devices. Fortunately, you were able to use the potentiometer with the ADC using the SPI bus to control several LEDs as a power meter simulation.

You saw a number of new things in this chapter, including the ADC, how to connect and set up an SPI device, how to read a potentiometer, and finally, how to use the debug feature to write out statements to the output window. While the project itself is rather simplistic, the code clearly was not.

The next chapter continues our set of example IoT projects that you can use to learn more about building IoT solutions. You learn how to use another sensor to measure ambient light. Indeed, you will be making a cool, automatic night-light.

■ ■ ■

## CHAPTER 11

Project 2: Measuring Light

The projects that you've encountered thus far have used GPIO pins to turn LEDs on or off. But what if you wanted to control the brightness of an LED? Given that you can only turn pins on (high: 3.3V) or off (low: 0V), you have no way to send less power to the LED. Recall the lower the power, the dimmer the LED. If you consider solutions (or features) of devices that are sensitive to ambient light—such as a backlight keyboard—you may have noticed it changes brightness as the room becomes darker. In this project, you're going to explore how such a feature works. That is, you're going to build a fancy LED night-light.

Unlike the last project, this project is a bit more complex in concept. You must measure the ambient light in the room, and then calculate how much power to send to the LED using a technique called *pulse-width modulation* (PWM). You will also use the same analog to digital converter (ADC) and SPI interface from the last project, but you will implement it in a slightly different way. Let's get started.

■ **Note** Although this project is written in C#, other than the syntax and mechanics of building, the concepts of using a class are the same in C++.

Overview

You will design and implement an LED night-light using a light-dependent resistor (LDR, also called a *photocell* or *photoresistor*)[1] read from an ADC (the LRD is an analog component like a potentiometer). The LDR is a special resistor that changes resistance depending on the intensity of ambient light.

You need to use the ADC because you want to read analog values and the Raspberry Pi (and many other boards) does not have analog-to-digital logic. That is, the GPIO pins are digital only. The ADC acts like a "bridge" between digital and analog devices. You will use the MCP3008 that you used in the last chapter.

The real trick to this solution is you must use a special library (namespace) in order to access the pulse-width modulation features of your Raspberry Pi. The library you will use is a contributed library and not part of the standard Visual Studio built-in libraries for Windows 10 IoT Core. However, the library isn't difficult to install and you will reuse the class you built in the last chapter as the basis for implementing a class for controlling an LED (fading), as well as the ADC. The library is named the Microsoft IoT Lightning Provider (also called the Lightning software development kit or SDK).

■ **Note** In order to use the Lightning SDK, you need Windows 10 IoT Core Version 10.0.10586.0 or later.

---

[1]See https://en.wikipedia.org/wiki/Photoresistor.

© Charles Bell 2016
C. Bell, *Windows 10 for the Internet of Things*, DOI 10.1007/978-1-4842-2108-2_11

Since the LDR responds to light and the ambient light can vary, you will also use a simple user interface to allow you to adjust the sensitivity of the light sensor by restricting its range. That is, you will be able to set the minimum and maximum values for the range. Thus, you will be able to tune the solution to fit the ambient light levels.

Let's look at the components that you need, and then see how to wire everything together.

Required Components

The following lists the components that you need. You can find these components in either of the kits mentioned in Chapter 9 or you can purchase the components separately from Adafruit (www.adafruit.com), SparkFun (www.sparkfun.com), or any electronics store that carries electronic components. Since this solution is a headed application, you will also need a monitor, a keyboard, and a mouse. If you want to run this application headless, you can but I recommend experimenting with the user interface to help tune the values that you need to make the fade effect work properly.

- (1) LED (any color)

- (1) 10K ohm resistor

- (1) 150 ohm resistors (or appropriate for your LEDs)

- (1) light-dependent resistor (photocell)

- MCP3008 ADC chip

- Jumper wires: (5) male-to-male, (8) male-to-female

- Breadboard (full size recommended but half size is OK)

- Raspberry Pi 2 or 3

- Power supply

- Monitor

- Keyboard and mouse

Set up the Hardware

Once again, in order to help get everything connected correctly, you will make a plan for how things should connect. To connect the components to the Raspberry Pi, you need four pins for the ADC, one for the LED, one for the LDR, and one each for power and ground. You will also need to make a number of connections on the breadboard to configure the ADC chip and connect the LDR to the ADS. Table 11-1 shows the map I designed for this project. I list the physical pin numbers in parenthesis for the named pins. You will use male/female jumper wires to make these connections.

Table 11-1. *Connection Map for Power Meter Project*

GPIO	Connection	Function	Notes
3.3V (1)	Breadboard power rail	Power	
5V (2)	10K resistor	Power to LDR	
GND (6)	Breadboard ground rail	GND	
MOSI (19)	SPI	MCP3008 pin 4	
MISO (21)	SPI	MCP3008 pin 4	
SCLK (23)	SPI	MCP3008 pin 3	
CC0 (24)	SPI	MCP3008 pin 5	
27	LED	Resistor for LED	

Next, you need to make a number of connections on the breadboard. For these, you will use male-to-male jumpers. Table 11-2 shows the connections needed on the breadboard.

Table 11-2. *Connections on the Breadboard*

From	To	Notes
Breadboard GND	LDR	
Breadboard power	ADC VDD (pin 16)	
Breadboard power	ADC VREF (pin 15)	
Breadboard GND	ADC AGND (pin 14)	
Breadboard GND	ADC GND (pin 9)	

▪ **Tip**   See Chapter 9 for details on how to determine the correct resistor for your LED.

Once again, I am cheating a bit by using the Adafruit GPIO Reference Card for Raspberry Pi 2 or 3 (www.adafruit.com/products/2263), which makes locating GPIO pins much easier than counting pin numbers.

Figure 11-1 shows all the connections needed. Take a close look at this drawing, because some connections may not be obvious (e.g., sending 5V power to the LDR).

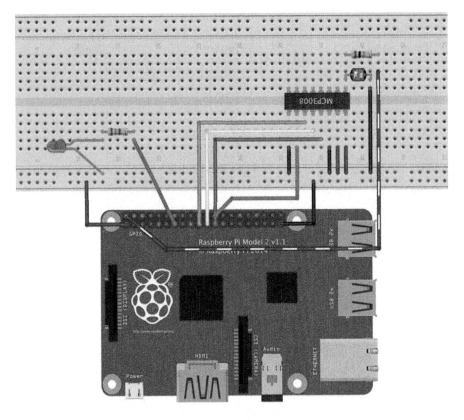

Figure 11-1. *Connections for the night-light project*

Notice how the LDR and 10k resistor are configured. This is not a mistake. You want the resistor to connect to the second leg of the LDR and the LDR connecting to the channel 0 pin of the ADC as a pullup resistor. Also note how power to the LDR comes from the 5V pin (2) on the GPIO. The SPI connections are the same as the last project.

WHAT IS PULSE-WIDTH MODULATION?

IoT devices such as the Raspberry Pi can switch between exactly two voltages on their GPIO pins (0.0V or 3.3V). The problem is that many components can operate on a range of voltage. For example, a fan can rotate at different speeds based on the current fed to it. The higher the current, the faster the fan spins. The same is true for other components, such as LEDs, servos, and motors.

It is possible to simulate sending different current levels to a component using a technique called *pulse-width modulation* (PWM). The process works by rapidly switching a pin (line) on and off at different frequencies (hence the pulse) for a given very short period of time (hence the width). The end effect is the device appears to be operating at lower voltages. For example, if you want only 3.0V from 3.3V, you must set the pin to operate at 90% or the pin will be on (through pulsing) at most 90% of the time.

This may sound a little strange and you may expect to see an LED fed by PWM flicker. The fact is, it is flickering, but it is doing so very quickly—too fast for humans to detect. In fact, many of the lights in your home or office cycle in a similar manner. The following diagram (see `https://en.wikipedia.org/wiki/Pulse-width_modulation`)[2] is an illustration of how PWM generates current.

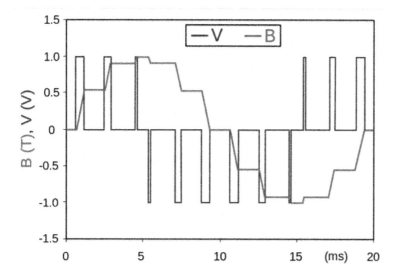

This shows a voltage source (V) modulated as a series of pulses creating a sine-like wave. The result is a sine-like current in the inductor (B). The current waveform is the integral of the voltage waveform. While this sounds all scientific/mathematical, just remember PWM is a technique for simulating different current values by turning the pin on and off rapidly in a specific pattern.

If you are following along with this chapter working on the project, go ahead and make the hardware connections now. Don't power on the board yet, but do double- and triple-check the connections.

Write the Code

Now it's time to write the code for our example. Since you are working with several new concepts as well as a new component, I will introduce the code for each in turn. I've decided to use C#, but you could implement this project in C++.

■ **Note** Since you have learned all the basics of creating projects in Visual Studio, including how to build and deploy applications, I omit the details of the common operations for brevity.

The project uses an LDR to measure ambient light read via an ADC (MCP3008), the value read is used to control the percentage of current sent to an LED via PWM. Unfortunately, while the IoT devices you have for Windows 10 IoT Core have PWM support, Windows 10 IoT Core has only rudimentary support for PWM. Fortunately, there is an add-on library called the Microsoft IoT Lightning Providers that get you what you need to create a PWM output on a GPIO pin.

---

[2]Creative Commons Attribution-Share Alike 3.0 Unported License

I like to use modularization to help make the code easier to write and maintain. It also helps to focus your efforts by concentrating on a specific concept to model. You will see this as you walk through the code. You will create a class for the ADC and a class for the PWM control of the LED. Finally, you use the DispatcherTimer class to periodically check the LDR and set the PWM percentage for the LED.[3]

Let's talk about starting a new project and adding the resources that you need first (e.g., the Microsoft IoT Lightning Providers), and then we will walk through the entire code.

New Project

You will use the same project template as the last project—the Blank App (Universal Windows) template. Use the name NightLight for the project name. You can save the project wherever you like or use the default location. Once the project opens, double-click the MainPage.xaml.cs file. There are two namespaces you need to include. Go ahead and add those now, as shown next.

```
using System.Diagnostics;              // add this for debugging
using Microsoft.IoT.Lightning.Providers;   // add for Lightning inteface
```

Next, you need to add some variables and constants. First, you create a constant for the GPIO pin for the LED (27) and an instance of one of the new classes that you create named LED_Fade. You see this class in a later section.

```
// Constants and variables for pin
private const int LED_PIN = 27;
private LED_Fade fader = new LED_Fade(LED_PIN);
```

Next, you create an instance for the timer.

```
// Timer to refresh the LED brightness
private DispatcherTimer refreshTimer;
```

Next, you add an instance of the ADC class, which you name **MCP3008** to distinguish it from the class you created in the last project. You will see this described in a later section.

```
// Add the new ADC Chip class
private MCP3008 adc = new MCP3008();
```

Finally, you need some variables and a constant for the LDR. You use these variables in conjunction with the user interface to adjust the lower and upper bounds for the light sensitivity. This is really nice because it allows you to experiment with the project while it runs rather than using constants or variables that you must modify, compile, and then rerun to test.

```
// Channel to read the photocell with initial min/max settings
private const int LDR_CHANNEL = 0;
private int max_LDR = 300;   // Tune this to match lighting
private int min_LDR = 100;   // Tune this to match lighting
```

The code in the MainPage() function initializes the components, the new LED fade class (fader), and the new MCP3008 class (adc). Each of these classes has an Initialize() method. There is also a method to initialize the lightning providers that must be called first. You will see this method in the next section.

---

[3]There are other ways to do this, but this uses a technique you've seen previously.

You also set up the timer. In this case, you use a value of 500, which is one-half a second. If you adjust this lower, the solution will be a bit more responsive. You can try this once you've got it to work and tested. Listing 11-1 shows the complete MainPage() method.

Listing 11-1. MainPage Method

```
public MainPage()
{
    InitializeComponent();        // init UI
    InitLightningProvider();      // setup lightning provider
    fader.Initialize();           // setup PWM
    adc.Initialize();             // setup ADC

    // Add code to setup timer
    refreshTimer = new DispatcherTimer();
    refreshTimer.Interval = TimeSpan.FromMilliseconds(500);
    refreshTimer.Tick += refreshTimer_Tick;
    refreshTimer.Start();
}
```

You need two references added to the solution. You need the Windows 10 IoT Extensions from the project property page. You do this by right-clicking the **References** item under the project in the Solution Explorer. You also need to install the Microsoft IoT Lightning Providers library.

Lighting Providers

The Microsoft IoT Lightning Provider is a set of providers to interface with GPIO, SPI, PWM, and I2C support via a direct memory access driver on the device. The lightning providers are implemented as a set of classes you can use to interface with components. In order to use the library, you must first turn on the Lightning driver on your device via the Device Portal. Figure 11-2 shows the **Devices** tab and the drop-down menu that you use to change the driver.

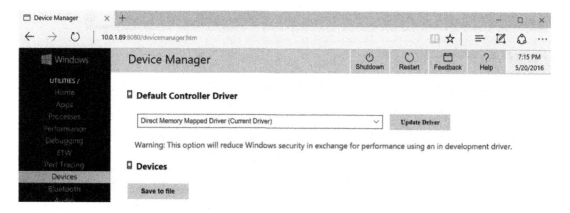

Figure 11-2. Setting the Lighting Driver on the Device Portal

When you select the direct memory mapped driver from the drop-down menu and click **Update Driver**, the portal warns you that the driver is a development-level pre-release and may not perform at peak efficiency. For our purposes, you can ignore the warning, as it has no bearing on the project. The device needs to reboot to complete the change.

■ **Caution**   You want to reverse this step once you have finished the project and before starting a new project that does not use the lightning library.

To add the lightning library, use the **NuGet Package Manager from the Tools ➤ NuGet Package Manager ➤ Manage NuGet Packages for Solution**... menu. Click the **Browse** tab and then type **lightning** in the search box. After a moment, the list updates. You should see a list with names that include **lightning**, as shown in Figure 11-3.

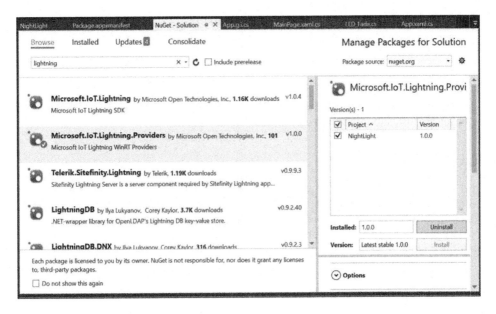

Figure 11-3. *NuGet Package Manager*

Select the entry named **Microsoft.IoT.Lightning.Providers** in the list, tick the project name (solution) in the list on the right, and finally, click **Install**. The installation starts. Visual Studio downloads a number of packages and then asks you permission to install them. Go ahead and let the installation complete. A dialog box tells you when the installation is complete.

■ **Tip**   You can also download this library and install it manually if the NuGet manager fails. See https:// developer.microsoft.com/en-us/windows/iot/win10/lightningproviders for more details.

Since this library isn't integrated with Visual Studio or the Windows UWP libraries, you must make a small alteration to the project files to grant permission to use the library (reference) in the project. Failure to do this step leads to some strange execution issues (but it compiles without errors or warnings). More specifically, the InitLightingProvder() method fails because the LightningProvider. IsLightningEnabled property is False.

In order to use the library, you need to manually update the `Package.appxmanifest` file to reference the Lightning device interface. To edit the file, right-click it in the Solution Explorer and then choose **View Code**. In the code editor, change the code at the top of the file as follows. I have noted the required changes in bold. The first line is a capability setting that enables the application to access custom devices and the second line is the globally unique identifier (GUID) for the lightning interface. Finally, you add **iot** to the namespaces list.

```
<Package
  xmlns="http://schemas.microsoft.com/appx/manifest/foundation/windows10"
  xmlns:mp="http://schemas.microsoft.com/appx/2014/phone/manifest"
  xmlns:uap="http://schemas.microsoft.com/appx/manifest/uap/windows10"
  xmlns:iot="http://schemas.microsoft.com/appx/manifest/iot/windows10"
  IgnorableNamespaces="uap mp iot">
...
```

Next, scroll to the bottom of the file and make the following changes. This adds the capabilities to match the changes at the top of the file.

```
...
  <Capabilities>
    <Capability Name="internetClient" />
    <iot:Capability Name="lowLevelDevices" />
    <DeviceCapability Name="109b86ad-f53d-4b76-aa5f-821e2ddf2141"/>
  </Capabilities>
</Package>
```

Now you're ready to add code to initialize the lightning providers. Since you are using the lightning providers, you must initialize the provider library in the `MainPage` class. The following method shows how to initialize the lightning provider.

```
private void InitLightningProvider()
{
    // Set the Lightning Provider as the default if Lightning driver is enabled on the
target device
    if (LightningProvider.IsLightningEnabled)
    {
        LowLevelDevicesController.DefaultProvider = LightningProvider.
GetAggregateProvider();
    }
}
```

The bulk of the code for the SPI and PWM are contained in the new classes you will add shortly. But first, let's implement the user interface and complete the `MainPage` class.

User Interface

The user interface is deliberately simple consisting of only a few labels and two sliders. The sliders control the minimum and maximum values for the range of sensitivity for the LDR. One of the labels displays the value read from the LDR. You can use this value to help tune the fade effect.

That is, you can set the lower value to the value read during ambient light levels. This effectively sets the percent power to 0 where the LED is off (or very dim). You can use the higher value slider to set the maximum value for the scale. You can find out what that value is by placing your hand over the LDR and reading the value. Thus, you can tune the night-light effect to match ambient light readings.

Listing 11-2 shows the code for the user interface. Open the `MainPage.xaml` file and make the changes as shown. This code uses a stacked panel to contain the controls. Note the lower value slider. You add a reference to a method to be called when the slider is updated. This helps you control the slider to avoid a nasty surprise should the lower value exceed the higher value. Speaking of which, I have added code to restrict the range to 0.0–1.0.

Listing 11-2. User Interface (XAML) Code

```xaml
<Page
    x:Class="NightLight.MainPage"
    xmlns="http://schemas.microsoft.com/winfx/2006/xaml/presentation"
    xmlns:x="http://schemas.microsoft.com/winfx/2006/xaml"
    xmlns:local="using:NightLight"
    xmlns:d="http://schemas.microsoft.com/expression/blend/2008"
    xmlns:mc="http://schemas.openxmlformats.org/markup-compatibility/2006"
    mc:Ignorable="d">

    <Grid Background="{ThemeResource ApplicationPageBackgroundThemeBrush}">
        <StackPanel Width="400" Height="400">
            <TextBlock x:Name="title" Height="60" TextWrapping="NoWrap"
                    Text="Night Light Experiment" FontSize="28" Foreground="Blue"
                    Margin="10" HorizontalAlignment="Center"/>
            <TextBlock Height="60" TextWrapping="NoWrap"
                    Text="High Value" FontSize="28" Foreground="Blue"
                    Margin="10" HorizontalAlignment="Center"/>
            <Slider x:Name="ldrHigh" Width="400" Value="350"
                    Orientation="Horizontal" HorizontalAlignment="Left"
                    Maximum="1023"/>
            <TextBlock Height="60" TextWrapping="NoWrap"
                    Text="Low Value" FontSize="28" Foreground="Blue"
                    Margin="10" HorizontalAlignment="Center"/>
            <Slider x:Name="ldrLow" Width="400" Value="125"
                    Orientation="Horizontal" HorizontalAlignment="Left"
                    Maximum="1023"
                    ValueChanged="ldrLow_ValueChanged"/>
            <TextBlock x:Name="status" Height="60" TextWrapping="NoWrap"
                    Text="LDR Value = 0" FontSize="28" Foreground="Blue"
                    Margin="10" HorizontalAlignment="Center"/>
        </StackPanel>
    </Grid>
</Page>
```

Figure 11-4 shows an example of the user interface.

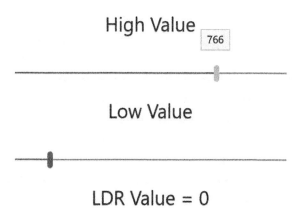

Figure 11-4. Sample User Interface

The sliders can be moved independently. This presents a problem if the user slides the lower value to a setting greater than the higher value. Thus, you want to control the sliders so that the lower value never exceeds the higher value. To do this, you complete the code for the ldrLow_ValueChanged() method as follows. This illustrates a common technique to manage the range of user input.

```
private void ldrLow_ValueChanged(object sender, Windows.UI.Xaml.Controls.Primitives.
RangeBaseValueChangedEventArgs e)
{
    // Make sure the low value doesn't exceed the high value
    if (ldrLow.Value > ldrHigh.Value)
    {
        ldrHigh.Value = ldrLow.Value + 1;
    }
}
```

■ **Tip**   This code needs one extra step to ensure that you stay within the limits of the slider. See if you can spot it yourself. Hint: What happens when the high value is already at the highest setting? Although not essential for our experiment, you should strive to improve error handling to include the fringe values and conditions.

Now let's see the code to read from the value of the LDR and fade the LED.

Controlling the LED

You are using a timer to refresh the project. This method, named refreshTimer_Tick(), first reads the value of the LDR from the MCP3008 class, and then calculates a percentage based on the minimum and maximum value scale as set via the user interface. You may have noticed that I used some constants values to set the initial range. Listing 11-3 shows the completed refreshTimer_Tick() method.

Listing 11-3. Timer Code

```csharp
private void refreshTimer_Tick(object sender, object e)
{
    float valRead = 0;
    float brightness = 0;

    Debug.WriteLine("Timer has fired the refreshTimer_Tick() method.");

    // Read value from the ADC
    valRead = adc.getValue(LDR_CHANNEL);
    Debug.Write("Val read = ");
    Debug.WriteLine(valRead);
    status.Text = "LDR Value = " + valRead;

    // Get min, max from sliders
    min_LDR = (int)ldrLow.Value;
    max_LDR = (int)ldrHigh.Value;
    Debug.WriteLine("Min LDR = " + min_LDR);
    Debug.WriteLine("Max LDR = " + max_LDR);

    // Calculate the brightness
    brightness = ((valRead - min_LDR) / (max_LDR - min_LDR));
    // Make sure the range stays 0.0 - 1.0
    if (brightness > 1)
    {
        brightness = (float)1.0;
    }
    else if (brightness < 0)
    {
        brightness = (float)0.0;
    }
    Debug.Write("Brightness percent = ");
    Debug.WriteLine(brightness);

    // Set the brightness
    fader.set_fade(brightness);

    // For extra credit, show the voltage returned from the LDR
    // convert the ADC readings to voltages to make them more friendly.
    float ldrVolts = adc.ADCToVoltage((int)valRead);

    // Let us know what was read in.
    Debug.WriteLine(String.Format("Voltage for value read: {0}, {1}v",
                                  valRead, ldrVolts));
}
```

Wow, that's a lot of code. For the most part, it is similar to the timer event from the last project. However, the calculation for the percentage to use for the PWM is new. Here the percentage represents the brightness of the LED. The higher the percentage calculated means the faster the LED pulses and therefore appears brighter.

I have added an interesting bit of data at the end of the method. This demonstrates how to calculate the voltage that is flowing through the LDR. You may need this technique for any projects where you are reading analog voltage.

Take some time to read through the code. You will see a number of additional debug statements that you can use when testing the project in debug mode. There is code to update the label in the user interface with the value read and the voltage calculated. If you are curious, you can use a multimeter to measure the voltage on the ADC side of the LDR (positive lead on the LDR and negative lead on ground). You should see the same voltage on the multimeter as shown in the user interface.

Now let's see how to put all of these pieces together.

Completing the Main Class

Now that you've seen all the code for the main code file, let's look at how the methods are placed in the class. Listing 11-4 shows the skeleton of the MainPage class. Use this as a guide to place the code in the right places. I omit the details of the methods for brevity.

Listing 11-4. MainPage Code Layout

```
...

namespace NightLight
{
    public sealed partial class MainPage : Page
    {
...
        public MainPage()
        {
...
        }

        private void InitLightningProvider()
        {
...
        }

        private void refreshTimer_Tick(object sender, object e)
        {
...
        }

        private void ldrLow_ValueChanged(object sender,
                Windows.UI.Xaml.Controls.Primitives.RangeBaseValueChangedEventArgs e)
        {
...
        }
    }
}
```

Now let's see how to create the code for the two new classes.

Code for the MCP3008

The code for the ADC class is very similar to the class you used in the last project. The difference is you will use the lightning providers to communicate with the ADC rather than the Windows 10 IoT extension. This is because the Windows 10 IoT extension and the lightning provider are incompatible. That is, you cannot use the same code from the last project together with the PWM code from the lightning provider. Fortunately, the lightning provider has SPI support.

The methods that you use in this class are same as you had in the last project. The following summarizes the methods that you need. The last method is a bonus method that you may need in the future if you reuse this code for more advanced projects.

- Constructor – MCP3008(): Log instantiation with a debug statement

- Initialize(): Initialize the SPI interface

- getValue(): Read a value from the ADC on a specified channel

- ADCToVoltage(): Return the voltage read for a specific value

Recall from Chapter 10 that to add a new class, right-click the project and choose **Add ➤ New Item...**. Name the class MCP3008 to distinguish it from the similar class in the last project. When you create a class, you create a new file, as shown earlier. Inside this new code file, you see a familiar skeleton. Of note is the same namespace as the main code and a list of namespaces that you will use. Open the file now and add the following namespaces. The need for each is shown in the comments.

```
using Windows.Devices.Spi;     // add this for SPI communication
using System.Diagnostics;      // add this for debugging
```

Next, you need some variables and constants. You need a variable for the SPI device and constants for the MCP3008. In this case, you set constants for helping you read data.

```
// SPI controller interface
private SpiDevice mcp_var;
const int SPI_CHIP_SELECT_LINE = 0;
const byte MCP3008_SingleEnded = 0x08;
const byte MCP3008_Differential = 0x00;
```

You also add a variable to store the reference voltage for the value read. You also store the maximum value of the component read from the ADC (1023). Of note is how I converted the constant 5.0 to a float by adding F to the end. This is a common shortcut that you can use instead of casting with (float). Also, be sure to change the reference voltage if you use a different voltage to power the components connected to the ADC channels. Recall from the wiring layout in Figure 11-1, you're using 5V on the LDR.

```
// These are used when you calculate the voltage from the ADC units
float ReferenceVoltage = 5.0F;
private const int MAX = 1023;
```

The constructor is simple; you just announce that you've instantiated the class via a debug statement. You will see this in the class layout later in the section. The really interesting changes appear in the Initialize() method. Listing 11-5 shows the updated method that uses the lightning provider class methods.

Listing 11-5. The Initialize() Method

```
// Setup the MCP3008 chip
public async void Initialize()
{
    Debug.WriteLine("Setting up the MCP3008.");
    try
    {
        // Settings for the SPI bus
        var SPI_settings = new SpiConnectionSettings(SPI_CHIP_SELECT_LINE);
        SPI_settings.ClockFrequency = 3600000;
        SPI_settings.Mode = SpiMode.Mode0;

        SpiController controller = await SpiController.GetDefaultAsync();
        mcp_var = controller.GetDevice(SPI_settings);
        if (mcp_var == null)
        {
            Debug.WriteLine("ERROR! SPI device may be in use.");
            return;
        }
    }
    catch (Exception e)
    {
        Debug.WriteLine("EXEPTION CAUGHT: " + e.Message + "\n" + e.StackTrace);
        throw;
    }
}
```

This method is very similar to the class that you used in the last project. The biggest difference is you can get the SPI device with a single class to the provider. This code is a pattern for the other providers in the Lightning Providers library. In fact, you see very similar code in the PWM class.

The getValue() method is the same as the class used in the previous class. You can simply copy it from the last project and past it in the new class. I omit the details from the list for brevity.

Now, let's see how all of these pieces fit together. Listing 11-6 shows the class layout for the MCP3008 class.

Listing 11-6. The MCP3008 Class Layout

```
...
namespace NightLight
{
    class MCP3008
    {
...

        public MCP3008()
        {
            Debug.WriteLine("New class instance of MCP3008 created.");
        }

        // Setup the MCP3008 chip
        public async void Initialize()
        {
```

```
...
        }
        public int getValue(byte whichChannel)
        {
...
        }

        // Utility function to get voltage from ADC
        public float ADCToVoltage(int value)
        {
            return (float)value * ReferenceVoltage / (float)MAX;
        }
    }
}
```

Notice the last method. This is a helper method that you can use to calculate the voltage read from the channel. You used this in the MainPage class to display the voltage when the value is read from the LDR.

■ **Tip** If you want to copy a class from another project, you can copy the file into the project folder and optionally rename the file. To add the class to the solution, right-click the project and choose **Add ➤ Existing Item...** and select the class file. If you renamed the class file, you may also have to rename the class itself as well as change the namespace to match the current project/solution.

Now let's look at the code for the PWM class.

Code for the PWM

The code to implement PWM on a GPIO pin is very straightforward and models the code in the MCP3008 class. The following summarizes the methods that you need. The last method is a bonus method that you may need in the future if you reuse this code for more advanced projects.

- Constructor – LED_Fade(): Log instantiation with a debug statement

- Initialize(): Initialize the PWM interface

- set_fade(): Set the duty cycle (percentage of time the LED is on) for the GPIO (LED) pin

Recall from Chapter 10, to add a new class, right-click the project and choose **Add ➤ New Item...**. Name the class LED_Fade. When you create a class, you create a new file, as shown. Inside this new code file, you will see a familiar skeleton. Of note is the same namespace as the main code and a list of namespaces that you will use. Open the file now and add the following namespaces. The need for each is shown in the comments.

```
using System.Diagnostics;    // add for Debug.Write()
using Windows.Devices.Pwm;   // add for PWM control (10586 and newer)
using Microsoft.IoT.Lightning.Providers;  // add for Lightning driver for Pwm
```

There are only two variables required: one for the GPIO (LED) pin and another for the PWM class instance from the lightning provider. You use this variable to communicate with the PWM features.

```
// Variables for controlling the PWM class
private int LED_pin;
private PwmPin Pwm;
```

The constructor sets the pin number as a parameter and prints a debug statement indicating the instance has been created. The code is shown next.

```
public LED_Fade(int pin_num = 27)
{
    LED_pin = pin_num;    // GPIO pin
    Debug.WriteLine("New class instance of LED_Fade created.");
}
```

The Initialize() method is very similar to the same method in the MCP3008 class. In this case, you are setting up the PWM provider. Listing 11-7 shows the complete Initialize() method.

Listing 11-7. The Initialize() Method

```
// Initialize the PwmController class instance
public async void Initialize()
{
    try
    {
        var pwmControllers = await
            PwmController.GetControllersAsync(LightningPwmProvider.GetPwmProvider());
        var pwmController = pwmControllers[1];    // the device controller
        pwmController.SetDesiredFrequency(50);

        Pwm = pwmController.OpenPin(LED_pin);
        Pwm.SetActiveDutyCyclePercentage(0);    // start at 0%
        Pwm.Start();
        if (Pwm == null)
        {
            Debug.WriteLine("ERROR! Pwm device {0} may be in use.");
            return;
        }
        Debug.WriteLine("GPIO pin setup for Pwm.");
    }
    catch (Exception e)
    {
        Debug.WriteLine("EXEPTION CAUGHT: " + e.Message + "\n" + e.StackTrace);
        throw;
    }
}
```

This code is a bit different than the SPI code. In this case, you need to open a pin for controlling the PWM and set the starting duty cycle (percent). In this case, you set it to 0, which effectively turns off the LED. Finally, you start the PWM class.

To control the fading effect, you provide a method named set_fade() that takes a floating point value (percent) that you use to change the duty cycle of the PWM. The following shows the complete code.

```
// Set percentage of brightness (or how many cycles are pulsed) where
public void set_fade(float percent)
{
    if (Pwm != null)
    {
        Pwm.SetActiveDutyCyclePercentage(percent);
        Debug.WriteLine("Pwm set.");
    }
    else
    {
        Debug.WriteLine("Cannot trigger Pwm.");
    }
}
```

Now, let's see how all of these pieces fit together. Listing 11-8 shows the class layout for the LED_Fade class.

Listing 11-8. The LED_Fade Class Layout

```
...
namespace NightLight
{
    class LED_Fade
    {
...
        public LED_Fade(int pin_num = 27)
        {
            LED_pin = pin_num;    // GPIO pin
        }

        // Initialize the PwmController class instance
        public async void Initialize()
        {
...
        }

        // Set percentage of brightness (or how many cycles are pulsed) where
        // 100 is fast or "bright" and 0 is slow or "dim"
        public void set_fade(float percent)
        {
...
        }
    }
}
```

That's it! The code is complete and ready for compilation. Be sure to check the prior listings to ensure that you have all the code in the right place. Once you have entered all the code, you should then attempt to compile the code. Correct any errors that you find until the code compiles without errors or warnings.

Deploy and Execute

Now it is time to deploy the application! Be sure to fix any compilation errors first. Like you have with other applications, you want to compile the application in debug first (but you can compile in release mode if you'd prefer) and you must turn on the debugger on your board. You do this with the device portal.

Go ahead and power on your board. Be sure to connect a monitor, mouse, and keyboard so you can use the user interface. When ready, connect to the board to run the device portal, turn on the debugger, and then open the project properties to target the device and run with the remote debugger.

Recall from the Chapter 9, you must modify two settings: the **Remote machine** name and the **Authentication Mode**. Set the **Remote machine** name to the IP address of your device with the port specified by the remote debugger when you started it from the device portal. For example, my device gave me the port number 8116 so I use 10.0.1.89:8116. Set the **Authentication Mode** to **None**.

Now you can deploy your application. Go ahead and do that now. You can run the deployment from the **Debug** menu and when you do, you see the debug statements in the output window. Listing 11-9 shows an excerpt of the data you should see. When you start the debugger, the output window automatically opens and displays the debug output. If it does not, you can open the window manually and select debug from the drop-down menu.

Listing 11-9. Example Debug (Output Window)

```
...
New class instance of MCP3008 created.
New class instance of LED_Fade created.
Setting up the MCP3008.
GPIO pin setup for Pwm.
...
Timer has fired the refreshTimer_Tick() method.
commandBuf = 11280
readBuf = 00239
Val read = 239
Min LDR = 235
Max LDR = 384
Brightness percent = 0.02684564
Pwm set.
Voltage for value read: 239, 1.168133v
Timer has fired the refreshTimer_Tick() method.
commandBuf = 11280
readBuf = 00239
Val read = 239
Min LDR = 235
Max LDR = 384
Brightness percent = 0.02684564
Pwm set.
Voltage for value read: 239, 1.168133v
Timer has fired the refreshTimer_Tick() method.
commandBuf = 11280
readBuf = 00238
Val read = 238
Min LDR = 235
Max LDR = 384
Brightness percent = 0.02013423
Pwm set.
Voltage for value read: 238, 1.163245v
```

Of course, you can run the application by starting it on your device using the Apps pane. You use the small triangle or arrow next to the application name to start it and the square icon to stop it or the trashcan icon to delete the application.

If everything worked correctly, you should be able to place your hand over the LDR and see the LED brighten. If it doesn't (and there is a good chance it won't), don't worry. Conduct an experiment by clearing away anything that can cast a shadow on the LDR and take note of the value in the user interface. Let it run for a few minutes and note the values. You should see it bounce around a bit but should settle to within about 20–30 of normal. For example, in my office, the ambient light level value was about 220. Next place your hand over the LDR and note the change in value. Let the values settle and note the value range. In my office, the value was about 450.

Now, adjust the high value slider to the value when your hand was over the LDR and the low value to the value for ambient light. Once set, you can experiment with the LED brightness by slowly placing your hand over the LDR. The LED should go from a very dim setting to full (or nearly) brightness. It may take some additional runs to fine tune the high and low values but once set, the project acts as a light-sensing night-light. How cool is that?

Summary

IoT solutions often require using analog components. Moreover, you may need to control analog components by adjusting the voltage to the component. This often requires using additional libraries and features that may not be in the standard libraries. In this case, you had to use a special library named the Microsoft IoT Lightning Providers to get access to PWM and SPI interfaces.

You saw a number of new things in this project, including another way to use an ADC with an SPI interface, how to read values from an LDR, and how to use PWM to control the brightness of an LED. While the project itself is rather simplistic, the code clearly was not and it was a bit more code than you've seen so far.[4]

The next chapter continues the set of example IoT projects that you can use to learn more about building IoT solutions. You learn how to write code to read more sensors. In Chapter 12, you create a project to report weather data from several sensors, and then in Chapter 13, you learn how to collect and store the data generated.

---

[4]This pattern of ever-increasing complexity continues in the next chapters—so strap in and hang on!

CHAPTER 12

■ ■ ■

Project 3: Using Weather Sensors

IoT solutions often employ a number of sensors to observe the world around us and while you've explored a project with a simple sensor, you have yet to see how to work with more sophisticated sensors, such as those available as breakout boards. One of the more popular choices of sensors includes those you use to observe weather. In this case, you'll start out with a sensor that measures temperature, barometric pressure, and calculates the altitude based on sea level pressure. There's a lot you learn with only those two measurements.

The sensor that you will use is the BMP280 I2C or the SPI barometric pressure and altitude sensor breakout board from Adafruit (www.adafruit.com/products/2651). This sensor comes with Microsoft IoT Pack for Raspberry Pi (but you can buy it separately). Although the board can be used with I2C or SPI, you will use the sensor with the I2C interface.

■ **Note**   Some newer releases of the Microsoft IoT Pack for Raspberry Pi come with a BME280 sensor instead of the BMP280. In which case, you cannot use it with this project because the library that we use only works with the BMP280 (but it can be modified to work; see the data sheet for the BME280). However, the project in the next chapter permits either the BMP280 or the BME280.

This project also represents another escalation of complexity from the last project. While you will learn how to use the I2C interface, you will also discover how to work with sensors and breakout boards that have a complex communication protocol. Recall from our work with the ADC chip, you had to interpret the data read from the chip. The same is true for most breakout boards and sensors. In the case of the BMP280, the protocol is quite a bit more complex. Fortunately, you are able to make use of code made by others and thus reduce the burden of having to figure it out for yourself.

Finally, you take a different tactic in this project and implement the project in Visual C++. However, since the third party code was written in C#, you also discover how to build a Windows 10 IoT solution that uses mixed languages. While this sounds like unnecessary complexity, it allows you to master leveraging one of the most impressive features of Visual Studio—mixing C# and C++ in the same solution. With so much to work on and learn about, let's take a moment to examine the goals of the project and what you need to get started.

© Charles Bell 2016
C. Bell, *Windows 10 for the Internet of Things*, DOI 10.1007/978-1-4842-2108-2_12

Overview

Weather projects can be a lot of fun since they are not only easy to wire up because the sensors are packaged in breakout boards, but also because they are more practical than projects that teach techniques alone. While the projects thus far have been mainly in that category[1], this project is something you can build and use. You may even want to take it a bit further and add additional sensors and features.

To make it more practical, this project is written with a basic user interface, so you can adapt it to a desktop solution or even a wall-mounted weather station. For example, you could use the Raspberry Pi 7" Touch LCD Panel or similar LCD to display the weather conditions.

The hardware for this project is very simplistic. All you need are the BMP280 sensor and four wires to connect it to your device. What isn't simplistic is the code needed to read data from the BMP280. Fortunately, there are two solutions available for you to use. More specifically, you do not have to write any code to use the BMP280 with these solutions (but you have to modify it slightly). One is a C# class that you can get from a tutorial from Adafruit and the other is a library that you can download and install from NuGet. Both of which are written in/for C#.

You use C++ to build this project to not only learn how to build Windows 10 IoT solutions in C++ but also to learn how to consume code and libraries built for other languages in the process. But don't be alarmed. The process you need to use to enable this cross language solution is not difficult (but requires a precise set of steps and settings to get correct).

This project is part one of two for writing a weather solution. In this chapter, you use one of the solutions for the BMP280 and in the next chapter you use the other solution. In fact, you will see a pure C# implementation in the next chapter. For this chapter, you use the code class solution so that you can see how to modify it slightly for use in a C++ solution. Fortunately, the changes are minor and help understand what you may need to do to use a C# code library in other C++ applications.

I demonstrate all the code necessary and more in the following sections. But first, let's talk briefly about the hardware that you need for this project.

OTHER WEATHER IOT SOLUTIONS

You can find a number of weather themed example projects on the Internet. In fact, there are three that you may want to look at to see how to use other sensors or libraries to build a weather project. The following are three of the better sample projects that offer the most interesting and complete solutions.

- https://www.hackster.io/windows-iot/weather-station-67e40d

- https://github.com/ms-iot/samples/tree/develop/WeatherStation

- https://developer.microsoft.com/en-us/windows/iot/win10/samples/weatherstationlightning

If you decide to explore these solutions, be sure to download the sample code rather than building it from scratch. In fact, the projects on GitHub do not have very clear documentation. Those projects use a different sensor too (see www.adafruit.com/products/1899) but are written in Visual Basic, C#, C++, and Python so they show a wide array of possible implementations. The last one uses the Lightning driver you saw in the last chapter.

---

[1]But still fun, I think.

Required Components

The following lists the components that you need. You can get the BMP280 sensor from Adafruit (www.adafruit.com) either in the Microsoft IoT Pack for Raspberry Pi or purchased separately, SparkFun (www.sparkfun.com), or any electronics store that carries electronic components. However, if you use a sensor made by someone other than Adafruit, you may need to alter the code to change the I2C address. Since this solution is a headed application, you also need a monitor, a keyboard, and a mouse.

- Adafruit BMP280 I2C or SPI barometric pressure and altitude sensor
- (4) jumper wires: male-to-female
- Breadboard (full size recommended but half size is OK)
- Raspberry Pi 2 or 3
- Power supply
- Monitor
- Keyboard and Mouse

Setup the Hardware

Although there are only four connections needed for this project, you will make a plan for how things should connect, which is good practice to hone. To connect the components to the Raspberry Pi, you need four pins for the BMP280 sensor, which requires only power, ground, and 2 pins for the I2C interface. Table 12-1 shows the map I designed for this project. I list the physical pin numbers in parenthesis for the named pins. You use male-to-female jumper wires to make these connections.

Table 12-1. *Connection Map for Weather Sensor Project*

GPIO	Connection	Function	Notes
5V (2)	BMP280 VIN	Power to breakout board	
GND (6)	BMP280 GND	GND on breakout board	
SDA1 (3)	SDI on breakout board	I2C	
SCL1 (5)	SCK on breakout board	I2C	

Figure 12-1 shows all the connections needed.

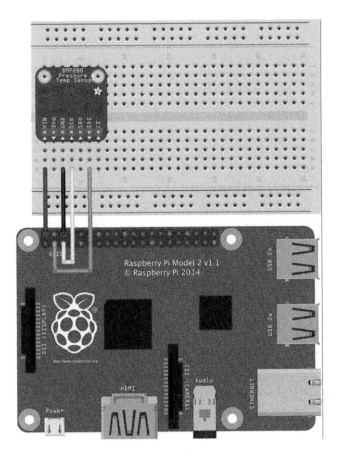

Figure 12-1. *Connections for the Weather Sensor project*

So what is this I2C interface? I2C[2] (pronounced "eye-two-see" or the less popular "eye-squared-see") is a fast digital protocol that uses two wires (plus power and ground) to read data from circuits, sensors, (or devices). One pin is used to read the data and the other is used as a clock to control how data is sent. There are several forms of how to read data from I2C, but fortunately the Windows 10 IoT Core has libraries that you can use to communicate with I2C devices. The IoT boards in this book all support the interface.

If you are following along with this chapter working on the project, go ahead and make the hardware connections now. Don't power on the board yet, but do double and triple check the connections.

Write the Code

Now it's time to write the code for our project. Since you are working with several new concepts as well as a new component (BMP280), I introduce the code for each in turn. I've decided to use a C++ project that consumes a C# project, which encapsulates a class library for the BMP280. You'll use a timer to read the data and update the user interface.

---

[2]See https://en.wikipedia.org/wiki/I%C2%B2C.

Note You may read that it is impossible to use a C# library in a C++ project. Sadly, that is a bit of misinformation. You can use a C# library in a C++ project provided they are compatible (same platform, etc.). However, you should implement the project in this chapter strictly as described. For example, if you use the wrong project template for the C# project, the solution will build and deploy, but may throw an exception or it will simply not work.

Recall that our project uses code to read the BMP280 from a tutorial written by someone else (Adafruit). In fact, the tutorial is one of the sample projects that Adafruit includes as an introduction to their Microsoft IoT Pack for the Raspberry Pi. Rather than simply present their solution, I present a C++ implementation that reuses their code for the BMP280. The code that you need can be downloaded from https://github.com/ms-iot/adafruitsample/tree/master/Lesson_203 /FullSolution. In this case, you only need the file named BMP280.cs. If you are following along with this chapter writing the code for the project, go ahead and download the sample code now then extract it. You will need the BMP280.cs shortly.

Once built, reading the temperature and pressure from the sensor is really easy. Rather than display the data using a debug interface, you will use a simple user interface to present the data. Yes, that means you will write some XAML code in our C++ project.

You are going to write the code in a slightly different order than previous projects. You first create the new projects in the same solution file then work on the C# code to get it correct then you'll return to the C++ project to fill in the user interface and code to read the sensor. As you will see, there isn't much code to write but there are a few things you must do in order to get everything working together. Let's start with the first project.

New Project

Since you are using C++ for the active project (called the startup project), you use the Blank App (Universal Windows) template under the **C++ ➤ Windows ➤ Universal** category. Use the name **Weather** for the project name. You can save the project wherever you like or use the default location.

Caution If you have just finished the project from Chapter 11, be sure to switch the driver back to the default controller driver otherwise, your project may not execute properly.

Add a C# Runtime Component Project

With the Weather C++ project open, right-click the solution and choose **Add ➤ New Project....** For this project, you want a C# Runtime Component project. Use the **Other Languages ➤ C# ➤ Windows ➤ Universal ➤ Windows Runtime Component** template. Name the project **WeatherSensor**. You can save the project wherever you like or use the default location.

What you should see at this point is one solution entitled Solution 'Weather' with two projects: Weather and Weather Sensor. The Weather project is the startup project. Figure 12-2 shows an example of what you should see in the Solution Explorer. If the Weather project is not the startup project, you can right-click it and set it as the startup project.

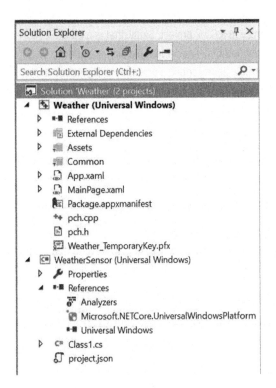

Figure 12-2. Solution Explorer with Weather and WeatherSensor projects

At this point, you have the projects created but you still must add the code for the BMP280 class, add references for the projects, the user interface, and code to read the sensor and display it in the user interface.

BMP280 Class

You are going to use the Adafruit code in our project. If you haven't downloaded the sample project, do that now. You need the file named BMP280.cs. To add it to the project, right-click the **WeatherSensor** project and choose **Add ➤ Existing Item...** then browse for the file and select it. Visual Studio copies the file to your project folder.

You will have to edit this file because it is not in the correct state for use in a runtime component project. But first, you can delete the Class1.cs file that the template provided. You do not need it. Simply right-click the file and choose **Delete**.

The modifications you need to make include changes to the class decorations and some of the methods. You want to allow the code to read from the sensor to run as a task, however, the code as written won't work. You must start the tasks from inside the project rather than from other code as it was originally written. You will change the following things. I go through each of them in turn. As you will see, the changes aren't complex but they are required for the C# project to be used with the C++ project.

- Class declarations must be sealed

- The Initialize() method must run the set up code as a task

- The Read* methods must be normal methods that run the Begin() code as a task

■ **Note** The following listings show unified difference output generated from diff.[3] Essentially, each line marked with a minus sign is replaced with the line immediately below marked with a plus sign (but you delete the plus sign).

Let's begin with changes to the classes. Open the BMP280.cs file. You see two classes: BMP280_CalibrationData and BMP280. Change the class declaration to a sealed class, as shown in Listing 12-1. This is an excerpt of the affected code for brevity and adjacent lines to show context.

Listing 12-1. Changes to the Class Declarations

```
...
 using Windows.Devices.Gpio;
 using Windows.Devices.I2c;
-namespace Lesson_203
+namespace WeatherSensor
 {
-    public class BMP280_CalibrationData
+    public sealed class BMP280_CalibrationData
     {
         //BMP280 Registers
         public UInt16 dig_T1 { get; set; }

...

         public Int16 dig_P9 { get; set; }
     }
-    public class BMP280
+    public sealed class BMP280
     {
         //The BMP280 register addresses according the the datasheet: http://www.adafruit.
com/datasheets/BST-BMP280-DS001-11.pdf
         const byte BMP280_Address = 0x77;
...
```

■ **Caution** Notice that you changed the namespace. If you skip this or forget to change it, your code will compile but fail to link or deploy presenting a strange and confusing set of errors regarding names missing from the template. If you see such, check your namespace declaration to make sure that it is correct.

Next, you must change the Initialize() method to call a new Setup() method as a task. Listing 12-2 shows the changes needed for that method.

Listing 12-2. Changes to the Initialize() Method

```
         //Method to initialize the BMP280 sensor
-        public async Task Initialize()
+        public void Initialize()
```

---

[3]A utility that is very familiar to Linux and Mac developers. See http://gnuwin32.sourceforge.net/packages/diffutils.htm for a Windows port of this tool.

```
      {
          Debug.WriteLine("BMP280::Initialize");
+         Task t_setup = Task.Run(() => Setup());
+         t_setup.Wait();
+
+     }
+
+
+     private async Task Setup()
+     {
+         Debug.WriteLine("BMP280::Setup");
...
```

You are splitting the set up code out as a separate method decorated to run as an asynchronous task. This is necessary to ensure that the code does not run in the same thread as the user interface (it can't).

Next, you must change the ReadTemperature() method to remove the async decorator and to call the Begin() method as a task. This protects the thread and ensures that the code executes outside of the user interface thread. Listing 12-3 shows the changes needed.

Listing 12-3. Changes to the ReadTemperature() Method

```
...
-     public async Task<float> ReadTemperature()
+     public float ReadTemperature()
      {
          //Make sure the I2C device is initialized
-         if (!init) await Begin();
+         if (!init)
+         {
+             Task t_begin = Task.Run(() => Begin());
+             t_begin.Wait();
+         }
...
```

You must change the ReadPressure() method in the same manner as ReadTemperature(). However, since this method must call ReadTemperature() and it is no longer a task, you can call the method directly. Listing 12-4 shows the changes needed.

Listing 12-4. Changes to the ReadPressure() Method

```
...
-     public async Task<float> ReadPreasure()
+     public float ReadPreasure()
      {
          //Make sure the I2C device is initialized
-         if (!init) await Begin();
+         if (!init)
+         {
+             Task t_begin = Task.Run(() => Begin());
+             t_begin.Wait();
+         }
          //Read the temperature first to load the t_fine value for compensation
          if (t_fine == Int32.MinValue)
          {
```

```
-              await ReadTemperature();
+              ReadTemperature();
        }
...
```

Finally, you must change the ReadAltitude() method in the same manner. Listing 12-5 shows the changes needed.

Listing 12-5. Changes to the ReadAltitude() Method

```
...
-        public async Task<float> ReadAltitude(float seaLevel)
+        public float ReadAltitude(float seaLevel)
        {
            //Make sure the I2C device is initialized
-            if (!init) await Begin();
+            if (!init)
+            {
+                Task t_begin = Task.Run(() => Begin());
+                t_begin.Wait();
+            }
            //Read the pressure first
-            float pressure = await ReadPreasure();
+            float pressure = ReadPreasure();
            //Convert the pressure to Hectopascals(hPa)
            pressure /= 100;
...
```

That's it! Not too bad, eh? Remember, the complete source code is available for download from the Apress web site so if you don't want to make these changes yourself, you can download the sample code and get the completed file. However, I do recommend you build the solution yourself otherwise so you can learn how to mix C++ and C# projects in the same solution.

You use a couple of tasks in the code. You do this because the initialization of the libraries (for example, the Wait() method) should not run in the UI thread. This is a very common technique when working with libraries and other assemblies.

But wait, how does this class work to read the data from the sensor? I mentioned the code needs to read multiple bytes from the sensor and decode them. The authors of the donor class have done their homework well and have implemented the code based on the recommendations from the manufacturer's data sheet. Take a moment while you are in the code to look around. There is a whole lot more code in there than what you've seen in the last few code listings!

■ **Tip**   The data sheet for the BMP280 can be found at https://cdn-shop.adafruit.com/datasheets/BST-BMP280-DS001-11.pdf.

So, how did the authors know what to do? Figure 12-3 shows an excerpt from the data sheet that outlines how to read data from the sensor.

5.2.1 I²C write

Writing is done by sending the slave address in write mode (RW = '0'), resulting in slave address 111011X0 ('X' is determined by state of SDO pin. Then the master sends pairs of register addresses and register data. The transaction is ended by a stop condition. This is depicted in Figure 7.

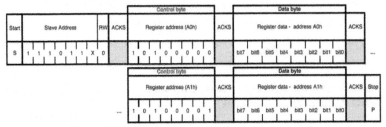

Figure 7: I²C multiple byte write (not auto-incremented)

5.2.2 I²C read

To be able to read registers, first the register address must be sent in write mode (slave address 111011X0). Then either a stop or a repeated start condition must be generated. After this the slave is addressed in read mode (RW = '1') at address 111011X1, after which the slave sends out data from auto-incremented register addresses until a NOACKM and stop condition occurs. This is depicted in Figure 8, where two bytes are read from register 0xF6 and 0xF7.

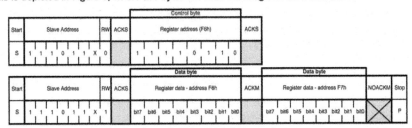

Figure 8: I²C multiple byte read

Figure 12-3. *Excerpt from BMP280 data sheet*

Clearly, there is a lot going on here! What complicates the code are features in the BMP280 to provide calibration data, which is used to correct the values read. Given you must read multiple bytes and interpret them, the code is quite complex. Fortunately, all the hard work has been done for us and you can simply use the code provided. This is often the case for complex breakout boards like the BMP280 from Adafruit. If you ever have to implement your own code for such a component, be sure to read and study the data sheet. You should also consider searching the Internet for solutions. Often times you can find a solution that may not be written in your language of choice but you can either rewrite or as you've seen in this chapter, use the power of Visual Studio to combine C# and C++ in the same solution.

Now, let's add the user interface code.

User Interface

The user interface code is very simple—just four labels: one for the title, and one for each of the values read (temperature, pressure, and altitude). Listing 12-6 shows the XAML code for the interface. Open the MainPage.xaml file in the Weather project and edit the file as shown. The new lines you need to add are shown in bold.

Listing 12-6. User Interface Code

```
...
    <Grid Background="{ThemeResource ApplicationPageBackgroundThemeBrush}">
        <StackPanel Width="400" Height="400">
            <TextBlock x:Name="Title" Height="60" TextWrapping="NoWrap"
                    Text="Weather Station" FontSize="28" Foreground="Blue"
                    Margin="10" HorizontalAlignment="Center"/>
            <TextBlock x:Name="Temp" Height="60" TextWrapping="NoWrap"
                    Text="Initializing..." FontSize="28" Foreground="Blue"
                    Margin="10" HorizontalAlignment="Center"/>
            <TextBlock x:Name="Press" Height="60" TextWrapping="NoWrap"
                    Text="Initializing..." FontSize="28" Foreground="Blue"
                    Margin="10" HorizontalAlignment="Center"/>
            <TextBlock x:Name="Alt" Height="60" TextWrapping="NoWrap"
                    Text="Initializing..." FontSize="28" Foreground="Blue"
                    Margin="10" HorizontalAlignment="Center"/>
        </StackPanel>
    </Grid>
...
```

Figure 12-4 shows an example of what the user interface will look like when the application runs on the Raspberry Pi.

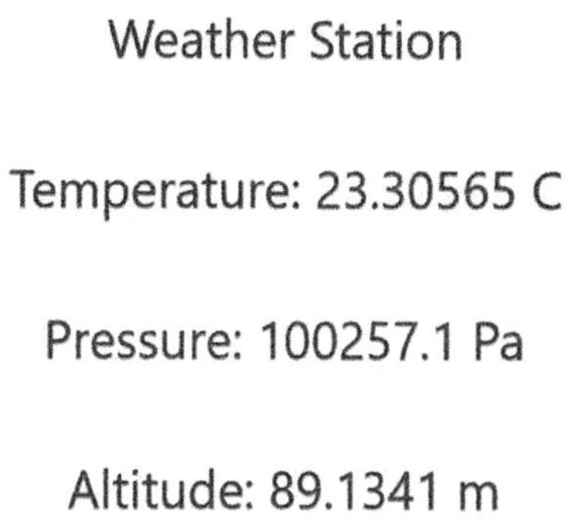

Weather Station

Temperature: 23.30565 C

Pressure: 100257.1 Pa

Altitude: 89.1341 m

Figure 12-4. *Example user interface*

Add References

There are only two references needed—one for each project. Let's start with the WeatherSensor project. Right-click **References** in the project tree in the Solution Explorer and add the **Windows 10 IoT Extensions** reference. You also need to add a reference in the Weather project that references the WeatherSensor project. Right-click **References** in the project tree in the Solution Explorer and add the WeatherSensor project, as shown in Figure 12-5. Be sure to tick the box before clicking **OK**.

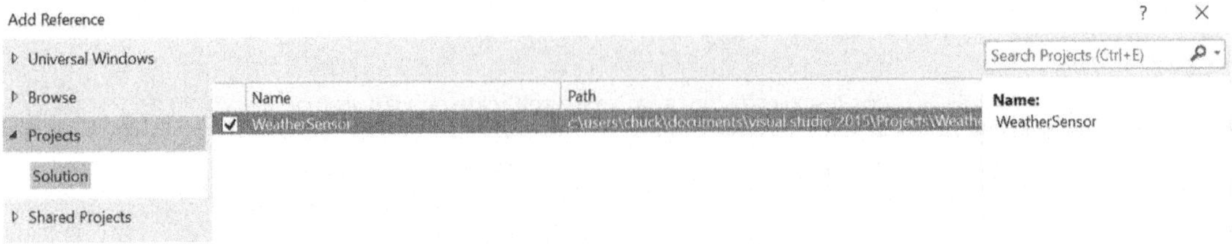

Figure 12-5. *Adding the WeatherSensor project reference*

You should now see the WeatherSensor project in the Weather project references and the Windows 10 IoT extensions in the WeatherSensor project references. If you encounter strange build issues or errors, be sure to check the references.

You may be presented with a dialog during project creation or when you add references to the project. The dialog asks you to determine the minimum and required versions for the IoT extensions. You should choose the **Build 10586** entries for both. Figure 12-6 shows an example of the dialog that you may see.

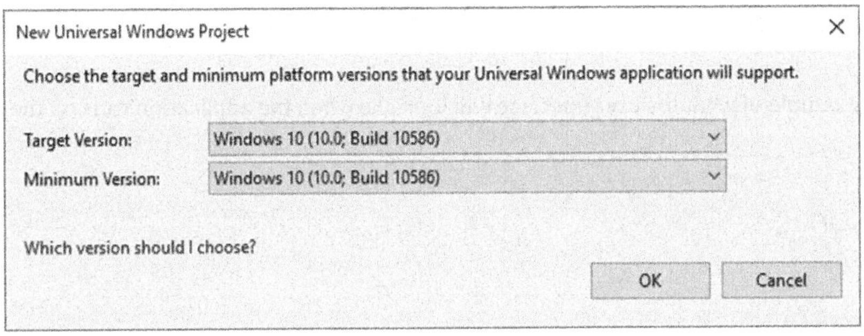

Figure 12-6. *Target version dialog*

You're almost there! You just need to add the code to use the BMP280 class and update the user interface. Let's do that now.

Reading the Weather Data

Now let's add the code to read the temperature, pressure, and altitude from the BMP280 class and display it in the user interface you implemented. Since you are working with C++, you have a separate header file that you put the variables and declarations for methods (called functions in C++). Open the MainPage.xaml.h file and make the following changes. Listing 12-7 shows the entire code for the file.

Listing 12-7. The MainPage.xaml.h Header File

```
#pragma once

#include "MainPage.g.h"

using namespace WeatherSensor;   // Add for sensor C# project

namespace Weather
{
```

```
public ref class MainPage sealed
{
public:
        MainPage();
private:
        // Reference to BMP280 C# class
        BMP280 ^bmp280 = ref new BMP280();

        // Variable for the timer
        Windows::UI::Xaml::DispatcherTimer ^timer_;

        // Timer on tick method
        void OnTick(Platform::Object ^sender, Platform::Object ^args);
};
}
```

Notice you added a namespace that references the BMP280 class (the WeatherSensor namespace), and then added a variable reference to the class. You add these to the private section by convention. You also add a variable for a timer that you will use to read and update the user interface and the method, which is called on each "tick" of the timer event.

Next, you edit the MainPage.xaml.cpp file and add the code to read the data. But first, you need to add a namespace for the timer code. In this case, you need the currency namespace.

Next, you edit the constructor code to call the Initialize() method on the BMP280 instance and then set up the timer code. You use a small update (delay) but you can increase this once you are comfortable the solution is working correctly. If you have been following along with the projects in the previous projects, this code should look familiar in a not-C#-but-similar manner.

Next, you add the code for the OnTick() method. You use this method to read the data and update the user interface. You first read each of the values and then update the corresponding label in the user interface. Listing 12-8 shows an excerpt of the file with all the code needed for the constructor and the new method.

Listing 12-8. The MainPage.xaml.cpp Source File

```
...
using namespace Windows::UI::Xaml::Navigation;
using namespace concurrency; // Add for timer

MainPage::MainPage()
{

        InitializeComponent();

        // Initialize the sensor
        bmp280->Initialize();

        // Setup timer
        timer_ = ref new DispatcherTimer();
        TimeSpan interval;
        interval.Duration = 1000 * 1000 * 10;
        timer_->Interval = interval;
        timer_->Tick += ref new EventHandler<Object ^>(this, &MainPage::OnTick);
        timer_->Start();
}
```

```
void MainPage::OnTick(Object ^sender, Object ^args)
{
        // Initialize the variables
        float temp = 0;
        float pressure = 0;
        float altitude = 0;

        // Create a constant for pressure at sea level.
        // This is based on your local sea level pressure (Unit: Hectopascal)
        // To find the sea level pressure for your area, go to:
        // weather.gov and enter your city then read the pressure from the
        // history.
        const float seaLevelPressure = 1013.25f;

        // Read samples of the data every 10 seconds
        temp = bmp280->ReadTemperature();
        pressure = bmp280->ReadPreasure();
        altitude = bmp280->ReadAltitude(seaLevelPressure);

        Temp->Text = "Temperature: " + temp + " C";
        Press->Text = "Pressure: " + pressure + " Pa";
        Alt->Text = "Altitude: " + altitude + " m";
}
```

You supply a constant for the sea level pressure. This value may be different for your location. Check the weather web sites for your location to get the latest calculated sea level pressure. Substitute your value in the code to get more accurate altitude calculations. That is, the altitude is the result of a calculation based on the barometric pressure and sea level pressure values. Thus, it may not be accurate but should be pretty close if you use the correct value for your area.

That's it! The code is complete and ready for compilation. Be sure to check the preceding listings to ensure that you have all the code in the right place. Once you have entered all the code, you should now attempt to compile the solution. You can compile the projects individually, but if you do so, start with the WeatherSensor project. Once that is compiling correctly, you can compile the weather project. Or you can just build the entire solution. Correct any errors you find until the code compiles without errors or warnings.

■ **Caution**   Be sure to choose the Any CPU platform before you build the solution. This ensures that the correct platform (WinRT) is built for the C# project.

Deploy and Execute

Now it is time to deploy the application! Be sure to fix any compilation errors first. Like you have with other applications, you want to compile the application in debug first (but you can compile in release mode if you'd prefer) and you must turn on the debugger on your board. You do this with the device portal.

Go ahead and power on your board. Be sure to connect a monitor, mouse, and keyboard so you can use the user interface. When ready, connect to the board to run the Device Portal, turn on the debugger, and then open the project properties to target the device and run with the remote debugger.

Recall from the Chapter 9, you must modify two settings: the **Remote machine** name and the **Authentication Mode**. Set the **Remote machine** name to the IP address of your device with the port specified by the remote debugger when you started it from the device portal. For example, my device gave me the port number 8116 so I use 10.0.1.89:8116. Set the **Authentication Mode** to **None**. Do this for the C++ startup project (Weather) and not the C# project (WeatherSensor).

Now you can deploy your application. Go ahead and do that now. You should see the user interface appear on the device after a few moments and then the screen update as the timer tick event fires.

If you do not see the user interface or you see the box with the X in it, then the default application reloads, chances are there is something wrong in your code. In this case, you should run the code in debug mode and observe the exception thrown. There are a number of things that can go wrong. The following summarizes some of the things (mistakes) I've encountered in designing this project.

- Ensure that the correct project is marked as the startup project (should show name in bold in Solution Explorer).

- Ensure that you are compiling the entire solution for *any* CPU.

- Ensure that the namespace of the BMP280 class is set correctly if you copied it from the Adafruit sample.

- Double-check your references to ensure they are correct.

- Ensure that the BMP280 sensor is wired correctly to your device.

- If you connected the sensor with 3.3V by mistake, it should still work, but double-check the GND pin is connected correctly.

- Make sure that the namespaces are specified for concurrency and the WeatherSensor (and in the correct files).

- Ensure that the debugger is started on your device before deploying.

- Use the Device Portal to start and stop the application after deployment and do not set the application to run as a startup until you have successfully run and debugged the code.

Once the project is running correctly, enjoy it for a while then put the hardware aside. You will need it again for the next chapter.

Summary

Weather IoT solutions can be a lot of fun to develop and unlike an experiment that teaches techniques, it can be a very practical solution. Indeed, you can show it to your friends and family and possibly get more than a non-committal "that's nice" accolade.[4]

In this chapter, you've seen a depth of complexity that presented a number of advanced tools and techniques from using a code library written by someone else to building a C++ headed application to incorporating a C# and C++ project in the same solution. Combining all of these together makes this project the most complex in the book.[5]

In the next chapter, you'll step back a bit from escalating the complexity of this project by seeing a pure C# implementation. To make it interesting, you will add the ability to record the data read to a MySQL database. Adding a database storage feature makes the project much more interesting and useful for post-event analysis. That is, you cannot perform any form of analysis without having stored the data collected over time. You could do this with a log file (and many solutions do), but having the ability to use a powerful query language like SQL can give your solution a great deal of sophistication and capability for free.

---

[4]You do show off your projects don't you? Well, this time you can show them something they can actually use!
[5]But you see less complex yet more diverse projects in the following chapters.

■ ■ ■

## CHAPTER 13

Project 4: Using MySQL to Store Data

IoT solutions, by definition if not implementation, can generate a lot of data. Indeed, most IoT solutions observe the world in one or often several ways. Those observations generate data at whatever rate the solution specifies (called a *sample rate*). To make the data most useful for historical or similar analytics, you have to save the data for later processing. Database systems provide a perfect solution for storing IoT data and making it available for later use.

With the exception of the last project, you wouldn't consider storing the data from the example projects you've seen thus far for any length of time. That changed when we started using sensors in our projects. The project from Chapter 12 is a perfect example; it observes weather data that could be useful for later analysis.

The project in this chapter takes the same project goals from the last chapter and implements them in a different way. There are two methods for connecting to an I2C device. You used one method in the last project; you will use the other in this project. You will also add a database element storing the data generated in a MySQL database. I demonstrate how to retrieve that data.

■ **Tip** I cover incorporating database systems and specifically using MySQL for the IoT in my book *MySQL for the Internet of Things* (Apress, 2015). If you want to explore the theory and application of database systems in the IoT, this book will get you started, even if you know very little about IoT or MySQL.

Before we get into the project design and how to implement the code, let's take a moment to discover MySQL. If you already have a lot of experience with MySQL, you may want to skim the section as a refresher. If you have never used MySQL, the following brief primer gets you started and covers everything you need to know in order to complete this project.

What Is MySQL?

MySQL is the world's most popular open source database system for many excellent reasons. First and foremost, it is open source, which means anyone can use it for a wide variety of tasks for free.[1] Best of all, MySQL is included in many platform repositories, making it easy to get and install. If your platform doesn't include MySQL in the repository (such as aptitude), you can download it from the MySQL web site (http://dev.mysql.com).

---

[1]According to GNU (www.gnu.org/philosophy/free-sw.html), "free software is a matter of liberty, not price. To understand the concept, you should think of "free" as in "free speech," not as in "free beer.""

© Charles Bell 2016
C. Bell, *Windows 10 for the Internet of Things*, DOI 10.1007/978-1-4842-2108-2_13

Oracle Corporation owns MySQL. Oracle obtained MySQL through an acquisition of Sun Microsystems, which acquired MySQL from its original owners, MySQL AB. Despite fears to the contrary, Oracle has shown excellent stewardship of MySQL by continuing to invest in the evolution and development of new features as well as faithfully maintaining its open source heritage. Although Oracle also offers commercial licenses of MySQL—just as its prior owners did in the past—MySQL is still open source and available to everyone.

WHAT IS OPEN SOURCE? IS IT REALLY FREE?

Open source software grew from a conscious resistance to the corporate-property mindset. While working for MIT, Richard Stallman, the father of the free software movement, resisted the trend of making software private (closed) and left MIT to start the GNU (GNU Not Unix) project and the Free Software Foundation (FSF).

Stallman's goal was to reestablish a cooperating community of developers. He had the foresight, however, to realize that the system needed a copyright license that guaranteed certain freedoms. (Some have called Stallman's take on copyright "copyleft," because it guarantees freedom rather than restricts it.) To solve this, Stallman created the GNU Public License (GPL). The GPL, a clever work of legal permissions that permits the code to be copied and modified without restriction, states that derivative works (the modified copies) must be distributed under the same license as the original version without any additional restrictions.

There was one problem with the free software movement. The term *free* was intended to guarantee freedom to use, modify, and distribute; it was not intended to mean "no cost" or "free to a good home." To counter this misconception, the Open Source Initiative (OSI) formed and later adopted and promoted the phrase *open source* to describe the freedoms guaranteed by the GPL license. For more information about open source software, visit `www.opensource.org`.

MySQL runs as a background process (or as a foreground process if you launch it from the command line)[2] on your system. Like most database systems, MySQL supports Structured Query Language (SQL). You can use SQL to create databases and objects (using data definition language [DDL]), write or change data (using data manipulation language [DML]), and execute various commands for managing the server.

To issue these commands, you must first connect to the database server. MySQL provides a client application that enables you to connect to and run commands on the server. The application is named mysql and is known as the mysql *client* (previously the mysql monitor). Note that Oracle has stopped using the older name so that it is not confused with the MySQL Enterprise Monitor, which is a premium product provided to customers who purchase an Enterprise license.

Listing 13-1 shows examples of each type of command in action using the mysql client:

Listing 13-1. Commands Using the mysql Client

```
$ mysql -uroot -pXXXX
Welcome to the MySQL monitor.  Commands end with ; or \g.
Your MySQL connection id is 4
Server version: 5.7.10-log MySQL Community Server (GPL)

Copyright (c) 2000, 2015, Oracle and/or its affiliates. All rights reserved.

Oracle is a registered trademark of Oracle Corporation and/or its
affiliates. Other names may be trademarks of their respective
```

---

[2]And use the `--console` command-line option on Windows systems.

owners.

```
Type 'help;' or '\h' for help. Type '\c' to clear the current input statement.

mysql> CREATE DATABASE testme;
Query OK, 1 row affected (0.00 sec)

mysql> CREATE TABLE testme.table1 (sensor_node char(30), sensor_value int, sensor_event
timestamp);
Query OK, 0 rows affected (0.00 sec)

mysql> INSERT INTO testme.table1 VALUES ('living room', 23, NULL);
Query OK, 1 row affected (0.00 sec)

mysql> SELECT * FROM testme.table1;
+-------------+--------------+---------------------+
| sensor_node | sensor_value | sensor_event        |
+-------------+--------------+---------------------+
| living room |           23 | 2013-02-04 20:30:13 |
+-------------+--------------+---------------------+
1 row in set (0.00 sec)

mysql> SET @@global.server_id = 111;
Query OK, 0 rows affected (0.00 sec)

mysql>
```

In this example, you see DDL in the form of the CREATE DATABASE and CREATE TABLE statements, DML in the form of the INSERT and SELECT statements, and a simple administrative command to set a global server variable. Next you see the creation of a database and a table to store the data, the addition of a row in the table, and finally retrieval of the data in the table.

A great many commands are available in MySQL. Fortunately, you need master only a few of the more common ones. The following are the commands that you use most often. The portions enclosed in <> indicate user-supplied components of the command, and [. . .] indicates that additional options are needed:

■ **Tip** If you use the mysql client, you must terminate each command with a semicolon (;) or \G.

- CREATE DATABASE <database_name>: Creates a database.

- USE <database>: Sets the default database.

- CREATE TABLE <table_name> [...]: Creates a table or structure to store data.

- INSERT INTO <table_name> [...]: Adds data to a table.

- UPDATE [...]: Changes one or more values for a specific row.

- DELETE FROM <table_name> [...]: Removes data from a table.

- SELECT [...]: Retrieves data (rows) from the table.

Although this list is only a short introduction and nothing like a complete syntax guide, there is an excellent online reference manual that explains each and every command (and much more) in great detail. You should refer to the online reference manual whenever you have a question about anything in MySQL. You can find it at http://dev.mysql.com/doc/.

If you are thinking that there is a lot more to MySQL than a few simple commands, you are absolutely correct. Despite its ease of use and fast startup time, MySQL is a full-fledged relational database management system (RDBMS). There is much more to it than you've seen here. For more information about MySQL, including all the advanced features, see the reference manual.

MYSQL: WHAT DOES IT MEAN?

The name MySQL is a combination of a proper name and an acronym. SQL is Structured Query Language. The *My* part isn't the possessive form—it is a name. In this case, My is the name of the founder's daughter. As for pronunciation, MySQL experts pronounce it "My-S-Q-L" and not "my sequel." Indeed, the mark of a savvy MySQL user is in their pronunciation of the product. There is a corollary with Mac OS X: Is it "Mac O-S Ex" or "Mac O-S Ten"? Check to see.

Getting Started with MySQL

Now that you know what MySQL is and how it is used, you need to know a bit more about RDBMSs and MySQL in particular before you start building your first database server. This section discusses how MySQL stores data (and where it is stored), how it communicates with other systems, and some basic administration tasks required in order to manage your new MySQL server.

WHAT'S A RELATIONAL DATABASE MANAGEMENT SYSTEM?

An RDBMS is a data storage-and-retrieval service based on the Relational Model of Data as proposed by E. F. Codd in 1970. These systems are the standard storage mechanism for structured data. A great deal of research is devoted to refining the essential model proposed by Codd, as discussed by C. J. Date in *The Database Relational Model: A Retrospective Review and Analysis*.[3] This evolution of theory and practice is best documented in *The Third Manifesto*.[4]

The relational model is an intuitive concept of a storage repository (database) that can be easily queried by using a mechanism called a *query language* to retrieve, update, and insert data. The relational model has been implemented by many vendors because it has a sound systematic theory, a firm mathematical foundation, and a simple structure. The most commonly used query mechanism is SQL, which resembles natural language. Although SQL is not included in the relational model, it provides an integral part of the practical application of the relational model in RDBMSs.

The data are represented as related pieces of information (attributes or *columns*) about a certain event or entity. The set of values for the attributes is formed as a *tuple* (sometimes called a *record* or a *row*). Tuples are stored in tables that have the same set of attributes. Tables can then be related to other tables through constraints on keys, attributes, and tuples.

Tables can have special mappings of columns called *indexes* that permit you to read the data in a specific order. Indexes are also very useful for fast retrieval of rows that match the value(s) of the indexed columns.

---

[3]C. J. Date, *The Database Relational Model: A Retrospective Review and Analysis* (Reading, MA: Addison-Wesley, 2001).
[4]C. J. Date and H. Darwen, *Foundation for Future Database Systems: The Third Manifesto* (Reading, MA: Addison-Wesley, 2000).

How and Where MySQL Stores Data

The MySQL database system stores data via an interesting mechanism of programmatic isolation; it is called a *storage engine*, which is governed by the handler interface. The handler interface permits the use of interchangeable storage components in the MySQL server so that the parser, the optimizer, and all manner of components can interact in storing data on disk using a common mechanism. This is also referred to as a *pluggable storage engine.*[5]

■ **Note** MySQL supports several storage engines. Most are designed to write data to disk by default. However, the MEMORY storage engine stores data in memory but is not persistent. That is, when the computer is rebooted, the data is lost. You can use the MEMORY storage engine for fast lookup tables. Indeed, one optimization technique is to create copies of lookup tables at startup using the MEMORY storage engine.

What does this mean to you? It means you have the choice of different mechanisms for storing data. You can specify the storage engine in the table CREATE statement shown in the following code sample. Notice the last line in the command; this is how a storage engine is specified. Leaving off this clause results in MySQL using the default storage engine.

■ **Tip** The default storage engine was changed from MyISAM to InnoDB in MySQL version 5.6.

```
Create Table:

CREATE TABLE `books` (
  `ISBN` varchar(15) DEFAULT NULL,
  `Title` varchar(125) DEFAULT NULL,
  `Authors` varchar(100) DEFAULT NULL,
  `Quantity` int(11) DEFAULT NULL,
  `Slot` int(11) DEFAULT NULL,
  `Thumbnail` varchar(100) DEFAULT NULL,
  `Description` text
) ENGINE=MyISAM;
```

Great! Now, what storage engines exist on MySQL? You can discover which storage engines are supported by issuing the following command. As you see, there are a lot to choose from. I cover a few that may be pertinent to planning sensor networks:

```
mysql> SHOW STORAGE ENGINES \G
*************************** 1. row ***************************
      Engine: InnoDB
     Support: YES
     Comment: Supports transactions, row-level locking, and foreign keys
Transactions: YES
          XA: YES
  Savepoints: YES
```

---

[5]If you would like to know more about storage engines and what makes them tick, see my book *Expert MySQL* (Apress, 2012).

```
*************************** 2. row ***************************
     Engine: MRG_MYISAM
    Support: YES
    Comment: Collection of identical MyISAM tables
Transactions: NO
         XA: NO
 Savepoints: NO
*************************** 3. row ***************************
     Engine: BLACKHOLE
    Support: YES
    Comment: /dev/null storage engine (anything you write to it disappears)
Transactions: NO
         XA: NO
 Savepoints: NO
*************************** 4. row ***************************
     Engine: CSV
    Support: YES
    Comment: CSV storage engine
Transactions: NO
         XA: NO
 Savepoints: NO
*************************** 5. row ***************************
     Engine: MEMORY
    Support: YES
    Comment: Hash based, stored in memory, useful for temporary tables
Transactions: NO
         XA: NO
 Savepoints: NO
*************************** 6. row ***************************
     Engine: FEDERATED
    Support: NO
    Comment: Federated MySQL storage engine
Transactions: NULL
         XA: NULL
 Savepoints: NULL
*************************** 7. row ***************************
     Engine: ARCHIVE
    Support: YES
    Comment: Archive storage engine
Transactions: NO
         XA: NO
 Savepoints: NO
*************************** 8. row ***************************
     Engine: MyISAM
    Support: DEFAULT
    Comment: Default engine as of MySQL 3.23 with great performance
Transactions: NO
         XA: NO
 Savepoints: NO
8 rows in set (0.00 sec)
```

As of version 5.6, MySQL uses the InnoDB storage engine by default. Previous versions used MyISAM as the default. InnoDB is a fully transactional, ACID[6] storage engine. A *transaction* is a batch of statements that must all succeed before any changes are written to disk. The classic example is a bank transfer. If you consider a system that requires deducting an amount from one account and then crediting that amount to another account to complete the act of moving funds, you would not want the first to succeed and the second to fail or vice versa!

Wrapping the statements in a transaction ensures that no data is written to disk until and unless all statements are completed without errors. Transactions in this case are designated with a BEGIN statement and concluded with either a COMMIT to save the changes or a ROLLBACK to undo the changes. InnoDB stores its data in a single file (with some additional files for managing indexes and transactions).

The MyISAM storage engine is optimized for reads. MyISAM has been the default for some time and was one of the first storage engines available. In fact, a large portion of the server is dedicated to supporting MyISAM. It differs from InnoDB in that it does not support transactions and stores its data in an indexed sequential access method format. This means it supports fast indexing. You would choose MyISAM over InnoDB if you did not need transactions and you wanted to be able to move or back up individual tables.

Another storage engine that you may want to consider, especially for sensor networks, is Archive. This engine does not support deletes (but you can drop entire tables) and is optimized for minimal storage on disk. Clearly, if you are running MySQL on a small system like a Raspberry Pi, small is almost always better! The inability to delete data may limit more advanced applications, but most sensor networks merely store data and rarely delete it. In this case, you can consider using the Archive storage engine.

There is also the CSV storage engine (where CSV stands for *comma-separated values*). This storage engine creates text files to store the data in plain text that can be read by other applications, such as a spreadsheet application. If you use your sensor data for statistical analysis, the CSV storage engine may make the process of ingesting the data easier.

So where is all this data? If you query the MySQL server and issue the SHOW VARIABLES LIKE "datadir"; command, you see the path to the location on disk that all storage engines use to store data. In the case of InnoDB, this is a single file on disk located in the data directory. InnoDB also creates a few administrative files, but the data is stored in the single file. For most other storage engines except NDB and MEMORY, the data for the tables is stored in a folder with the name of the database under the data directory. Listing 13-2 shows how to determine the location of the data directory (it is typically in a protected folder).

Listing 13-2. Finding Where Your Data Is Located

```
mysql> SHOW VARIABLES LIKE 'datadir';
+---------------+-----------------------+
| Variable_name | Value                 |
+---------------+-----------------------+
| datadir       | /usr/local/mysql/data/ |
+---------------+-----------------------+
1 row in set (0.00 sec)
```

If you navigate to the location (path) shown and (with elevated privileges) issue a directory listing command, you can see the InnoDB files identified by the ib and ibd prefixes. You may also see a number of directories, all of which are the databases on this server.

■ **Tip** If you want to copy data from one server to another by copying files, be sure to copy the .frm files as well! This is easy for MyISAM and Archive but much harder with InnoDB. In the case of InnoDB, you have to copy all the database folders and the InnoDB files to make sure that you get everything.

---

[6]See http://en.wikipedia.org/wiki/ACID.

Although it is unlikely that you would require a transactional storage engine for your IoT solutions, MySQL 5.6 has one, and it's turned on by default. A more likely scenario is that you would use the MyISAM or Archive engine for your tables.

For more information about storage engines and the choices and features of each, please see the online MySQL Reference Manual section "Storage Engines" (http://dev.mysql.com/doc/).

The MySQL Configuration File

The MySQL server can be configured using a configuration file, similar to the way you configure other Windows applications. On Windows, the MySQL configuration file is named my.ini and located in the program data folder under the MySQL server version (e.g., C:\ProgramData\MySQL\MySQL Server 5.7). This file contains several sections, one of which is labeled [mysqld]. The items in this list are key-value pairs: the name on the left of the equal sign is the option, and its value on the right. The following is a typical configuration file (with many lines suppressed for brevity):

```
[mysqld]
port=3306
datadir=C:/ProgramData/MySQL/MySQL Server 5.7\Data
character-set-server=utf8
default-storage-engine=INNODB
sql-mode="STRICT_TRANS_TABLES,NO_AUTO_CREATE_USER,NO_ENGINE_SUBSTITUTION"
log-output=FILE
general-log=0
general_log_file="CHUCKSURFACE.log"
slow-query-log=1
slow_query_log_file="CHUCKSURFACE-slow.log"
long_query_time=10
log-error="CHUCKSURFACE.err"
server-id=1
```

As you can see, this is a simple way to configure a system. This example sets the TCP port, base directory (the root of the MySQL installation, including the data as well as binary and auxiliary files), data directory, and server ID (used for replication, as discussed shortly) and turns on the general log (when the Boolean switch is included, it turns on the log). There are many such variables you can set for MySQL. See the online MySQL reference manual for details concerning using the configuration file.

How to Get and Install MySQL

The MySQL server is available for a variety of platforms, including most Linux and Unix platforms, Mac OS X, and Windows. To download MySQL server, visit http://dev.mysql.com/downloads/ and click **Community** and then **MySQL Community Server**. This is the GPLv2 license of MySQL. The page automatically detects your operating system. If you want to download for another platform, you can select it from the drop-down menu.

Oracle has provided a special installation packaging for Windows named the Windows Installer. This package includes all the MySQL products available under the community license, including MySQL Server, Workbench, Utilities, Fabric, and all the available connectors (program libraries for connecting to MySQL). This makes installing on Windows a one-stop, one-installation affair. How cool is that? Figure 13-1 shows the download page for the Windows installer.

Figure 13-1. *Download page for Windows installer*

To install MySQL, begin by choosing either the Windows installer 32- or 64-bit installation package that matches your Windows version. Once the file is downloaded, click the file to begin installation. Note that some browsers, such as the new Edge browser, may ask you if you want to launch the installation. You may need to reply to a dialog permitting the installation.

The installation is fully automated with a series of dialogs to help you configure your installation. Like most Windows installation packages, you can choose what you want to install as well as where to install it. I recommend accepting the defaults since these are optimized for typical use cases. You can even choose to automatically start MySQL. You may also be given a temporary password for the root user account. Be sure to write that down and change it later (with the SET PASSWORD command; see the following) .

How to Start, Stop, and Restart MySQL

While working with your databases and configuring MySQL on your Windows system, you may need to control the startup and shutdown of the MySQL server. The default mode for installing MySQL is to automatically start on boot and stop on shutdown, but you may want to change that, or you may need to stop and start the server after changing a parameter. In addition, when you change the configuration file, you need to restart the server to see the effect of your changes.

You can start, pause (stop), and restart the MySQL server with the *Services* application in Windows. Simply type **Services** in the search box and choose the application. You can then scroll down, choose MySQL from the list and start, stop, or restart the server using the links provided. Figure 13-2 shows an example of the application with the commands highlighted.

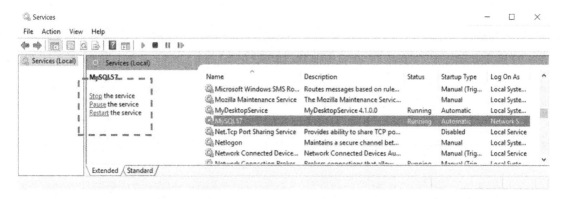

Figure 13-2. *Services application*

Creating Users and Granting Access

You need to know about two additional administrative operations before working with MySQL: creating user accounts and granting access to databases. MySQL can perform both of these with the GRANT statement, which automatically creates a user if one does not exist. But the more pedantic method is first to issue a CREATE USER command followed by one or more GRANT commands. For example, the following shows the creation of a user named sensor1 and grants the user access to the database room_temp:

```
CREATE USER 'sensor1'@'%' IDENTIFIED BY 'secret';
GRANT SELECT, INSERT, UPDATE ON room_temp.* TO 'sensor1'@'%';
```

The first command creates the user named sensor1, but the name also has an @ followed by another string. This second string is the host name of the machine with which the user is associated. That is, each user in MySQL has both a user name and a host name, in the form user@host, to uniquely identify them. That means the user and host sensor1@10.0.1.16 and the user and host sensor1@10.0.1.17 are not the same. However, the % symbol can be used as a wildcard to associate the user with any host. The IDENTIFIED BY clause sets the password for the user.

A NOTE ABOUT SECURITY

It is always a good idea to create a user for your application that does not have full access to the MySQL system. This is so you can minimize any accidental changes and also to prevent exploitation. For sensor networks, it is recommended that you create a user with access only to those databases where you store (or retrieve) data. You can change MySQL user passwords with the following command:

```
SET PASSWORD FOR sensor1@"%" = PASSWORD("secret");
```

Also be careful about using the wildcard % for the host. Although it makes it easier to create a single user and let the user access the database server from any host, it also makes it much easier for someone bent on malice to access your server (once they discover the password).

Another consideration is connectivity. If you connect a database to your network and the network is in turn connected to the Internet, it may be possible for other users on your network or the Internet to gain access to the database. Don't make it easy for them—change your root user password, and create users for your applications.

The second command allows access to databases. There are many privileges that you can give a user. The example shows the most likely set that you would want to give a user of a sensor network database: read (SELECT), add data (INSERT), and change data (UPDATE). See the online reference manual for more about security and account access privileges.

The command also specifies a database and objects to which to grant the privilege. Thus, it is possible to give a user read (SELECT) privileges to some tables and write (INSERT, UPDATE) privileges to other tables. This example gives the user access to all objects (tables, views, and so on) in the room_temp database.

As mentioned, you can combine these two commands into a single command. You are likely to see this form more often in the literature. The following shows the combined syntax. In this case, all you need to do is add the IDENTIFIED BY clause to the GRANT statement. Cool!

```
GRANT SELECT, INSERT, UPDATE ON room_temp.* TO  "sensor1 "@'%" IDENTIFIED BY "secret ";
```

Now that you know more about MySQL, let's talk about the project and how you will develop the database portion.

Overview

This project is a rewrite of the Weather project from the last chapter. The project uses a BMP280 sensor breakout board with an I2C interface. Unlike the last project, you will not be writing a user interface. Similarly, you will implement the project in C# rather than C++.

You will write this project purely in C# so that you can take advantage of a NuGet package to read data from the sensor using an I2C interface. This is because the library you want is written in C# and only works with C# applications. Interestingly, the library you will use provides a bit more data than the previous code library producing some different units for barometric precision. The extra data are extrapolated from the BMP280 data and may be interesting if you're a real weather buff.

Otherwise, the project goals are the same as the previous project. You want to read data from the sensor but instead of displaying it for an instant and replacing it with new values, you will store the data in a MySQL database.

You will use the same hardware, as you did in the previous project except you won't need a monitor, a keyboard, or a mouse since this project will run headless.

I demonstrate all the code necessary and more in the following sections. But first, let's discover how to set up the MySQL database for the project.

Set up the Database

If you have not installed MySQL, you should consider doing that before you complete the code. You do not have to run MySQL on your laptop. You could run it on another machine, across the Internet, or wherever. You just need to have your IoT device configured so that it can connect to the MySQL server via TCP/IP.

I normally set up the database I want to use as a first step. That is, I design the database and table(s)[7] before I write the code. In this case, I've done the design work for you. To keep things simple, you will store data in a single table one row at a time. That is, as you read data from the sensor, you will store it together as a single entry (via an INSERT statement).

In order to keep the rows unique, you will use an auto increment column, which is a surrogate key technique that generates an increasing integer value. This allows you to refer to a row by the key for faster access (via indexing). You will also use a special column called a *timestamp* that the server will automatically supply the current date and time you inserted the row. This is a great technique and a best practice for generating database tables that store data that may be used for historical analysis.

Finally, you add a column for each of the data elements you collect. This includes temperature (in both Celsius and Fahrenheit - the library converts it for you), and bars, hectopascals, and atmospheres for barometric pressure. Listing 13-3 shows the commands you issue to create a database named weather with a single table named history.

Listing 13-3. SQL Commands for Weather Database

```
mysql> CREATE DATABASE weather;
Query OK, 1 row affected (0.01 sec)

mysql> USE weather;
Database changed
mysql> CREATE TABLE history (
    ->    id int not null auto_increment primary key,
    ->    dateRecorded timestamp,
    ->    degreesCelsius float,
    ->    degreesFarenheit float,
```

---

[7]And any additional database objects that I need.

```
    ->    bars float,
    ->    hectopascals float,
    ->    atmospheres float
    -> );
Query OK, 0 rows affected (0.10 sec)
```

There is just one more step to accomplish. It is always a good practice to create a new user account that has rights to update the database (and only the one database). This allows you to ensure that casual intrusion via your application limits exposure of other data on your system. The following command creates a user named w_user with a specific password and access only to the weather database. You can use a different password if you'd like, just be sure to remember it when you write the code.

```
mysql> CREATE USER w_user@'%' IDENTIFIED BY 'secret';
Query OK, 0 rows affected (0.04 sec)

mysql> GRANT ALL ON weather.* TO w_user@'%';
Query OK, 0 rows affected (0.03 sec)
```

■ **Note**   Depending on how your MySQL server is set up, you may be able to issue only the GRANT command to create the user, but I like to use the CREATE command first.

Now that you have the database table set up, let's review the hardware you need for this project. I repeat the data from the last chapter for clarity.

Required Components

The following lists the components that you need. You can get the BMP280 sensor from Adafruit (www.adafruit.com) either in the Microsoft IoT Pack for Raspberry Pi or purchased separately, SparkFun (www.sparkfun.com), or any electronics store that carries electronic components. However, if you use a sensor made by someone other than Adafruit, you may need to alter the code to change the I2C address. Since this solution is a headless application, you do not need a monitor, a keyboard, or a mouse.

- Adafruit BMP280 I2C or SPI barometric pressure and altitude sensor

- Jumper wires: (4) male-to-female

- Breadboard (full size recommended but half size is OK)

- Raspberry Pi 2 or 3

- Power supply

Set up the Hardware

Although there are only four connections needed for this project, you will make a plan for how things should connect, which is good practice to hone. To connect the components to the Raspberry Pi, you need four pins for the BMP280 sensor, which requires only power, ground, and 2 pins for the I2C interface. Refer to Chapter 12 for how to wire the sensor. I include Figure 13-3 as a reminder of how things are connected.

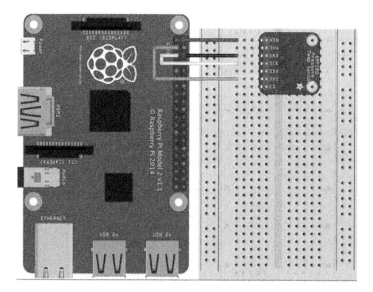

Figure 13-3. *Connections for the Weather Sensor Project*

If you are following along with this chapter working on the project, go ahead and make the hardware connections now. Don't power on the board yet, but do double and triple check the connections.

Write the Code

Now it's time to write the code for the project. You will use the same BMP280 sensor from the last project, but you will use a different library that you can download from NuGet. You will also be using the Connector/Net database connector from Oracle to connect to your MySQL server. You'll use a timer to read the data and update the database. Once you launch the new project, you can monitor its progress by running a SELECT query on the MySQL server.

Let's begin with setting up the new project, and then you'll see how to add the new resources and complete the code for both reading from the sensor and writing the data to the database.

New Project

You will use a C# project template for this project—the Blank App (Universal Windows) template. Use the name WeatherDatabase for the project name. You can save the project wherever you like or use the default location. Once the project opens, double-click the MainPage.xaml.cs file. There are a number of namespaces you need to include. Go ahead and add those now, as shown next.

```
using System.Threading.Tasks;          // Add for Task.Run()
using MySql.Data.MySqlClient;          // Add for MySQL connection
using System.Diagnostics;              // Add for debugging
using Glovebox.IoT.Devices.Sensors;    // Add for BMP280 (or BME280)
```

Here you added namespaces for threading so you can run a task (for the BMP280 library), the MySQL database client, the diagnostics for writing debug statements to the log, and finally, the sensor library from Glovebox. You'll see how to add the MySQL and Glovebox references in the next sections.

Next, you need to add some variables. You add a variable for the `DispatcherTimer` class, a variable for the `MySQLConnection` class, and a variable for the BMP280 class. The following shows the correct code for defining these variables. These are placed in the `MainPage` class.

```
private DispatcherTimer bmpTimer;     // Timer
private MySqlConnection mysql_conn;   // Connection to MySQL
private BMP280 tempAndPressure;       // Instance of BMP280 class
```

Next, you need to add some constants for the database. First, you add a connection string. This string contains several parts that you need to modify to match your systems. I highlight each in bold italics. These are the IP or hostname of the server, the user account and password that you want to use to connect to the MySQL server (you can use what is shown if you issued the preceding commands to create the user), the port (3306 is the default), the default database (can be omitted), and the SSL mode (which must be set to None because the RT version of the connector does not support SSL). Next, you add a formatted string so that you can issue the INSERT statement using parameters for the values.

```
// String constants for database
private string connStr = "server=10.0.1.18;user=w_user;password=secret;" +
                         "port=3306;database=weather;sslMode=None";
private string INSERT = "INSERT INTO weather.history VALUES (null, null, " +
                         "{0}, {1}, {2}, {3}, {4})";
```

You may notice that the first two columns for the INSERT statement are null. You do this to tell the database server to use the default values for the specific columns. In this case, the auto increment column named Id and the timestamp column named dateRecorded. I should note that you could have used a different form of the INSERT statement by specifying the columns by name but passing null is a bit easier if not a bit lazy.

The code in the `MainPage()` function initializes the components, the BMP280 class and call a new method imaginative named `Connect()` to connect to the database server. You make this a new method to keep things easier to write (and the code easier to understand). You'll see this new method shortly. You also set up the timer. In this case, you use a value of 5000, which is 5 seconds. You may want to consider making value greater if you plan to use the project for practical use cases. Listing 13-4 shows the complete `MainPage()` method.

Listing 13-4. MainPage Method

```
public MainPage()
{
    this.InitializeComponent();

    // Instantiate a new BMP280 class instance
    tempAndPressure = new BMP280();

    // Connect to MySQL. If successful, setup timer
    if (this.Connect())
    {
        this.bmpTimer = new DispatcherTimer();
        this.bmpTimer.Interval = TimeSpan.FromMilliseconds(5000);
        this.bmpTimer.Tick += BmpTimer_Tick;
        this.bmpTimer.Start();
    }
    else
    {
        Debug.WriteLine("ERROR: Cannot proceed without database connection.");
    }
}
```

Notice that if the Connect() method returns false, you issue a debug statement stating you cannot connect to the MySQL server. Thus, you should run this project with debug to ensure that you are connecting to the server properly (see the "Why Can't I Connect?" sidebar).

Now you need three references added to the solution. You need the Glovebox.IoT.Devices library from NuGet, which in turn requires the Units.NET library from NuGet. You add the reference to the Connector/Net library. Finally, you need the Windows 10 IoT Extensions library. Let's discuss each of these in turn.

Glovebox.IoT.Devices

Glovebox.IoT.Devices is a library that supports a host of sensors—each presented as a separate class to make it easy to use for specific sensors, like the BMP280. It also takes care of the I2C communication for you. All you need to do is instantiate the class and call the methods to read the data.

To add the Glovebox.IoT.Devices library, use the NuGet Package Manager from the **Tools ➤ NuGet Package Manager ➤ Manage NuGet Packages for Solution...** menu. Click the **Browse** tab and then type **glovebox** in the search box, as shown in Figure 13-4.

Figure 13-4. *NuGet Package Manager, results for Glovebox*

Select the entry named **Glovebox.IoT.Devices** in the list, tick the project name (solution) in the list on the right, and finally, click **Install**. The installation starts and Visual Studio downloads a number of packages and then asks you permission to install them. Go ahead and let the installation complete. A dialog box will open to tell you that the installation is complete.

■ **Caution**   If you have just finished the project from Chapter 11, be sure to switch the driver back to the default controller driver; otherwise, your project may not execute properly.

Unlike the lightning library you used in Chapter 11, there are no additional steps needed to use the library, however, the library requires another library—the Units.NET library, also available via NuGet.

Units.NET

Glovebox.IoT.Devices requires another library that provides helper methods for converting units of measure. The notes state you must have the Units.NET library installed.

To add the Units.NET library, use the NuGet Package Manager from the **Tools ➤ NuGet Package Manager ➤ Manage NuGet Packages for Solution...** menu. Click the **Browse** tab and then type **units** in the search box, as shown in Figure 13-5.

Figure 13-5. *NuGet Package Manager, results for Units*

Select the entry named Units.NET in the list, tick the project name (solution) in the list on the right, and finally, click **Install**. The installation starts and Visual Studio downloads a number of packages and then ask you permission to install them. Go ahead and let the installation complete. A dialog box opens to tell you when the installation is complete.

Now, let's install the database connector library.

Connector/Net

Connector/Net is a Microsoft .NET database library for connecting to a MySQL database server. You can download the connector from Oracle at `http://dev.mysql.com/downloads/connector/net/`.

■ **Note**   If you are building your project for the MinnowBoard Max–compatible board, you can use the normal .Net connector and these steps are unnecessary.

You need to download this library to use with the project. You can download the Windows (x86 or 64-bit) installer separately but if you installed MySQL with the MySQL Windows Installer, it is already installed on your system. However, you need a different version of the library to run on the Raspberry Pi.

■ **Tip** Check your system before you download the connector. If you see the directory C:\Program Files (x86)\MySQL\MySQL Connector Net 6.9.8\Assemblies\RT, you already have the library you need. You can skip ahead to adding it to your project but use this path instead.

In this case, you need the RT version of the connector. You can download it from the URL listed earlier, but instead of downloading the Windows version, click the drop-down menu and choose the **.NET & Mono** version. Figure 13-6 shows an excerpt from the download page with the correct version selected. This link downloads a compressed file named mysql-connector-net-6.9.8-noinstall.zip (or similar if using a different version).

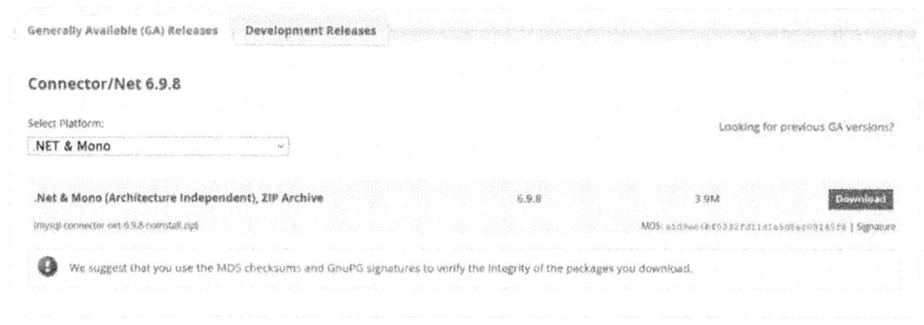

Figure 13-6. *Selecting the .NET & Mono version of Connector/Net for download*

Once the file has downloaded, extract it and find it at <downloads folder>\mysql-connector-net-6.9.8-noinstall\RT\MySql.Data.RT.dll, replacing the path with your downloads folder.

Now, return to Visual Studio and right-click the References item in the Solution Explorer and choose **Add Resource...**. When dialog opens, choose the **Browse** tab, click the **Browse** button, locate the MySql.Data.RT.dll file, and select it. This adds the library to your project.

WHAT ABOUT C++ OR PYTHON?

You may be wondering if you can access MySQL from an IoT project written in C++ or Python. It turns out you can, but not easily.

For C++, you need to use the Connector/C++ from Oracle (http://dev.mysql.com/downloads/connector/cpp/) but it takes some work. Since there is no compiled version for Microsoft RT, you must set up a project to compile the connector along with your own code but since the connector requires some parts of the server code, the process is not trivial (but still possible in principle).[8]

For Python, you can use the Connector/Python from Oracle (http://dev.mysql.com/downloads/connector/python/) by adding the connector to a Python environment that you set up in the project inside Visual Studio. Sadly, the only platform supported is the x86-based MinnowBoard Max–compatible boards. However, it is only a matter of time before someone figures out how to deploy the connector on ARM-based platforms.

---

[8]I have yet to try this myself. I encourage you to try it yourself and blog about it!

Windows 10 IoT Extensions

Finally, you must add the reference to the Windows 10 IoT Extensions from the project property page. You do this by right-clicking the **References** item on the project in Solution Explorer. Once you have loaded the three references, Solution Explorer should show the resources shown in Figure 13-7.

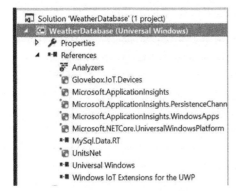

Figure 13-7. *References for the WeatherDatabase project*

Now that you have all the resources you need, let's see the code for connecting to MySQL.

Connecting to MySQL

Connecting to MySQL with Connector/Net is pretty easy. You simply need to instantiate a MySQLConnection object passing it the connection string defined earlier, and then you open the connection. You wrap the Open() call in a try block so that you can detect whether the connection fails (the library throws an exception). You add some debug statements to help diagnose the problem—just remember to run the project in debug to see the messages. Listing 13-5 shows the Connect() method.

Listing 13-5. Connecting to MySQL

```
private Boolean Connect()
{
    mysql_conn = new MySqlConnection(connStr);
    try
    {
        Debug.WriteLine("Connecting to MySQL...");
        mysql_conn.Open();
        Debug.WriteLine("Connected to " + mysql_conn.ServerVersion + ".");
    }
    catch (Exception ex)
    {
        Debug.Write("ERROR: ");
        Debug.WriteLine(ex.ToString());
        return false;
    }
    return true;
}
```

■ **Tip**   For complete documentation on Connector/Net, see http://dev.mysql.com/doc/connector-net/en/.

Now, let's see the code for reading the weather data.

Reading the Weather Data

Now let's add the code to read the weather data from the BMP280. A timer is used to fire an event every 5 seconds to read (and save) the data. Thus, you need the event for the DispatcherTimer object defined earlier, named BmpTimer_Tick().

However, instead of putting the code to read the data in this event, you use another method and run it as a task (in a new thread). This is because the BMP280 object cannot run in the user interface thread.[9] Thus, you create a new method named getData() to read the data. You also write the data to MySQL in the same method, but you look at that in the next section.

The BMP280 class from the Glovebox.IoT.Devices library provides a variety of data from the sensor, including temperature in both Fahrenheit and Celsius, and barometric pressure in several units. You read all of these with the Temperature and Pressure attributes, as shown in Listing 13-6. This listing only shows the code for reading from the sensor. The code for writing the database is shown in the next section.

Listing 13-6.   Reading the Weather Data

```
private void BmpTimer_Tick(object sender, object e)
{
    var t = Task.Run(() => getData());
}

public void getData()
{
    var degreesCelsius = tempAndPressure.Temperature.DegreesCelsius;
    var degreesFahrenheit = tempAndPressure.Temperature.DegreesFahrenheit;
    var bars = tempAndPressure.Pressure.Bars;
    var hectopascals = tempAndPressure.Pressure.Hectopascals;
    var atmospheres = tempAndPressure.Pressure.Atmospheres;

    Debug.WriteLine(degreesCelsius);
    Debug.WriteLine(degreesFahrenheit);
    Debug.WriteLine(bars);
    Debug.WriteLine(hectopascals);
    Debug.WriteLine(atmospheres);
...
```

Writing the Data to the Database

Writing data to the database is also easy to do. You create a string constant that you can use to fill in the data as parameters. Thus, you use the INSERT constant with the String.Format() method and pass in the data you read. You must put these values in the same order as the table columns. This means that you supply temperature in Celsius, temperature in Fahrenheit, bars, hectopascals, and atmospheres for the pressure.

---

[9]You could overcome this by changing the project template, but I like to use the blank app template in case I ever decide to add a user interface.

Next, you create a new class instance for a MySQLCommand class passing the query you just formatted. Once the class is instantiated, you then call the method ExecuteNonQuery() to run the query. This method is the one you use if there are no results returned. Other methods are provided for reading data from the database server (see the documentation for more information).

You wrap all of this code in a try block so that you can capture any exceptions and display them using debug statements. Listing 13-7 shows the rest of the getData() method with the database code highlighted.

Listing 13-7. Writing Data to MySQL in the getData() Method

```
...
    try
    {
        // Format the query string with data read
        String insert_str = String.Format(INSERT, degreesCelsius, degreesFahrenheit,
            bars, hectopascals, atmospheres);
        // Create a new command and setup the query
        MySqlCommand cmd = new MySqlCommand(insert_str, mysql_conn);
        // Execute the query
        cmd.ExecuteNonQuery();
        Debug.WriteLine("Data inserted.");
    }
    catch (Exception ex)
    {
        Debug.Write("ERROR: ");
        Debug.WriteLine(ex.ToString());
    }
}
```

A debug statement was added to inform you that the data was inserted in the database. This could be helpful if you are debugging the code.

That's it! The code is complete and ready for compilation. Be sure to check the earlier listings to ensure that you have all the code in the right place. Once you have entered all the code, you should now attempt to compile the solution. Correct any errors you find until the code compiles without errors or warnings.

Deploy and Execute

Now it is time to deploy the application! Be sure to fix any compilation errors first. Like you have with other applications, you want to compile the application in debug first (but you can compile in release mode if you'd prefer) and you must turn on the debugger on your board. You do this with the device portal.

Go ahead and power on your board. When ready, connect to the board to run the device portal, turn on the debugger, and then open the project properties to target the device and run with the remote debugger.

Recall from Chapter 9 that you must modify two settings: the **Remote machine** name and the **Authentication Mode**. Set the **Remote machine** name to the IP address of your device with the port specified by the remote debugger when you started it from the device portal. Figure 13-8 shows the debug settings for the WeatherDatabase project.

Figure 13-8. WeatherDatabase project debug settings

Now you can deploy your application. Go ahead and do that now but run it in debug so that you can see any error messages. Run it a few times until you see the debug statements reporting the values read and the data being written to the database. If you encounter errors, stop the application and work through the problems.

You are most likely going to see problems making the connection. This is more likely if you use a MySQL server setup differently than the default, is managed by another entity, or you don't have the database server or user accounts set up properly. If this is the case, take the following advice and be sure to run all the steps in the previous sections on setting up the database for the project. Correct any errors until the code is running properly in debug.

WHY CAN'T I CONNECT?

One of the most frustrating experiences for those new to writing database applications is trying to discover why the code will not connect to the server. If this happens to you, you should resist the temptation to change your code or blame the connector or database server. Rather, you should perform the follow diagnostic steps to discover the problem.

- Use the `mysql` client tool with the values from your connection string to ensure that you have the correct user name, password, host, port, and so forth.

- If you can't connect, check your database server to ensure that the user account your using has privileges to connect to the server (remember, the user and host must match).

- If you can connect, but cannot access the database objects, check the user privileges in MySQL to ensure that everything is set correctly.

- In some cases, you may need to modify your MySQL configuration file to remove the `bind_address` setting or modify your firewall to ensure that the port for MySQL is not blocked.

- Finally, ensure that there are no networking issues from your device to your server (e.g., isolated LAN segment).

When you have isolated the problem, change your connection string or database server settings and try your code again.

If you run the application in debug and have it stating it is writing data to the database, you can issue the following query to see the results, as shown in Listing 13-8.

Listing 13-8. Example Weather Data

```
mysql> select * from weather.history \G
*************************** 1. row ***************************
           id: 33
 dateRecorded: 2016-05-30 21:09:06
degreesCelsius: 24.88
degreesFarenheit: 76.784
         bars: 1.01215
  hectopascals: 1012.13
   atmospheres: 0.998898
*************************** 2. row ***************************
           id: 34
 dateRecorded: 2016-05-30 21:09:10
degreesCelsius: 24.89
degreesFarenheit: 76.802
         bars: 1.01214
  hectopascals: 1012.14
   atmospheres: 0.998902
*************************** 3. row ***************************
           id: 35
 dateRecorded: 2016-05-30 21:09:15
degreesCelsius: 24.88
degreesFarenheit: 76.784
         bars: 1.0121
  hectopascals: 1012.1
   atmospheres: 0.998864
*************************** 4. row ***************************
           id: 36
 dateRecorded: 2016-05-30 21:09:21
degreesCelsius: 24.88
degreesFarenheit: 76.784
         bars: 1.01213
  hectopascals: 1012.13
   atmospheres: 0.99889
*************************** 5. row ***************************
           id: 37
 dateRecorded: 2016-05-30 21:09:25
degreesCelsius: 24.89
degreesFarenheit: 76.802
         bars: 1.01213
  hectopascals: 1012.13
   atmospheres: 0.998892
*************************** 6. row ***************************
           id: 38
 dateRecorded: 2016-05-30 21:09:30
degreesCelsius: 24.89
degreesFarenheit: 76.802
         bars: 1.01214
  hectopascals: 1012.14
   atmospheres: 0.998904
6 rows in set (0.00 sec)
```

There are several things in this listing to note. First, the id column is filled in for you (you passed null for this column in the insert statement) as well as the timestamp column. Not only can you see the values change, you can also see when the sample was taken. Thus, you can issue queries later, such as the following (and many more).

- Dates when the temperature exceeded a certain value

- Average temperature for a day, week, month, and so forth

- Plot temperature or pressure using a spreadsheet program

■ **Tip**   If you are interested in learning how to issue queries to find data for these or similar questions, see *MySQL Cookbook: Solutions for Database Developers and Administrators* by Paul DuBois (O'Reilly, 2014).

If you want to reset the data, you can issue the following command to empty the table thereby purging the data. Use this command with caution as a DELETE without a WHERE clause has ruined many database administrators' day. Once run, you can't undo it!

```
DELETE FROM weather.history;
```

Once you have the application working correctly, you can deploy it to your device and start or stop the application from the Apps tab on the Device Portal, as shown in Figure 13-9. The triangle icon is used to start the application and the square icon shows a running application that you can stop.

▤ **Apps**

	APP NAME	APP TYPE	STARTUP	
▷	BlinkPythonStyle	Background	Add to Startup	🗑
↺	IoTCoreDefaultApp	Foreground	Default App	🗑
▷	IoTUAPOOBE	Foreground	Set as Default App	🗑
▷	Purchase Dialog	Foreground	Set as Default App	🗑
▷	Work or school account	Foreground	Set as Default App	🗑
▷	ZWaveHeadlessAdapterApp	Background	Add to Startup	🗑
▷	Lightning	Foreground	Set as Default App	🗑
▷	NightLight	Foreground	Set as Default App	🗑
▷	PowerMeter	Foreground	Set as Default App	🗑
▷	WeatherStation	Background	Add to Startup	🗑
▷	App2	Foreground	Set as Default App	🗑
▷	BackgroundApplication4	Background	Add to Startup	🗑
▷	Examples	Background	Add to Startup	🗑
▷	App12	Foreground	Set as Default App	🗑
▷	App13	Foreground	Set as Default App	🗑
▷	Lesson_203	Foreground	Set as Default App	🗑
▷	Weather	Foreground	Set as Default App	🗑
☐	WeatherDatabase	Foreground	Set as Default App	🗑

Figure 13-9. *Starting and stopping apps using the Device Portal*

Now that you have the project working, rejoice! You have just completed the most sophisticated project in this book and have learning how to leverage off-device storage to keep your data in an organized, reliable storage medium—MySQL!

Summary

Now that you've seen how easy it is to save data to MySQL from your IoT projects, you can leverage the power and convenience of structured storage in a database in your own IoT projects. In fact, there are very few IoT projects that generate meaningful data that I would not consider using a database to store the data. Sure, there are plenty of projects that generate data that is only meaningful in the current context, but those IoT projects that I refer to are those that help you understand the world by drawing conclusions from the data. The best way to do that is to process the data using tools designed for such analysis and by putting the data in the database, you open the door to a host of tools that can read and process the data (including your own code!).

In this chapter, you saw how to add a database component to a C# project. As I mention in a sidebar, you can also use the Connector/Python database connector for the MinnowBoard Max–compatible boards (but not the ARM-based boards—yet). You now have an excellent template to use to write more IoT projects that store data in MySQL.

In the next chapter, you look at a different technique for IoT solutions—controlling hardware remotely via the Internet (Web). The project demonstrates what is possible with only a little bit of work.

■ ■ ■

Project 5: Using a Web Server to Control Hardware

Some of the more interesting IoT solutions are those that implement actionable devices or features that allow you to control hardware remotely. This could be controlling a remote control toy (car, truck, drone, moving a camera (pan, tilt, zoom), opening or closing a gate, and so forth, or simply controlling lights or locks remotely.

That's the real key and trick to the solution—making the solution available remotely. One of the best ways to do that is to write a web interface (web page) as the user interface. More specifically, you implement a web server that has one or more pages with controls that allow you to control the actionable devices, lights, locks, and so forth.

In this chapter, you will see one method for building IoT solutions that control hardware remotely using a web page. You have already discovered the tools and techniques for turning LEDs on/off, reading data from sensors, and implementing PWM-controlled devices.

Unlike previous projects, the code for this project is not overly complicated, but it is different from the techniques you've seen. That is, you will use HTML and Python to implement a simple web server to remotely control hardware. I've kept the hardware portion simple so that you can focus on the implementation. There is an optional part that requires some soldering and cutting or drilling to assemble making the hardware for this project a bit more complicated than previous projects. Regardless, you can take the concepts presented and make all manner of interesting solutions. Let's get started.

Overview

If you work in a cube farm[1] or have a similar office arrangement, you may enjoy this project. You're going to implement an out of office sign that you can use to inform your coworkers whether you are in your office or not. To make it interesting, you will use a servo to raise or lower a flag for IN or OUT. You will combine that with a set of LEDs that you can use to indicate why you are out. More specifically, you will use the following states.

- *Available*: You are available.

- *Do not disturb*: You do not want to be interrupted.

- *Out to lunch*: You are away from your desk on lunch break.

- *Be back later*: You are away from your desk but expect to return.

- *Gone for the day*: You are not returning to your desk until the next business day.

---

[1]See http://dilbert.com/search_results?terms=Cube+Farming.

© Charles Bell 2016
C. Bell, *Windows 10 for the Internet of Things*, DOI 10.1007/978-1-4842-2108-2_14

A servo operates with pulse-width modulation (PWM).[2] That is, the faster the pulses you send it, the more (further) it will rotate. Typically, you would choose specific patterns to move the servo to one of several positions. Servos are used in all manner of solutions from mechanical movements in toys, robots, remote control planes, cars, and even 3D printers. Basically, if you need a small motor to move a lever, rotate something, steer, or move something in a precision manner, you may want to use a servo.

The servo that you use in this project is a typical miniature servo that you can find at most online electronics stores, including Adafruit (www.adafruit.com/products/169), SparkFun (www.sparkfun.com/products/9065), and online auction sites. Figure 14-1 shows a typical micro hobby servo.

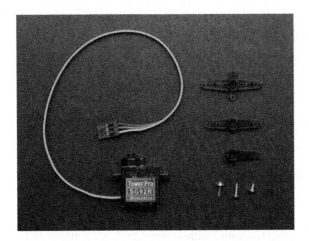

Figure 14-1. *Micro servo (courtesy of Adafruit)*

As you can see in the photo, servos typically come with a number of arms you can use to connect a thin wire or rod from the servo to another component that you want to move. Servos have a range of motion of 90 degrees. A special form of servo—called a *continuous rotation servo*—can rotate 360 degrees. For this project, the normal 90-degree range of motion is all that you need.

When you use a servo, you must discover the positions you want for the feature. For example, in this project, you simply want to raise a flag so you need to know only two positions: one for when the flag is down and one for when the flag is raised. To do this, you need to know how to use PWM in your projects. Fortunately, you have seen an example of how to do this. Recall that you saw how to use PWM in Chapter 11. However, you're going to cheat a little and use a nifty new product from SparkFun called a *servo trigger* (www.sparkfun.com/products/13118).

The servo trigger is a breakout board that has three small potentiometers that you can use to control the fully counterclockwise position (or "off") and fully clockwise position (or "on"/triggered) as well as the speed at which the servo turns. This is perfect for this project because you only need the two positions. Furthermore, since the board is designed to use a simple trigger event, you can use a single GPIO pin to toggle the board. How cool is that? Figure 14-2 shows the servo trigger with the connections highlighted.

---

[2]See https://en.wikipedia.org/wiki/Pulse-width_modulation.

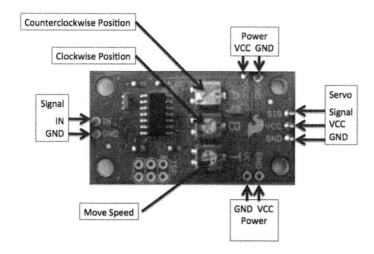

Figure 14-2. Servo Trigger from SparkFun

On the left is the signal connector with only two wires - a ground and signal. On the right is the connector for the servo. On the top and bottom are power connectors for the servo. This allows you to power the servo using 5V (most servos can operate on 3-9V). There are two connectors so that you can wire several servo triggers in series so you can use less wire for connecting power.

While the servo trigger makes using a servo for this project really easy, it does not come with any headers soldered. Thus, you will have to solder headers to the board. At a minimum, you need male headers on the signal, one of the power connections, and the servo connector. Soldering is not difficult but if you do not know how to solder, ask a friend to help you. If nothing else, you now have a really good reason to learn how to solder!

▪ **Note** The servo trigger is customized via the programming interface. See the product site on SparkFun for a complete guide (called the *hookup guide*).

Servos have three wires: one for the signal, one for ground, and another for power. These are normally colored brown, orange, and yellow respectfully. However, not all servos have the same colored wires. Table 14-1 shows some of the possible color schemes that vendors use (but there are many such variations. Just make sure that you don't connect it backward! It is always best to check with your vendor on how to wire your servo.

Table 14-1. Servo Wires

Ground (pin 1)	Power (pin 2)	Signal (pin 3)
Black	Red	White
Black	Brown	Yellow
Black	Brown	White
Black	Red	Yellow
Brown	Red	White
Brown	Orange	Yellow

■ **Tip**   To learn more about servos, see jameco.com/jameco/workshop/howitworks/how-servo-motors-work.html.

You will also use a number of LEDs to indicate a message for your coworkers. That is, when the LED is on, it indicates the current state. For example, if the flag is set to IN and the LED for "Do not disturb" is turned on, you are in the office but working on things that require your attention (such as a phone call or similar meeting) and do not want to be interrupted.

Finally, you will use HTML to write a small web page that you can use to control the hardware. To keep things as simple as possible, you will also use Python to write the code and use the Python web server library to implement a very basic web server. You will also keep the HTML code simple.

What you should gain from this project is how to write small applications to control hardware via a web page, how to use a servo (with help from SparkFun), and how to create specialized web-based solutions in Windows 10 IoT Core.

The really fun part of this project is building the sign itself. You have two options: (1) build the circuit on the breadboard, which allows you to explore all the basic concepts without the extra work, or (2) build the solution in an enclosure using a simple cardboard box that you can mount the LEDs, servo, and flag. You can hang this box on your cubical wall so that visitors can see your office status at a glance. However, this option does require a bit of soldering.

I present both options; thus, I recommend doing the first (on a breadboard) and then later build the solution in an enclosure. But first, let's look at the hardware needed for this project.

Required Components

The following lists the components that you need. You can get the LEDs and servo from Adafruit (www.adafruit.com), SparkFun (www.sparkfun.com), or any electronics store that carries electronic components. The Servo Trigger board is available from SparkFun (www.sparkfun.com/products/13118). Since this solution is a headless application, you need a monitor, a keyboard, and a mouse.

- SparkFun Servo Trigger (www.sparkfun.com/products/13118)
- Servo (www.adafruit.com/products/169) or (www.sparkfun.com/products/9065)
- (4) red LEDs
- (1) green LED
- (5) 150 ohm resistors
- jumper wires: (10) male-to-female, (1) male-to-male
- breadboard (full size recommended but half size is OK)
- Raspberry Pi 2 or 3
- power supply

■ **Tip**   The servo mount and flag are available in the source code download. They are 3D printer STL files that you can download and print or have printed for you using a 3D printing service.

Set up the Hardware

There are a number of connections needed for this project, and as usual, you will make a plan for how things should connect. To connect the components to the Raspberry Pi, you need five pins for the LEDs, one pin for the servo trigger, 5V power, ground, and four jumpers to complete the connections to the servo trigger breakout board. Table 14-2 shows the map I designed for this project.

Table 14-2. *Connection Map for Out of Office Project*

GPIO	Connection	Function	Notes
5V (2)	Power for Servo Trigger	Power to breakout board	
GND (6)	Ground for Servo Trigger, LEDs	GND on breakout board	
20	Servo Trigger Signal	Engage Servo	
21	Green LED	LED on	
22	Red LED #1	LED on	
23	Red LED #2	LED on	
24	Red LED #3	LED on	
25	Red LED #4	LED on	

Next, you need to make a number of connections on the breadboard for joining the ground rails together and connecting the servo trigger. Table 14-3 shows the connections needed.

Table 14-3. *Connections on the Breadboard*

From	To	Notes
Breadboard GND Rail#1	Breadboard GND Rail#2	
Breadboard Power	Servo Trigger VCC	
Breadboard GND	Servo GND (power in)	
Breadboard GND	Servo GND (signal)	
Servo	Servo Trigger servo header	

Figure 14-3 shows all the connections needed.

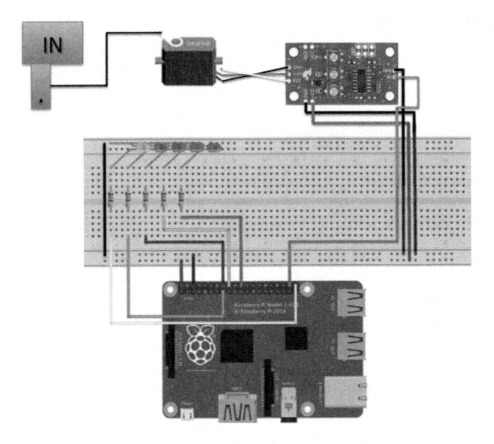

Figure 14-3. *Connections for the Out of Office project*

Here you see the breadboard implementation of the project. Recall from earlier discussions, this is how most projects start out—as a set of circuits implemented on a breadboard before being transferred to an enclosure, a printed circuit board designed, and so forth.

There are GPIO connections for each of the LEDs with the negative lead plugged into the breadboard for each LED. You also see a GPIO pin for the servo trigger as well as power connections for the servo trigger and ground for the LEDs. Lastly, you see the servo depicted connected to the servo trigger and an arm connected to the flag to raise and lower it on command.

If you are following along with this chapter working on the project, go ahead and make the hardware connections now. Don't power on the board yet, but do double and triple check the connections.

COOL GADGET

SparkFun has a really neat adapter for working with the GPIO header on a Raspberry Pi with a breadboard. It's called a Raspberry Pi Wedge (www.sparkfun.com/products/13717). It comes with a 40-pin ribbon cable that you connect to your Raspberry Pi thereby allowing you to position the Raspberry Pi farther away from the breadboard or simply reposition the device without disturbing the breadboard circuits.

What I like most about it is it plugs into the breadboard along the DIP channel allowing you to connect more than one jumper to the pins, which is really helpful for connecting power or ground. The following is a photo of the Pi Wedge.

If you use this device, you can save yourself some time looking for the correct GPIO pin since they are clearly marked on the board (but they are not in the same order as the GPIO header on the board but, in my opinion, grouped together a bit more logically).

Write the Code

Now it is time to write the code for the project. Since it is written in Python, it is very easy to follow however the web server portion is not as intuitive as some of the previous projects. Let's begin with creating the project and then walk through each of the parts that you need to implement.

New Project

Begin by opening a new project template. Choose **Python ➤ Windows IoT Core in the tree and the Background Application (IoT)** template in the list. This template creates a new solution with all the source files and resources you need for a UWP headed application. Figure 14-4 shows the project template that you need. Use the project name **OutOfOffice** for the project name.

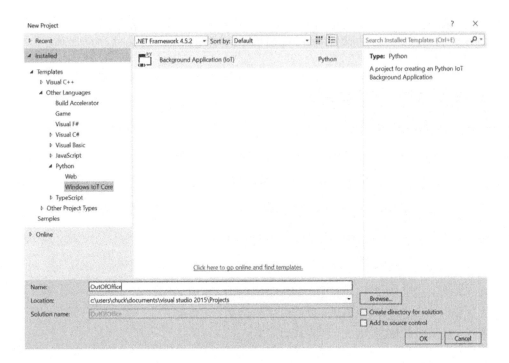

Figure 14-4. *New Project dialog: Background Application (IoT)*

The first thing you need to do is add a reference to the GPIO Python module. This is contained in the pywindevices.zip file from CPython UWP SDK. To add the reference, right-click the References in the Solution Explorer and choose **Add Reference...**. Once the dialog opens, click **Browse** and then navigate to the folder where you extracted the files. In that folder are subfolders for ARM or x86 builds. Choose the folder that matches your board (e.g., ARM for Raspberry Pi). Figure 14-5 shows the dialog that you can use to add the reference.

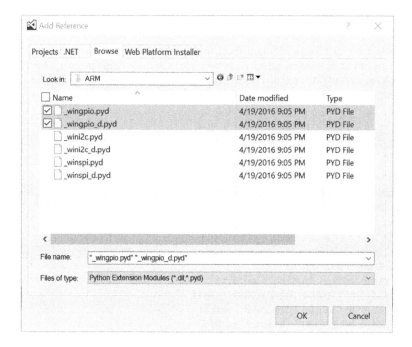

Figure 14-5. *Adding the Python module references*

Recall from Chapter 7, the module you want is named _wingpio. There are two versions of the reference: one for release builds (_wingpio.pyd), and one for debug builds (_wingpio_d.pyd). I recommend using the debug version until your application runs correctly. Once selected, you see all the available modules.

Now you can return to the code file and add the modules you want to use. You need two: the GPIO module and the http server module. You can add them as follows.

```
import http.server     # Add for html webpage support
import _wingpio as gpio  # Add for GPIO
```

Next, you can write the code to create some variables that you need and initialize the GPIO module. The code to do this is as follows. There are a number of ways you could write these constants. I chose to use GPIO pins 20-25 and constants for each pin along with an array to make it easier to initialize the pins in a loop.

```
# Constants for GPIO pins
SERVO_PIN = 20
AVAIL = 21
DND = 22
BBL = 23
LUNCH = 24
GFTD = 25
LED_PINS = [AVAIL, DND, BBL, LUNCH, GFTD]
```

Next, you'll add the HTML code for the web page.

Web Interface

To keep things as simple as possible, I decided to use a Python string to store the HTML code for the user interface. Further, I keep the code as simple as possible as not to detract from the goals of the project. That is, you want to learn how to interact with your IoT device via a web interface. I am certain those who know how to write HTML can improve the code, but it is fine to use as it is. Listing 14-1 shows the complete HTML code for the project.

Listing 14-1. HTML Code

```
#
# Create the HTML for the body of the page post (or get).
# This is a parameterized string to make it easy to modify
# without building the HTML code dynamically. Specifically,
# the following are the parameters in the code.
#
# 0 - "" or disabled to turn input "IN" on/off
# 1 - "" or disabled to turn input "OUT" on/off
# 2 - checked or "" to turn radio button Available on/off
# 3 - checked or "" to turn radio button Do Not Disturb on/off
# 4 - checked or "" to turn radio button Be Back Later on/off
# 5 - checked or "" to turn radio button Out to lunch on/off
# 6 - checked or "" to turn radio button Gone for the day on/off
#
# Initial states are:
#   OUT = disabled
#   Available = checked
#
HTML = """
<html>
  <title>Windows 10 Hardware Control Example</title>
  <body>
    <form method=POST><br>Chuck's Office Status Board: <br>
      <P><input type=submit Value='Set Status to IN ' name='in' {0}/><br>
      <P><input type=submit Value='Set Status to OUT' name='out' {1}/><br><br>
      <input type="radio" name="status" value="avail" {2}> Available<br>
      <input type="radio" name="status" value="dnd" {3}>Do not disturb <br>
      <input type="radio" name="status" value="bbl" {4}> Be back later<br>
      <input type="radio" name="status" value="lunch" {5}> At lunch <br>
      <input type="radio" name="status" value="gftd" {6}> Gone for the day <br>
    </form>
  </body>
</html>
"""
```

There are a number of integers (0–6) surrounded by {}. These are placeholders for variables that you can complete with the format() function. You'll see how to do that in the next section.

Also, the web page supports two input areas (like a button) that you use to choose whether you are in the office or out of the office. You also have a set of five radio buttons for choosing a status of your occupancy. Take a moment to read through the comments and code to understand the interface.

To make the code that updates the interface easier, I use an array of values to help populate the variables. In this case, you want to control which input areas are unavailable for clicking on (you use **disabled** to signal the state) and which radio button is checked (you use **checked** to turn it on). Thus, you create a variable called states, as follows. This is the default state for the input controls. More specifically, the **IN** input area is disabled and the available radio button is checked. Thus, the default starting state is **IN** and **Available**.

```
# List of states for the HTML code; each position is one of the controls.
states = ["disabled", "", "checked", "", "", "", ""]
```

Again, if you feel you want to make the interface a bit better (or simply change the name), I encourage you to do so. Now, let's look at the web server code.

Web Server Code

The web server code is a Python class derived from the http.server.BaseHTTPRequestHandler class named RequestHandler. There are several methods you need to implement: three required methods and one helper method. The following describes the purpose of each.

- do_HEAD(): This method is used to return the headers and is called by the web server class itself. It is used to set the headers for the web page.

- do_GET(): This method is called when the user connects to the web server to "get" the web page. Normally, this method is implemented to read an HTML file, such as index.html and send the contents to the client but for this project, you'll use a string constant.

- do_POST(): This method is called when the user submits input; in this case, when either of the input areas that you use as buttons.

- _set_headers(): This is a private helper method to send the headers rather than duplicating the code in the preceding methods.

This sounds pretty simple and it really is a bare bones, lightweight web server. All you are going to implement is a simple sending of the preceding HTML code with the variables filled in. Let's walk through what each of the preceding methods. Let's begin with the basic layout of the class. The following shows the skeleton for the class.

```
# Implement a class for the request handler
class RequestHandler(http.server.BaseHTTPRequestHandler):
    def _set_headers(self):
...

    def do_GET(self):
...

    def do_HEAD(self):
...

    def do_POST(self):
...
```

■ **Note**   I use a number of debug `print()` methods in the code. You can see this in the output window when interactively debugging the code.

The `_set_headers()` method sends the header back to the client. This is a standard mechanism and is coded as follows.

```
def _set_headers(self):
    # Prepare headers
    self.send_response(200)
    self.send_header('Content-type', 'text/html')
    self.end_headers()
```

The do_GET() method calls the _set_headers() method, and then prints a debug statement, and finally sends the HTML code to the client with the `self.wfile.write()` method. The following shows the code for this method.

```
def do_GET(self):
    # Prepare webpage
    self._set_headers()
    print("Setting states: {0}".format(",".join(states)))
    self.wfile.write(HTML.format(states[0], states[1], states[2], states[3],
                                 states[4], states[5], states[6]).encode())
```

Notice the HTML.format() call. Here you are formatting the HTML string populating it with the states values. The format() method replaces the variables in the string with the arguments passed (in order). In this case, you simply return the current status of the states.

The do_HEAD() method is simply a placeholder to call the _set_headers() method as follows.

```
def do_HEAD(self):
    # Send header
    self._set_headers()
```

The do_POST() method is where all the work for the functionality takes place. I walk through this code a section at a time, followed by a complete listing to show the code in context. This method is called when the user submits the form. For example, when the input area is clicked. To determine what the user chose as input, you have to read and parse the data sent from the client.

To do this, you read the data sent to the server (the posted data) using the self.rfile.read() method. But first, you have to know how much data to read. You get this information from the self.headers.get() method passing content-length as the option. This returns the number of bytes in text (all data is in text) that you convert to an integer and use that to tell how many bytes to read. The following shows the code for this part of the method.

```
def do_POST(self):
    # Get response and repost results
    content_len = int(self.headers.get('content-length', 0))
    post_body = self.rfile.read(content_len)
    option = None
```

Next, you reset the states array and all the GPIO pins to ensure that they are set correctly. This is not entirely necessary but helps when debugging the code to ensure that you know what the states are. It also helps to ensure that only the LED you want turned on is on at the end of the method. The following shows the code to accomplish these goals.

```
# reset states for safety
for i in range(0,7):
    states[i] = ""
# Turn all pins off
for i in LED_PINS:
    gpio.output(i, gpio.LOW)
```

Next, you check the data posted to see if the user clicked the IN or OUT input area. If it is the IN input area, the post data starts with "IN" and similarly, "OUT" if the OUT input area was clicked. We need to compare bytes not characters so you use the byte conversion shortcut.

If the IN input area was clicked, you trigger the servo to the LOW position, which rotates the servo to the default position. Note that it doesn't matter which you choose for the default—LOW or HIGH—provided you orient the servo and flag (and label it) correctly to display IN for the default. If the OUT input area was clicked, you trigger the servo to the HIGH position, which rotates the server so that the flag displays OUT.

You also set the states for the input areas to toggle between the IN and OUT input areas by making one disabled. More specifically, if IN is clicked, the OUT input area is enabled and the IN input area is disabled. The reverse occurs when OUT is clicked. The following shows how you can accomplish these goals.

```
# Engage the servo
print("post body = {0}".format(post_body))
if post_body.startswith(b"in"):
    gpio.output(SERVO_PIN, gpio.LOW)
    states[0] = "disabled"
    states[1] = ""
# Turn on the LEDS
elif post_body.startswith(b"out"):
    gpio.output(SERVO_PIN, gpio.HIGH)
    states[0] = ""
    states[1] = "disabled"
```

Notice the use of b before the strings. This is because the server class is expecting bytes rather than characters. Thus, you can convert a string to bytes using the shortcut b. You use these everywhere you work with strings in the method.

Next, you parse the body of the post to get which of LED switches (radio buttons) is checked. This information is passed in the post along with the name of the input area clicked. All you need to do is split the data return on the ampersand or "and" sign (&). Then, you can determine the option that was checked by splitting the second part (position 1) by the equal sign (=). This allows you to get the option (radio button) selected when the data was posted. Depending on which it is, you turn on that LED. The following shows the code needed.

```
# Need to parse to get the LED switches
parts = post_body.split(b"&", 1)
if len(parts) > 1:
    option = parts[1].split(b'=')[1]
self._set_headers()
```

```
if option:
    if option == b"avail":
        states[2] = "checked"
        gpio.output(AVAIL, gpio.HIGH)
    elif option == b"dnd":
        states[3] = "checked"
        gpio.output(DND, gpio.HIGH)
    elif option == b"bbl":
        states[4] = "checked"
        gpio.output(BBL, gpio.HIGH)
    elif option == b"lunch":
        states[5] = "checked"
        gpio.output(LUNCH, gpio.HIGH)
    elif option == b"gftd":
        states[6] = "checked"
        gpio.output(GFTD, gpio.HIGH)
```

Notice that you once again use the byte conversion shortcut to compare the option (radio button selected). You also set the state for the radio buttons so that only the radio button that is checked is displayed. You do this because you are sending a new HTML stream to the client thus overriding the client state.

Finally, you call the self.wfile.write() method and format the HTML string with the variables as determined by the previous code.

```
print("Setting states: {0}".format(",".join(states)))
self.wfile.write(HTML.format(states[0], states[1], states[2], states[3],
                            states[4], states[5], states[6]).encode())
```

Listing 14-2 shows the completed web server code with all the methods placed in the class for reference (notice the indentation).

Listing 14-2. Completed Web Server Code

```
# Implement a class for the request handler
class RequestHandler(http.server.BaseHTTPRequestHandler):
    def _set_headers(self):
        # Prepare headers
        self.send_response(200)
        self.send_header('Content-type', 'text/html')
        self.end_headers()

    def do_GET(self):
        # Prepare webpage
        self._set_headers()
        print("Setting states: {0}".format(",".join(states)))
        self.wfile.write(HTML.format(states[0], states[1], states[2],
                                    states[3], states[4], states[5],
                                    states[6]).encode())

    def do_HEAD(self):
        # Send header
        self._set_headers()
```

```python
def do_POST(self):
    # Get response and repost results
    content_len = int(self.headers.get('content-length', 0))
    post_body = self.rfile.read(content_len)
    option = None
    # reset states for safety
    for i in range(0,7):
        states[i] = ""
    # Turn all pins off
    for i in LED_PINS:
        gpio.output(i, gpio.LOW)
    # Engage the servo
    print("post body = {0}".format(post_body))
    if post_body.startswith(b"in"):
        gpio.output(SERVO_PIN, gpio.LOW)
        states[0] = "disabled"
        states[1] = ""
    # Turn on the LEDS
    elif post_body.startswith(b"out"):
        gpio.output(SERVO_PIN, gpio.HIGH)
        states[0] = ""
        states[1] = "disabled"
    # Need to parse to get the LED switches
    parts = post_body.split(b"&", 1)
    if len(parts) > 1:
        option = parts[1].split(b'=')[1]
    self._set_headers()
    if option:
        if option == b"avail":
            states[2] = "checked"
            gpio.output(AVAIL, gpio.HIGH)
        elif option == b"dnd":
            states[3] = "checked"
            gpio.output(DND, gpio.HIGH)
        elif option == b"bbl":
            states[4] = "checked"
            gpio.output(BBL, gpio.HIGH)
        elif option == b"lunch":
            states[5] = "checked"
            gpio.output(LUNCH, gpio.HIGH)
        elif option == b"gftd":
            states[6] = "checked"
            gpio.output(GFTD, gpio.HIGH)
    print("Setting states: {0}".format(",".join(states)))
    self.wfile.write(HTML.format(states[0], states[1], states[2],
                                 states[3], states[4], states[5],
                                 states[6]).encode())
```

Wow! That's a lot of code. Fortunately, the code isn't too difficult but could use a bit of refactoring. That is, I kept the code simple and limited to the four methods described, but you could move the code to detect the input radio button and act on it to a separate method. But that is something you can do as an exercise.

■ **Tip** In yet another effort to keep the code simple, there are no checks for validity of the radio button selections and the IN/OUT state. For example, it is possible to have the out to lunch button selected and the state IN. I leave this as an exercise for you to experiment with putting some practical limits on what can be selected. Hint: move the code for this to a new method to help keep the do_POST() method small.

Now, let's discuss the code to initialize the GPIO pins and start the web server class.

Initializing the GPIO and Starting the Web Server

The last part of the code is used to initialize the GPIO pins and start the web server. For the GPIO pins, I use a loop to set up the pins and set the pins to LOW (off). Note that since the Available state is on at the start, you turn that GPIO HIGH (on).

Starting the web server requires making a call to http.server.HTTPServer() where you pass in the port you want to use (8081) and the name of our class (RequestHandler). This returns an instance to the HTTP server (you use the variable name httpd) and simply calls the httpd.serve_forever() method to start the web server and run indefinitely. Listing 14-3 shows the initialization and web server startup code.

Listing 14-3. Initialization and Web Server Startup Code

```
# Initialize the pins, set to LOW (OFF)
for i in LED_PINS:
    gpio.setup(i, gpio.OUT, gpio.PUD_OFF, gpio.HIGH)
    gpio.output(i, gpio.LOW)
gpio.setup(SERVO_PIN, gpio.OUT, gpio.PUD_OFF, gpio.HIGH)
gpio.output(SERVO_PIN, gpio.LOW)

# default is "Available"
gpio.output(AVAIL, gpio.HIGH)

# Run the http server indefinitely
httpd = http.server.HTTPServer(("", 8081), RequestHandler)
print('Started web server on port %d' % httpd.server_address[1])
httpd.serve_forever()
```

Now let's look at all of this code in context.

Completing the Code

So far, you've seen each part of the code separately. Listing 14-4 shows a code skeleton for reference to help you understand the order in which you put the code.

Listing 14-4. Complete Code for Out of Office Project

```
#
# Out of Office sign with mechanical IN/OUT sign
#
# Windows 10 for the Internet of Things (Apress)
#
```

```python
# Dr. Charles Bell
#
import http.server     # Add for html webpage support
import _wingpio as gpio # Add for GPIO

# Constants for GPIO pins
...
#
# Create the HTML for the body of the page post (or get).
# This is a parameterized string to make it easy to modify
# without building the HTML code dynamically. Specifically,
# the following are the parameters in the code.
#
...
HTML = """
<html>
...
</html>
"""

# List of states for the HTML code
states = ["disabled", "", "checked", "", "", "", ""]

# Implement a class for the request handler
class RequestHandler(http.server.BaseHTTPRequestHandler):
    def _set_headers(self):
...

    def do_GET(self):
...

    def do_HEAD(self):
...

    def do_POST(self):
...

# Initalize the pins, set to LOW (OFF)
...

# default is "Available"
gpio.output(AVAIL, gpio.HIGH)

# Run the http server indefinitely
...
```

■ **Caution**   Compiling or building the code performs only the most basic of tests. It does not verify that the code is correct.

OK, now you're ready to deploy the application to your device. Go ahead, set everything up and power on your device.

Deploy and Execute

Once the Raspberry Pi has booted, you're ready to deploy the application to your Raspberry Pi (or other device). Recall from Chapter 7, you have to set up the debug settings to specify the IP address of your Raspberry Pi. Fortunately, you only have to change two items as indicated in Figure 14-6. Remember to choose ARM for the platform. Click **Apply** and then **OK** to close the dialog.

UWP Project Settings
Configuration: Active (Debug)
Platform: Active (ARM)
Debug Settings
Remote Device: 10.0.1.89
Remote Port: 5678

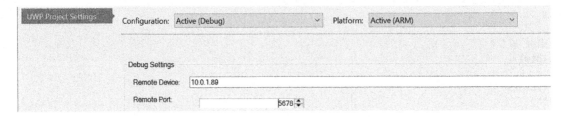

Figure 14-6. *Debug Settings for deployment*

Once these are set, you can deploy the application from the Build menu. Once the deployment is complete, you can start the application. We do this by visiting the Device Portal and selecting the application from the Apps panel. Figure 14-7 shows an excerpt from this page.

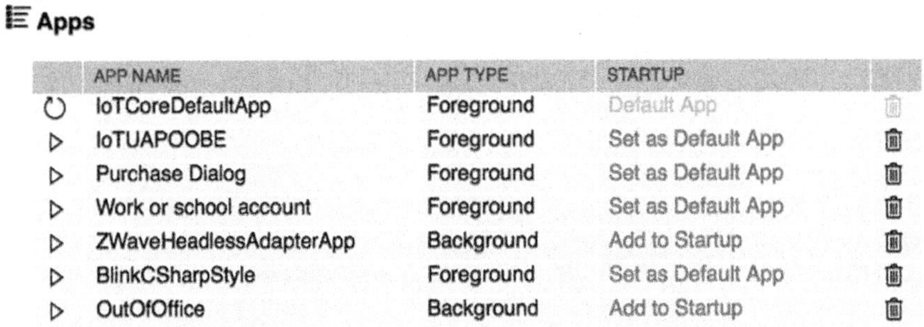

Apps
APP NAME
APP TYPE
STARTUP
IoTCoreDefaultApp
Foreground
Default App
IoTUAPOOBE
Foreground
Set as Default App
Purchase Dialog
Foreground
Set as Default App
Work or school account
Foreground
Set as Default App
ZWaveHeadlessAdapterApp
Background
Add to Startup
BlinkCSharpStyle
Foreground
Set as Default App
OutOfOffice
Background
Add to Startup

Figure 14-7. *Device Portal apps page: Starting the app*

Since this is a headless application, you must add it as a startup application in order to execute. To do so, click the **Add to Startup** link. This launches the application and in a moment you should see the triangle icon turn into a small square (toggles from start to stop) and the link changes to **Remove from Startup**. You cannot start the application with the triangle icon because this is a startup project. If you try to do this, you get a message that the application failed to start. Be sure to click the link instead. To stop the application, click the **Remove from Startup** link. Figure 14-8 shows how the link changes once the application starts.

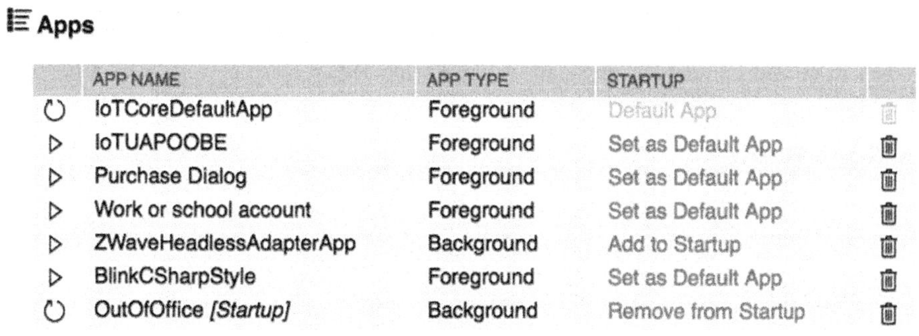

☰ Apps

	APP NAME	APP TYPE	STARTUP	
↻	IoTCoreDefaultApp	Foreground	Default App	🗑
▷	IoTUAPOOBE	Foreground	Set as Default App	🗑
▷	Purchase Dialog	Foreground	Set as Default App	🗑
▷	Work or school account	Foreground	Set as Default App	🗑
▷	ZWaveHeadlessAdapterApp	Background	Add to Startup	🗑
▷	BlinkCSharpStyle	Foreground	Set as Default App	🗑
↻	OutOfOffice *[Startup]*	Background	Remove from Startup	🗑

Figure 14-8. *Device Portal apps page: stopping the app*

Once the application has started, you can now connect to the device via the web server port 8081 using the following link. Be sure to substitute the IP address of your device.

`10.0.1.89:8081`

Once you connect, you should see the web page as shown in Figure 14-9. This is the default settings for the first run of the page and is generated from the preceding do_GET() method.

Chuck's Office Status Board:

Set Status to IN

Set Status to OUT

- ◉ Available
- ○ Do not disturb
- ○ Be back later
- ○ At lunch
- ○ Gone for the day

Figure 14-9. *Web interface: default state*

Now, play around with the input areas and radio buttons for a bit. You should see the servo trigger from one position to another as you change the IN/OUT state. Similarly, the LEDs should illuminate corresponding with the radio buttons you chose. Figure 14-10 shows one such state you could toggle.

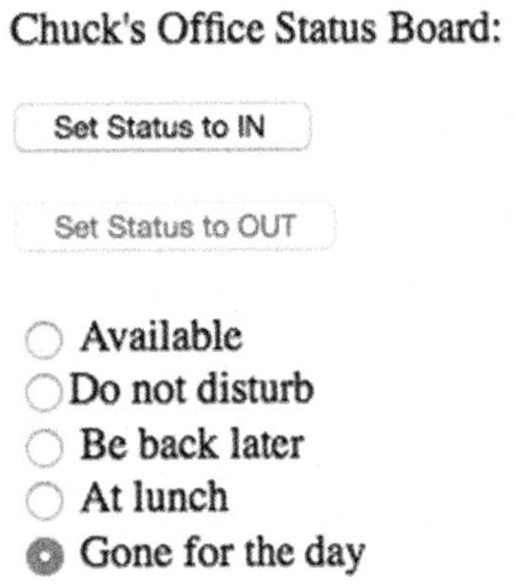

Chuck's Office Status Board:

[Set Status to IN]

[Set Status to OUT]

○ Available
○ Do not disturb
○ Be back later
○ At lunch
◉ Gone for the day

Figure 14-10. *Web interface: out of office status*

You should see the servo turn to one position when the IN input is clicked and another when the OUT input is clicked (and back again). This may not seem very interesting since you want to use the servo to raise and lower a flag, but you can use your imagination at least and see the wonder of remotely controlling mechanical devices. In fact, I show you how to prototype an enclosure complete with the mechanical flag apparatus in the next section.

■ **Tip** You do not have to complete the enclosure task. I provide it as a mild diversion to show what is possible. As you will see, it requires a bit of mechanical aptitude (must be able to use a hobby knife, scissors, etc.) as well as a bit of patience to get it going. If nothing else, it may be an interesting read that sparks your own ideas.

If you got this to work, congratulations! You have now been introduced to a whole new venue of IoT applications—making things move. If this interests you, I encourage you to look for more helpers and aides, such as the Servo Trigger board from SparkFun. There are an awful lot of cool projects you can imagine with just a servo or a dozen. But making things move in your IoT projects isn't limited to servos. There are stepper motors, continuous rotation motors, actuators and much more that you can explore.

If you are having trouble getting this project to run or deploy, be sure to revisit Chapter 4 and Chapter 7 for tips. If the application starts but doesn't respond correctly, double-check your code to ensure that it is the same as listed in this chapter. Remember, you can always run the interactive debugger and step through the code. If you decide to do this, I recommend placing breakpoints in each of the do_GET() and do_POST() methods so you can trace them to diagnose errors.

Now let's go over how to get started embellishing this project by prototyping an enclosure.

Prototyping the Out of Office Sign Enclosure

The following is a brief detour from your discovery of Windows 10 for the IoT. In this section, I present one possible way to take the preceding project and make it into something you could use every day. That is, one thing that separates an experiment from a practical application is how it is packaged.

You could leave your components plugged into a breadboard and use it, but some projects like the one in this chapter aren't very practical. In this case, it is because you have a mechanical element—the IN/OUT flag. Wouldn't it be nice to see this in an enclosure so you can see the flag move?

You may be wondering how to get started. Or perhaps you have doubts about how to build such a thing. Read on to see one approach to discovering how to build a permanent enclosure. You're going to use a technique called *prototyping*, where you build a solution with the intent to experiment on how best to solve the problem.

Most prototypes therefore are called *throwaway prototypes* because they often are of lower quality and makeshift at best. Moreover, they often have little or no aesthetic value. Finally, prototypes are often used to prove a concept. That is what you will do in this section. The goal is to see how to design an enclosure but you also build it so that you can dismantle it either to build a permanent solution or to recover the components for use in other projects.

As you can imagine, this task requires a bit of work and of course you need to power down your device and partially disassemble the project. You can leave the breadboard connections to the Raspberry Pi because you reuse those, but all others can be disconnected.

I try to keep it relatively easy by using a cardboard box and as few tools and additional components as possible. That said, the following lists some additional parts you need if you want to follow along with this exercise.

- Small cardboard box with or without lid

- Double-sided tape (e.g., 3M mirror mounting tape)

- Assorted M3 screws and nuts (or equivalent)

- Mini breadboard (optional, but helpful)

- Additional male/female jumper wires

- Small solder prototyping board (called a *protoboard*)

- Soldering iron and solder

- 6 to 8 inches of small gauge wire to connect servo to the flag

- 3D printed parts from the source download

OK, so that's a lot more than a few items. And, yes, you are going to be soldering a bit. You could skip this part, but you need some way to attach wires to the LED legs and resistors to the LEDs. As you will see, the board that you will build isn't complicated. If you do not know how to solder, ask a friend to help you—now is a great time to learn![3]

The 3D printed parts are pretty simple and can be found in the source code archive for the book from the Apress web site. The parts are shown in Figure 14-11 and include an example flag and a mount for the servo.

---

[3]Depending on the temperament of your friend, it could be a good time to watch someone solder. Be sure to take notes and pay attention so you can learn on your own.

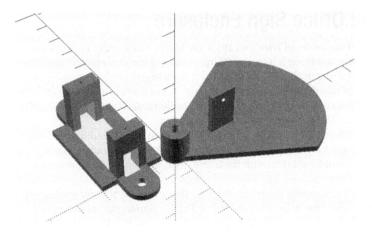

Figure 14-11. 3D Printed parts for enclosure

If you do not have a 3D printer or don't know anyone who does, you can visit a number of online 3D printing services, which you can upload design and (for a fee) they will print and ship them to you. I like shapeways.com as they have an excellent reputation and offer a variety of materials and finishes. However, you could also checkout freelance 3D printing services, such as 3D Hubs (www.3dhubs.com), which allows you to search in your area for someone willing to print parts for you—sometimes at a much reduced price or even for only the price of the weight of filament (plastic) used. Enthusiasts are cool like that.

The nice thing about these parts is you can either bolt them to the enclosure or in the case of the servo mount, glue them to the enclosure (if you don't want to see bolts through the face of the enclosure).

Of course, you could invent your own parts for mounting the servo and creating a flag. For example, you could use double-sided tape or glue to attach the servo to the box. For the flag, you could cutout another piece of cardboard or other material. The good news is this is a prototype and you are free to experiment at will with whatever ideas or materials you have at hand.

WHY USE A CARDBOARD BOX?

Cardboard boxes make excellent prototyping material. They are plentiful, cheap (free), easy to cut, you can write on them, and more. Best of all, if you mess one up, just throw it in the fire and start over with a new one! Once you get the enclosure the way you want, your cardboard box becomes a template you can use to transfer to another medium—you already know where (and where not) to put the holes.

Now, let's see how to assemble a circuit board for mounting the LEDs and resistors.

Assembling the Circuit Board

I chose to use a soldered prototyping board (called a *protoboard* for short) to mount the LEDs to make it easier to mount them in the enclosure. As I mentioned, you could simply solder the resistors to the LEDs and then solder a wire to them eliminating the need for a breadboard or circuit board, but having them on the board makes it easier to transfer the components from the prototype to a permanent enclosure.

In fact, you are going to use the solder protoboard to mount the LEDs and resistors just like you did on the breadboard. More specifically, you are going to solder all the ground legs together and each positive leg to its own resistor. You then use the legs of the resistor and one of the legs from the LEDs as mounting pins for your female jumper wires. Don't worry. It's not as crazy or difficult as it sounds.

Begin by placing the LEDs on the protoboard so that the legs are in separate solder pads. Take a look at the protoboard closely. You want to use the solder pads to hold the LED, but you need to make sure that you don't solder the legs to together. Go ahead and solder them in place. Figure 14-12 shows how the LEDs should look when soldered in place.

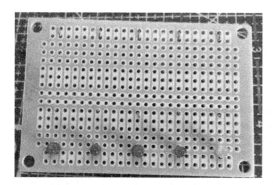

Figure 14-12. *LEDs soldered to protoboard*

I placed the green LED on the right side. I plan to use this as the "Available" indicator. It doesn't really matter which color you choose and it could be a bit easier to use the same color. That is, using a different color means having to orient the board a certain way. Just be sure to imagine the LEDs positioned on the reverse side so you can place them in the correct order if you are using different colors.

Next, place the resistor on the opposite side of the protoboard bending the LED leg so that it touches on side of the resistor. Take the other side and push it through and back out the backside of the protoboard. Now solder the LED leg to the resistor and the other side of the resistor to the protoboard solder pad.

While you are soldering on the reverse side of the protoboard, go ahead and bend the GND pins of the LEDs toward one edge (the long edge). This allows you to solder them all together saving the last one as a mounting point. This is easier to visualize in a photo. Figure 14-13 shows the completed reverse side of the protoboard.

Figure 14-13. *Completed protoboard (LED circuit board)*

Notice how I use the extra leg from the resistors as mounting points. Here I just bent them 90 degrees to the board. They were a bit long so I cut them off but be sure to leave at least 12-15 mm so you can fit the jumper wires. Look closely and you will see on the left side how I joined the GND legs of the LEDs together, terminating with one leg of the last LED used for mounting the jumper wire from the Raspberry Pi.

I have two 3-pin headers on the bottom edge of the board. I placed these here so that I can connect one VCC and one GND from the Raspberry Pi but I decided to not use these. You can safely ignore them.

Now that you have the LED circuit board, the mount for the servo, and flag, let's now see how to modify the cardboard box to house these components. In other words, here's where the prototype effort pays the bills.

Making the Enclosure

Let's begin by mounting the servo onto the servo mount. You can use two M1.5 screws or similar to mount the servo. The fit will be tight, but you can thread the wire leads through the opening in the side of the mount before you bolt it down. If the box has an attached lid or cover, you may want to cut it off to make it easier to work with.

Next, take a section of thin wire, such as the type used in auto and motorcycle racing. The wire I use is called *safety wire*, which is used to secure nuts and bolts from coming loose or worse falling out.[4] A length of about 6 to 8 inches should be enough. This wire is often called a *cable* or a *rod* when used with a servo. The small plastic piece that mounts to the servo is called the *servo arm*. Choose the longest of the arms and place it on the servo. Don't secure it yet. You need to make some modifications first. Take the wire and end it so that one end forms a vertical area that the servo can use to push and pull. The other end can be bent in a U shape for the same purpose when mounted to the flag. Finally, you need to put another bend in the center so that the arm can align with the flag (the servo is taller than the flag mount). Figure 14-4 shows an illustration of how the arm should be shaped.

Servo   Flag

Figure 14-14. *Forming the servo rod/cable*

Next, you need to place the servo and flag on the inside of the box. Be sure to orient the flag so that it is about 40mm above the servo. Go ahead and attach the rod to the servo and the flag. You may need to drill out the servo arm and the flag mount a bit depending on the thickness of the wire. Looser is better.

Without the servo connected and powered on, try moving the arm in a 90-degree arc to see where to place the flag. The placement will depend on the size of the box you are using and the length of the rod. For my box, a 6-inch rod (before bending) was a perfect fit. I aligned the center of the servo with the pivot mount for the flag. Take some time to experiment with this until you're convinced it will work. That's part of the fun—seeing where to place things.

Once you have the general placement of the servo and flag, you can start marking and cutting the box.

Marking the Box for Holes

Let's start with the LED circuit board. For this, you can place five 2.5mm holes spaced 13mm apart on the right side of the box. Press or drill out these holes, and then mount the LED circuit board with double-sided tape to the inside of the box.

---

[4]See https://en.wikipedia.org/wiki/Safety_wire.

Next, place the servo and flag inside the box and once again rotate the servo to ensure that there are no collisions. Once you figure out where to place the servo, mount it with double-sided tape or drill holes, and use bolts to mount the servo.

Next, place the flag in the box and mark where the hole for the bolt should be. Don't mount the flag yet, just drill the hole for it, and then mark a box approximately 22mm wide by 17mm high next to it. This is the window in which the IN or OUT values will show. Figure 14-15 shows the box with holes for the LEDs, flag, and flag mount.

Figure 14-15. *Box with holes drilled*

Before you mount the flag, place a piece of white paper over the flag, cut it to fit with some overlap, and then tape the ears down, as shown in Figure 14-16. This allows you to write on the flag from outside the box. Go ahead and mount the flag in the box with a long M3 bolt and nut or similar.

Figure 14-16. *Masking the flag*

Now you can put the rest of the components in the box.

Putting it all Together

Use double-sided tape to mount the Servo Trigger in the box away from the flag. Also, mount the mini breadboard inside the box to the right of the servo. Most mini breadboards come with double-sided tape. Now you can wire everything together. Use the mini breadboard to route the 5 LED wires, GND wires for the Servo Trigger and LEDs (you use the same row on the breadboard). Connect the servo to the Servo Trigger and finally GND and 5V power from the Servo Trigger to the breadboard. Figure 14-17 shows the completed wiring inside the box.

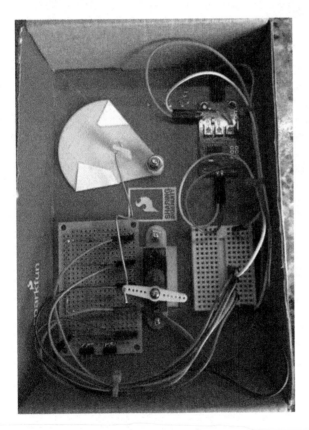

Figure 14-17. Components mounted

Be sure the Servo Trigger is unobstructed or at least positioned so that you can get to the three potentiometers on the board to tune it.

■ **Caution** Never rotate a servo when it is powered on. Also, never connect or disconnect a servo when the circuit is powered on.

Now you can connect your enclosure to the Raspberry Pi via the original breadboard. Once again, you wire the 5 LEDs, Servo Trigger, GND, and 5V power from the Raspberry Pi breadboard to the mini breadboard in the enclosure. Figure 14-18 shows the completed enclosure minus the Raspberry Pi.

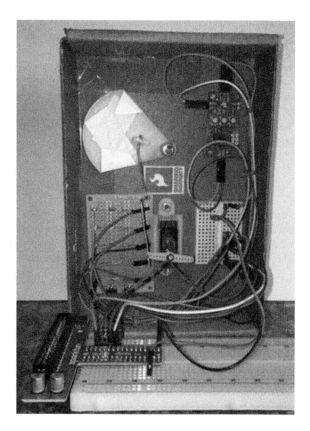

Figure 14-18. Raspberry Pi connected

I used a couple of zip ties to keep the unruly wires under control. Take a moment to examine the photo for ideas on where to place components in your box. Be sure there are no collisions in the movement of the servo and flag.

■ **Caution** Be sure to double-check and triple-check your wiring before powering on your device. There are a lot of wires in there and it is very easy to get things mixed up. Use the tables earlier in the chapter as a guide to trace each and every wire from the Raspberry Pi GPIO pin to the device. For example, ensure that GPIO 20 is connected to the Servo Trigger signal pin.

There's just one more step to perform. You need to tune the servo so that the flag moves correctly and it is shown in the window.

Tuning the Servo Trigger

The Servo Trigger has three small potentiometers on the board for adjusting the servo. Use these to change the off and on (triggered) positions of the servo so that your flag clearly shows in the cutout window and that it travels far enough to change from one value to another (from IN to OUT).

Use a small flat blade screwdriver to adjust the potentiometers. The A potentiometer adjusts the off position and the B potentiometer adjusts the on (triggered) position. If you find this to be the opposite, you can either roll with it that way or simply rotate the entire servo mount 180 degrees.

You may find that you have to experiment a bit to get this right. This is really the hardest part of the prototype and the reason you're using a temporary housing. You may have to adjust the wire that connects the servo arm to the flag shortening or bending it so that the action runs smooth. You may also have to detach and realign the servo arm to get it in the correct location.

Once you get the servo tuned correctly, take a moment to draw a small outline of the cutout on the flag in each position. You can then disassemble the flag and write IN and OUT in those locations.

OK, now you have your components assembled and the servo tuned. Let's test it out!

Testing the Prototype

OK, so now the real fun begins. Before you power on the device, if you haven't already done so, turn the box over and write the IN/OUT values on the flag and label the LEDs. I added my own comments here and there. Yes, my handwriting is terrible,[5] but since it's a prototype, all that matters is that it is legible.

Now, fire up the device and start the application. Your prototype should look very similar to the one in Figure 14-19.

Figure 14-19. *Chuck is in his office!*

If you're thinking it looks a little crude and very handmade, that's OK. It's a prototype and by definition (or practice), you don't worry about aesthetics. Rather, once it works, you can then decide to implement it using a different medium with nicer materials, better handwriting (or better still printed labels), and more.

---

[5]Thereby supporting the assumption that academics and medical doctors, however successful, have all failed basic handwriting and penmanship in elementary school. Guilty.

Now, try out your application and watch your sign change according to your commands. Isn't that cool? The sound of the servo cycling is a really powerful satisfying aspect of this prototype. Figure 14-20 shows the end state of my prototype for the day. I'm out of here!

Figure 14-20. *Chuck has left the building...*

Now that everything is working, I am sad to say the coolness and jubilation eventually wears off. Sure, you can impress your office mates with your prototype, but it eventually starts to look like a fifth grader made it. But remember, you built this as an exercise with the expectation you would either use it to design and build a better enclosure or simply dismantle the components for later reuse. Either way, bask in the delight of becoming a Maker and show it to all of your friends!

Taking it a Step Further

I chose to use a cardboard box for the enclosure for the project because it is easy to modify with a small hobby knife, you can write on it with a marker, and doesn't detract from the lessons learned in writing the code. However, you may want to consider implementing this project using a more robust enclosure.[6] Moreover, you could use an enclosure large enough to house your Raspberry Pi, too.

For example, you could purchase a plastic or aluminum enclosure and cut or drill the holes for the LEDS and flag mount. If you have access to a 3D printer, you could not only print the mounts and flag, as described earlier, or you can even design and print your own bezel. Another possibility is using a small piece of thin plywood to make the bezel and mount it in a deep photo frame or shadow box.

If you like this project and had as much fun as I did building it or better still if you want to take it to work and put it in service, I encourage you to consider taking the project a step further with a better enclosure. See Chapter 17 for ideas on where to publish your work. Be sure to reference this book as the origins of your idea.

---

[6]Or at least one that looks better!

Summary

I find those IoT projects (or any project) with mechanical elements really fun to design and implement. Mechanical movements allow you to bring a bit of whimsy and wonder to your projects. My interest in mechanical movements began as a child when I saw the early animatronics displays at Disney World and other amusement parks. Now that you have had a small taste of one method you can use to create such devices, you can begin to think about how to incorporate similar mechanisms in your projects.

In this chapter, you discovered how to build a nifty out of office sign with a mechanical flag and LEDs controlled from a web page. This represents the fundamental building blocks for other remote controlled IoT projects.

In the next chapter, you explore an exciting new venue for Windows 10 IoT development—the use of the widely popular Arduino microcontroller boards. Yes, that's right. You can even make use of the Arduino with Windows 10! How cool is that?

■ ■ ■

Project 6: Windows IoT and Arduino

If you are just starting to work with hardware and IoT solutions, you may not have encountered the world that is Arduino and microcontrollers. These are small boards with components that support GPIO pins with a limited processor (not a CPU) and memory that permits you to write small programs to control the hardware. In essence, it is a hardware development platform.

There are many such boards and the Arduino is perhaps the most popular with a community that spans the globe providing a vast assortment of sample libraries, code, blogs, books, and documentation. This makes the Arduino one of the most popular choices for hardware development. Some may say it is even more popular than the Raspberry Pi.

Microsoft recognized this dominance and has built three technologies that complement the Arduino platform: one allows Windows 10 IoT Core developers to use the vast library of Arduino code and examples running them in Windows 10; the two others allow Windows 10 IoT Core developers to create IoT solutions that incorporate their Arduino boards and thereby leverage their existing hardware and software complimented with Windows 10. These technologies are named Arduino Wiring and UWP Lightning Providers, Windows Remote Arduino, and Windows Virtual Shields for Arduino.[1]

In this chapter, you explore the Arduino platform along with the three Arduino technologies from Microsoft. You begin with a short tutorial on the Arduino and an in-depth look at using the Arduino Wiring libraries. You won't spend a lot of time on the latter two technologies as they have limited usability and, quite frankly, they are not nearly as polished as other Microsoft offerings. They suffer from insufficient documentation and few (or no) examples, and they haven't been updated in over a year. Sadly, the technologies have a lot of promise but clearly need more work.

Thus, you only look at those two technologies in a cursory manner to complete the coverage of the Arduino offerings. However, the Arduino Wiring and UWP Lightning Providers offer a profound and powerful tool to leverage the world of Arduino. Before you look at the Arduino Wiring library, let's discover what the Arduino is all about.

What Is an Arduino?

The Arduino is an open source hardware prototyping platform supported by an open source software environment. It was first introduced in 2005, designed with the goal of making the hardware and software easy to use and available to the widest audience possible. Thus, you do not have to be an electronics expert to use the Arduino.

The original target audience included artists and hobbyists who need a microcontroller to make their creations and designs more interesting. However, given its ease of use and versatility, the Arduino has quickly become the choice for a wider audience and a wider variety of projects, including the growing community of IoT developers.

---

[1]https://developer.microsoft.com/en-us/windows/iot/win10/arduinoandwindows10

© Charles Bell 2016
C. Bell, *Windows 10 for the Internet of Things*, DOI 10.1007/978-1-4842-2108-2_15

Another aspect that has helped the rapid adoption of the Arduino platform is the growing community of contributors to a wealth of information made available through the official Arduino web site (http://arduino.cc/en/). When you visit the web site, you find an excellent getting started tutorial, as well as a list of helpful project ideas and a full reference guide to the C-like language for writing the code to control the Arduino (called a *sketch*).

Arduino also provides an integrated development environment called the Arduino IDE. The IDE runs on your computer (called the *host*) where you can write and compile sketches, and then upload them to the Arduino via USB connections. The IDE is available for Windows, Linux, and Mac. It is designed around a text editor specially designed for writing code and a set of limited functions designed to support compilation and loading of sketches.

Sketches are written in a special format consisting of only two required methods—one that executes when the Arduino is reset or powered on and another that executes repeatedly. Thus, your initialization code goes in setup() and your code to control the Arduino goes in loop(). The language is C-like and you may define your own variables and functions. For a complete guide to writing sketches, see http://arduino.cc/en/Tutorial/Sketch.

Expansion of the software includes the capability to write libraries to encapsulate certain functionality like those required for networking, using memory cards, connecting to databases, mathematics, and more. Many of these libraries are included with the IDE. There are also some libraries written by others and contributed to Arduino.cc via the open source agreements—some of which have been bundled with the IDE.

The Arduino supports a number of analog and digital pins that you can use to connect to various devices and components and interact with them. The mainstream boards have a specific pin layout, or headers, that allow the use of expansion boards called *shields*. Shields allow you to add additional hardware capabilities, such as Ethernet, Bluetooth, or XBee[2] support to your Arduino. The physical layout of the Arduino and the shield allow you to stack shields. Thus, you can have an Ethernet shield as well as an XBee shield since each use different I/O pins.

In the next sections, you see examples of some Arduino boards available. There are many more boards available, but those that are compatible with the Microsoft technologies are the Uno and Mega boards.[3] As you will see, some of the examples in this chapter do not use the Arduino board, but you need to be familiar with the boards and how they operate in order to work with the Microsoft IoT projects in this chapter.

Arduino Models

There are a growing number of Arduino boards. Some are configured for special applications while others are designed with different processors and memory configurations. There are boards that are considered official Arduino boards because they are branded and endorsed by Arduino.cc. Since the Arduino is open source, anyone can build and even sell Arduino-compatible boards (often called an Arduino clone). In this section, you examine some of the more popular Arduino-brand boards.

The basic layout of an Arduino board consists of a USB connection, a power connector, reset switch, LEDs for power and serial communication, and a standard spaced set of headers for attaching shields. The official boards sport a distinctive blue colored printed circuit board (PCB) with white lettering. With the exception of one model, all the official boards can be mounted in a chassis (they have holes in the PCB for mounting screws). The exception is an Arduino designed for mounting on breadboards.

---

[2]http://www.digi.com/products/xbee-rf-solutions
[3]I have tried several boards myself and it appears some other boards may be compatible but officially, these are the two that Microsoft confirms support. I can tell you the Leonardo board works well.

Uno

The Uno board is the first standard Arduino board featuring a ATmega328 processor, 14 digital I/O pins, of which six can be used as PWM output, and 6 analog input pins. The Uno board has 32KB of flash memory and 2KB of SRAM.

The Uno is available either as a surface-mount device (SMD) or standard IC socket. The IC socket version allows you to exchange processors should you desire to use an external IC programmer to build custom solutions. Details and a full datasheet are available at http://arduino.cc/en/Main/ArduinoBoardUno. It has a standard USB type B connector and supports all shields. Figure 15-1 depicts the Arduino Uno revision 3 board.

Figure 15-1. *Arduino Uno (image courtesy of Arduino.cc)*

Mega 2560

The Arduino Mega 2560 is an older form of the Arduino Due board. It is based on the ATmega2560 processor (hence the name). Like the Due, the board supports a massive 54 digital I/O ports: 14 PWM outputs, 16 analog inputs, and 4 UARTs (hardware serial ports). It uses a 16MHz clock and has 256KB of flash memory.

The Mega 2560 is essentially a larger form of the standard Arduino (Uno, Duemilanove, etc.) and supports the standard shields. Figure 15-2 shows the Arduino Mega 2560 board.

Figure 15-2. *Arduino Mega (image courtesy of Arduino.cc)*

Interestingly, the Arduino Mega 256 is the board of choice for Prusa Mendel and similar 3D printers that require the use of a controller board named RepRap Arduino Mega Pololu Shield (RAMPS).

So, Which Do I Buy?

If you're wondering which Arduino to buy the answer depends on what you want to do. For most of the projects in this chapter, any Arduino Uno or Mega clone that supports the standard shield headers is fine.[4] You can build all the Arduino projects in this chapter with the smaller Arduino Uno board. However, if you want to build your own solutions and need more pins or more memory, you should consider buying the larger Arduino Mega board. I have built the solutions in this chapter with both boards.

WHAT ABOUT ARDUINO CLONES?

You may encounter Arduino boards that are not branded as Arduino, manufactured by other vendors, or simply imposters with all the normal lithography but not a true Arduino (they weren't made by `ardunio.cc`). Regardless, since the Arduino is open hardware, as long as the vendor is building the boards to the specification, you can normally rely on clone boards to work. Just keep in mind some may see the clone boards as inferior. This was warranted some years ago, but I haven't seen a bad clone board in some time.

Where to Buy

Due to the popularity of the Arduino platform, there are many vendors selling Arduino and Arduino clone boards, shields, and accessories. The Arduino.cc web site (`http://arduino.cc/en/Main/Buy`) also has a page devoted to approved distributors. If none of the resources listed here are available to you, you may want to check this page for a retailer near you.

Online Retailers

There are a growing number of online retailers where you can buy Arduino boards and accessories. The following lists a few of the more popular sites.

- *SparkFun*: From discrete components to their own branded Arduino clones and shields, SparkFun has just about anything you could possibly want for the Arduino platform. (`www.sparkfun.com`)

- *Adafruit*: Carries a growing array of components, gadgets, and more. They also have a growing number of products for the electronics hobbyist, including a full line of Arduino products. They also have an outstanding documentation library and wiki to support all products they sell. (`www.adafruit.com`)

- *Maker Shed*: The front store for *Make:* magazine and the Maker movement, Maker Shed has many products for the Arduino platform, including a growing number of custom project kits. From building your own bass guitar to building a robot, this store tells you that they have a lot to offer. (`www.makershed.com`)

---

[4]I found that most Uno clones, including the newer Leonardo board, worked well, but the newer boards with the newer processors, such as the Zero, did not work. If you have other boards, you can try them but the Uno is the best choice.

Retail Stores (USA)

There are also brick and mortar stores that carry Arduino products. While there aren't as many as online retailers and their inventories are typically limited, if you need a new Arduino board quickly you can find them in the following retailers. You may find additional retailers in your area. Look for popular hobby electronic stores.

- *Radio Shack*: While most stores are independently owned, most carry a modest array of Arduino boards, shields, and accessories, including Arduino-brand products as well as popular clone products. (`www.radioshack.com`)

- *Fry's*: An electronics superstore with a huge inventory of electronics, components, microcontrollers, computer parts, and more. If you have ever had the chance to visit a Fry's store, you should be prepared to spend some time there. They carry Arduino-brand boards, shields, and accessories, as well as products from Parallax, SparkFun, and many more. (`www.frys.com`)

- *Micro Center*: Micro Center is similar to Fry's, offering a huge inventory of products. However, most Micro Center stores have a smaller inventory of electronic components than Fry's. (`www.microcenter.com`)

Now that you have a better understanding of the hardware details and the variety of Arduino boards available, let's dive into how to use and program the Arduino. The next section provides a tutorial for installing the Arduino programming environment and programming the Arduino. You need this knowledge when building some of the solutions later in this chapter. If you already know how to use the Arduino IDE, you can skip ahead to the Arduino Wiring project.

Getting Started with Arduino

This section is a short tutorial on getting started using an Arduino. It covers obtaining and installing the IDE and writing a sample sketch. Rather than duplicate the excellent works that precede this book, you cover the highlights and refer readers who are less familiar with the Arduino to online resources and other books that offer a much deeper introduction to the Arduino. Also, the Arduino IDE has many sample sketches that you can use to explore the Arduino on your own. Most have corresponding tutorials on the arduino.cc site.

Learning Resources

There is a lot of information about the Arduino platform. If you are just getting started with the Arduino, Apress offers an impressive array of books covering all manner of topics concerning the Arduino ranging in topics from getting started using the microcontroller to learning the details of its design and implementation. The following is a list of the more popular books.

- *Beginning Arduino* by Michael McRoberts (Apress, 2010)

- *Practical Arduino: Cool Projects for Open Source Hardware (Technology in Action)* by Jonathan Oxer and Hugh Blemings (Apress, 2009)

- *Arduino Internals* by Dale Wheat (Apress, 2011)

The Arduino IDE

You can download the IDE for Windows at `https://www.arduino.cc/download_handler.php?f=/arduino-1.6.9-windows.exe`. Installing the IDE is straightforward. You omit the actual steps of installing the IDE for brevity but if you require a walkthrough of installing the IDE you can see the getting started link on the download page.

Once the IDE launches, you see a simple interface with a text editor area (a white background by default), a message area beneath the editor (a black background by default), and a simple button bar at the top. Figure 15-3 shows the Arduino IDE.

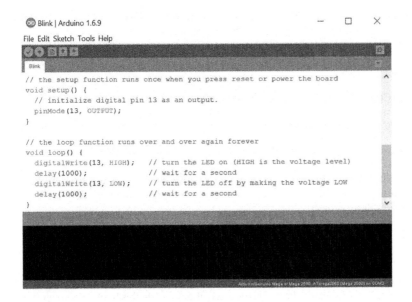

Figure 15-3. *The Arduino IDE*

The buttons are (from left-to-right): compile, compile and upload, new, open, and save. There is also a button to the right side called the *serial monitor*. You use the serial monitor to view messages from the Arduino sent (or printed) via the Serial library. You see this in action in the first project.

In the figure you see a sample sketch (called *blink*) and the result of a successful compile operation. Notice also at the bottom it tells use you are programming an Arduino Uno board on a specific serial port.

Due to the differences in processor and supporting architecture, there are some differences in how the compiler builds the program (and how the IDE uploads it). Thus, one of the first things you should do when you start the IDE is choose your board from the **Tools ➤ Board** menu. Figure 15-4 shows a sample of selecting the board.

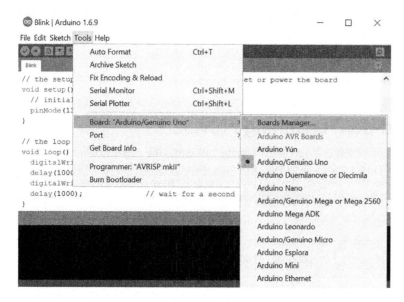

Figure 15-4. *Choosing the Arduino board*

Notice the number of boards available. Be sure to choose the one that matches your board. If you are using a clone board, check the manufacturer's site for the recommended setting to use for their boards. If you choose the wrong board, you typically get an error during upload but it may not be obvious that you've chosen the wrong board. Since I have so many different boards, I've made it a habit to choose the board each time I launch the IDE.

The next thing you need to do is choose the serial port that the Arduino board is connected. To do so, use the **Tools ➤ Port** menu option. Figure 15-5 shows an example of selecting the port.

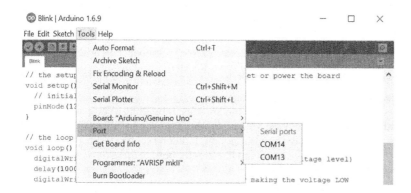

Figure 15-5. *Choosing the serial port*

To write a sketch for your Arduino board, you first plug the board into your USB port with the appropriate USB cable, and then open the IDE and select the board and port. From there, you can choose to load an example script from the dozens of built-in examples or start a new sketch.

Now that you know how to use the Arduino IDE, let's look at the three technologies Microsoft has provided for working with the Arduino and Windows 10.

Arduino Wiring and UWP Lightning Providers

The most exciting technology Microsoft has created concerning the Arduino is the ability to use existing Arduino code on Windows 10. Yes, that's right. You can use the same code that you would write for the Arduino in a Visual Studio Windows UWP project. The best part is you don't have to install a bunch of libraries, emulators, and the like. With just a few simple steps you can start writing Arduino sketches for Windows 10.

■ **Note** Arduino Wiring is not supported on DragonBoard 410c.

This means you can reuse the Arduino sketches and even the same Arduino libraries to access GPIO, I2C, and even SPI controllers and sensors. If you're not sure if that is a big deal, just do a casual search on the Internet for Arduino sample projects. You will find thousands of them from blogs to documentation to video demonstrations. Arduino is everywhere. Why not take advantage of that knowledge and run it in Windows 10?

While there are some minor changes needed to the Arduino code, for the most part you can use existing Arduino examples right on your Windows 10 device! There are two important changes needed in order to use existing Arduino libraries. First, and perhaps the most important, is to use the correct pin definitions. These definitions are documented and take the form of GPIO_XX and similar definitions that you can use to replace the numeric constants in the code. You will see these definitions used in a later section.

Next, you need to remove all calls to the serial library, as they will not work. Fortunately, you can replace them with the Log() method with very little effort. For a complete list of the changes needed as well as examples and troubleshooting tips, see https://developer.microsoft.com/en-us/windows/iot/win10/arduinowiringportingguide.

In order to use Arduino Wiring, you only have to install the Microsoft IoT Templates. To access the hardware, you use the Lightning Providers that you saw in Chapter 11. Recall that you installed the templates in Chapter 2, but in case you need to install them on another PC, the templates can be found by searching for Windows IoT Core Project Templates in the Visual Studio Gallery.

In this section, you see how easy it is to reuse existing Arduino libraries by building a color-sensing device using a Raspberry Pi and Arduino code. In this case, you use the color sensor from the Adafruit Microsoft Windows IoT Pack for Raspberry Pi along with an LCD for a standalone solution with a simple user interface triggered by a pushbutton. Better still, you use Arduino libraries for both of these devices without modification—right from Visual Studio!

The example project for this section incorporates a number of components and quite a few connections are needed to get the LCD and the color sensor wired to the Raspberry Pi. Let's look at the components and wiring before you get into the code.

Required Components

The following lists the components that you need. You can get the RGB Color Sensor (TCS34725) from Adafruit (www.adafruit.com) either in the Microsoft IoT Pack for Raspberry Pi or purchased separately, or from SparkFun (www.sparkfun.com), or any electronics store that carries similar electronic components.

However, if you use a sensor made by someone other than Adafruit, you may need to alter the code. The LCD module and other components are also available from Adafruit or SparkFun. Since this solution is a headless application, you do not need a monitor, keyboard, and mouse.

- Adafruit RGB Color Sensor with IR filter and White LED - TCS34725 Sensor
- 1602 (16 characters on 2 lines) LCD module
- Potentiometer (breadboard friendly)
- Pushbutton
- 10K ohm resistor (for the pushbutton)
- 220 ohm resistor (for the LCD)
- Jumper wires: (12) male-to-female, (12) male-to-male
- Breadboard (full size recommended)
- Raspberry Pi 2 or 3
- Power supply

Set up the Hardware

Since there are quite a number of connections, you will once again make a plan for how things should connect. Briefly, you need 12 connections to wire the LCD, 4 connections for the color sensor, 3 for the potentiometer, 2 for the pushbutton, and a few more to utilize the power rails on the breadboard. Table 15-1 shows the map that I designed for this project. I list the physical pin numbers in parenthesis for the named pins. You will use both male-to-female and male-to-male jumper wires to make these connections.

Table 15-1. *Connection Map for Power Meter Project*

GPIO	Connection	Function	Notes
3.3V (1)	Pushbutton Power	Power	
5V (2)	Breadboard power rail	5V power	
SDA1 (3)	SDI on breakout board	I2C	
SCL1 (5)	SCK on breakout board	I2C	
GND (6)	Breadboard ground rail	GND	
GPIO_4 (7)	Pushbutton signal	Button press	
GPIO_16 (36)	LCD	1602 pin E	
GPIO_17 (11)	LCD	1602 pin D5	
GPIO_18 (12)	LCD	1602 pin D4	
GPIO_19 (35)	LCD	1602 pin D6	
GPIO_20 (38)	LCD	1602 pin D7	
GPIO_21 (40)	LCD	1602 pin RS	

Next, you need to make a number of connections on the breadboard. For these, you will use male-to-male jumpers. Table 15-2 shows the connections needed on the breadboard. I also include the two resistors needed for the pushbutton and LCD.

Table 15-2. *Connections on the Breadboard*

From	To	Notes
Breadboard GND Rail #1	Breadboard GND Rail #2	
Breadboard Power Rail #1	Breadboard Power Rail #2	
TCS34725 INT	TCS34725 LED	
Breadboard GND	TCS34725 GND	
Breadboard Power	TCS34725 VDD	
TCS34725 GND	Pushbutton (shares ground)	
TCS34725 GND	10K ohm resistor	
Pushbutton	10K ohm resistor	
Breadboard GND	1602 pin GND	
Breadboard Power	1602 pin VCC	
Breadboard GND	Potentiometer pin 1	
1602 pin VEE	Potentiometer pin 2	
Breadboard Power	Potentiometer pin 3	
1602 pin BL+	220 ohm resistor	
Breadboard GND	220 ohm resistor	
Breadboard GND	1602 pin BL-	

Figure 15-6 shows all the connections needed. Take a close look at this drawing as some connections may not be obvious (e.g., sending 5V power to the pushbutton).

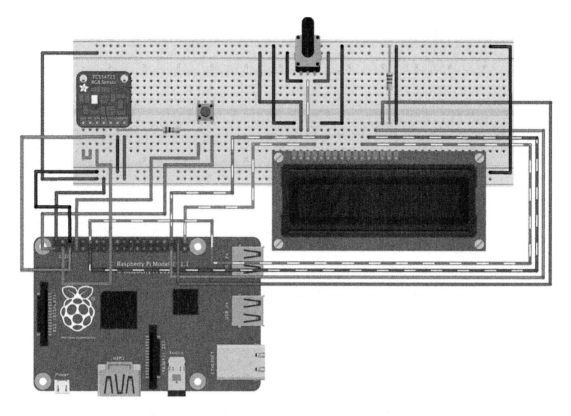

Figure 15-6. *Connections for the color sensor Arduino Wiring project*

Wow, that's a lot of connections! Be sure to use Figure 15-6, as well as Tables 15-1 and 15-2, to double-check your connections. If you are following along with this chapter working on the project, go ahead and make the hardware connections now. Don't power on the board yet, but do double and triple check the connections.

The potentiometer in the drawing is used to control the brightness of the LCD. If turned down too low (less power), the LCD will not show any characters; if turned up too high (more power), the character blocks become solid and thus hide the characters. Be sure to experiment with the potentiometer setting once you deploy and launch the application.

Now, let's write some Arduino code!

Write the Code

Now it's time to write the code for our project. Since you are working with several new concepts and two new components (the RGB color sensor and LCD), I introduce the code for each in turn. The user interface is an LCD, so you will look at how to write code to display data via the LCD. You are going to use Arduino libraries for both the sensor and LCD, so things will look a little odd at first (its Arduino Wiring code).

REMOVE THE ARDUINO IDE FOR VISUAL STUDIO PLUGIN

If you installed the plugin called Arduino IDE for Visual Studio, you must uninstall it before continuing. This plugin is incompatible with the Arduino Wiring feature and causes compilation errors and other strange warnings and errors when compiling your code. In fact, it attempts to compile using the Arduino machine files, which will create all manner of grief. Be sure to uninstall this plugin before you attempt to create the project because it creates links that will not work. If you created the project before uninstalling the plugin, you will have to re-create the project.

New Project

You will use a C++ project type for this project—the **Visual C++ ➤ Windows ➤ Windows IoT Core ➤ Arduino Wiring Application for Windows IoT Core** template. Use the name **WiringColorSensor** for the project name. You can save the project wherever you like or use the default location. Once the project opens, double-click the WiringColorSensor.ino file. There are a number of things that you need to include and there are a few variables and constants to define. Let's begin with the include files, as shown next.

```
#include <Wire.h>
#include "Adafruit_TCS34725.h"
#include <LiquidCrystal.h>
```

Here you include the headers for the LCD and Wire libraries. The Wire library is used for I2C communication on the Arduino. You also need to include the header for the Adafruit TCS34725 Arduino library. You will download this shortly.

Next, you need to create a few variables, including variables for the LCD pins, as shown next. These pins correspond to the wires in the layout in Figure 15-6. For more information about the LiquidCrystal library, see https://www.arduino.cc/en/Reference/LiquidCrystal.

```
int enablePin = GPIO16;
int registerSelectPin = GPIO21;
int dataPin11 = GPIO17;
int dataPin12 = GPIO18;
int dataPin13 = GPIO19;
int dataPin14 = GPIO20;
```

Next, you need an instance of the LCD class, an instance of the Adafruit_TCS34725 sensor library, and a variable for the button, as shown next. The parameters passed to the Adafruit_TCS34725 class set the integration timing and gain. These values are defined in the header file and are fine for most applications. See the header file or https://github.com/adafruit/Adafruit_TCS34725 for more information about the parameters.

```
//create a pointer to an instance of LiquidCrystal, yet to be created
LiquidCrystal *lcd;

Adafruit_TCS34725 colorSensor = Adafruit_TCS34725(TCS34725_INTEGRATIONTIME_700MS, TCS34725_
GAIN_1X);
const int buttonPin = GPIO4;
```

Next, you need to add the Lightning Providers and the Adafruit library to our solution.

■ **Note** Some versions of the project template add the Lightning Provider for you, but I found it best to uninstall it and install the latest version (1.0.4). Otherwise, you may get compilation errors.

Lighting Providers

In Chapter 11, you discovered the Microsoft IoT Lightning Provider is a set of providers to interface with GPIO, SPI, PWM, and I2C support through the Lightning direct memory access driver. In order to use the library, you must first turn on the Lightning driver on your device via the Device Portal. Figure 15-7 shows the **Devices** tab and the drop-down list you use to change the driver.

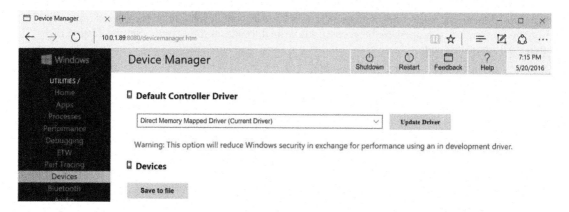

Figure 15-7. *Setting the Lighting driver on the Device Portal*

When you select the direct memory mapped driver from the drop-down list and click **Update Driver**, the portal warns you that the driver is a development-level pre-release and may not perform at peak efficiency. For our purposes, you can ignore the warning, as it has no bearing on the project. The device needs to reboot to complete the change.

■ **Caution** You want to reverse this step once you have finished the project and before starting a new project that does not use the lightning library.

To add the lightning library, use the NuGet Package Manager from the **Tools ➤ NuGet Package Manager ➤ Manage NuGet Packages for Solution...** menu. Click the **Browse** tab and type **lightning** in the search box. Choose the library and add it to your project.

■ **Caution** If you have version 1.0.2 installed, uninstall it, and then install version 1.0.4, which seems to work better than the 1.0.2 and 1.0.3 versions. I encountered compilation errors when using the latest version. If you get compilation errors, try a lower version.

Adding the Adafruit TCS34725 Library

We are going to use the Adafruit library for the TCS34725 RGB color sensor breakout board. You need to download the files from GitHub (https://github.com/adafruit/Adafruit_TCS34725). You can do so by clicking the **Clone or Download** drop-down list to the right side of the page, and then clicking the **Download Zip** button. Once downloaded, unzip the file and locate the Adafruit_TCS34725.h and Adafruit_TCS34725.cpp files.

Drag these files to your project folder. Be sure to put them in the same folder as the WireColorSensor.ino file. If you have your project open in Visual Studio, you may want to close the project first, perform the file copy, and then reopen the project so that the libraries load correctly when the project is opened.

Next, right-click your project in Solution Explorer, choose **Add ➤ Existing Item...**, and select the files. This adds the files to your project. You should now see them in Solution Explorer.

Setup Code

Arduino sketches require only two methods: a setup() method that is run once and used to set up class instances, initiate hardware connections, and so forth, and a loop() method that is executed repeatedly until the application is stopped. In this section, you write the code for the setup() method. Listing 15-1 shows the completed code.

Listing 15-1. Setup() Method for WireColorSensor Project

```
void setup()
{
    //create the LiquidCrystal instance with the pins as set
    lcd = new LiquidCrystal(registerSelectPin, enablePin, dataPin11,
                            dataPin12, dataPin13, dataPin14);

    // set up the LCD's number of columns and rows:
    lcd->begin(16, 2);

    lcd->clear();
    lcd->setCursor(0, 0);
    if (colorSensor.begin()) {
        lcd->print("Sensor found.");
        colorSensor.setInterrupt(true);
    }
    else {
        Log("ERROR: Cannot connect to LCD");
        lcd->print("NO SENSOR!");
        while (1);
    }
    lcd->setCursor(0, 1);
    lcd->print("Ready!");
}
```

The first thing you do is initiating a class instance of the LCD library (LiquidCrystal) passing in the select pin, enable pin, and data pins defined earlier. Next, you start the library by calling the begin() method, declaring the number of characters and rows. Yes, the library works for LCDs with more characters per line and more rows. If you choose to use a larger LCD, be sure to change the values to match your LCD. Next, you clear the LCD, set the cursor to column 0 and row 0.

The next section of code sets up the sensor library by calling its begin() method. Here if the method returns true, you print a statement on the LCD and then set the interrupt on (true), which turns off the LED on the sensor board (it is quite bright and only needed when reading the color value. If the LCD is not found or there are errors, you log the error with the Log() method, which writes to the output window in Visual Studio when run in debug mode. After that, you attempt to print a message and then run the code in the loop forever. If you find your application hangs without printing anything to the LCD, run it in debug mode and check the output window for the error.

Finally, you print "Ready" on the LCD to indicate everything started up OK and you are ready to read the color of an object (when the button is pressed).

Next, you create a method to read the values from the sensor and display them on the LCD.

Read the Data from the Sensor

You need to create a method to read data from the sensor to make the code easier to read and maintain. Here you simply call the methods in the Adafruit library to get the raw data (getRawData()) and then another helper method to get the red, green, and blue color values. There are other methods you can use to get even more data from the library, but the LCD is quite small so you stick to the color temperature and RGB values. If you would like to see what other data is available, see the library reference page (https://learn.adafruit.com/adafruit-color-sensors/program-it).

Once you read the data, you then display it on the LCD with the lcd->print() method. You use the clear() method to clear the LCD and the setCursor() method to place the cursor at the start of each line before you display the data. Listing 15-2 shows the completed code for the method.

Listing 15-2. read_colour() Method for WireColorSensor Project

```
void read_colour() {
    uint16_t red, green, blue, raw, colourTemperature;

    // Read the values amd calculate the colour and lumens values
    colorSensor.getRawData(&red, &green, &blue, &raw);
    colourTemperature = colorSensor.calculateColorTemperature(red, green, blue);

    // Display the results in the debug window.
    lcd->clear();
    lcd->setCursor(0, 0);
    lcd->print("Color Temp: ");
    lcd->print(colourTemperature);
    lcd->setCursor(0, 1);
    lcd->print(red);
    lcd->print(",");
    lcd->print(green);
    lcd->print(",");
    lcd->print(blue);
}
```

If you want to display the same data in the output window in Visual Studio for debugging, you can add the following code. I've taken a bit of a liberty (some would say sacrilege) by placing multiple commands on the same line to make the code shorter.[5]

---

[5]Most programmers would consider this bad form.

```
Log("Color temperature: "); Log(colourTemperature.ToString()->Begin());
Log("(K) Lumens: "); Log(lumens.ToString()->Begin());
Log(" RGB=("); Log(red.ToString()->Begin());
Log(","); Log(green.ToString()->Begin()); Log(","); Log(blue.ToString()->Begin());
Log(") Raw value: "); Log(raw.ToString()->Begin()); Log("\n");
```

This code reads all the data from the sensor, including the temperature, lumens, and the RGB values. This could be handy if you want to see more of what the sensor can do. Next, you complete the loop() method.

Loop Code

The last bit of code you need to complete is the loop() method, which executes repeatedly until the application is stopped, the device powered off, or (unlikely) an error occurs.

Here you implement code to read the button state. If it is high (button pressed), you call the interrupt on the sensor to turn on the LED, wait half a second, call the read_colour() method, and then wait another half second. This allows you to place an object over the sensor, press the button to read the data, display it, and then turn off the LED. Note that you need to press the button only until the LED comes on. Keeping it pressed fires the read over and over again. Listing 15-3 shows the complete code for the method.

Listing 15-3. loop() Method for WireColorSensor Project

```
void loop() {
    int buttonState = 0;

    delay(1000);
    buttonState = digitalRead(buttonPin);
    if (buttonState == HIGH) {
        lcd->clear();
        lcd->print("Reading sensor.");
        colorSensor.setInterrupt(false);
        delay(500);
        read_colour();
        delay(500);
    }
    else {
        colorSensor.setInterrupt(true);
    }
}
```

OK, now you're ready to compile the application and deploy it. You should change the platform to match your board and choose either debug or release mode. Be sure to compile the entire solution. It is OK if you get some warnings when the Adafruit code is compiled. Most of these are not a worry, but you may see a few warning messages similar to those shown in Listing 15-4.

Listing 15-4. Example Compilation Warnings for WireColorSensor Project

```
>------ Rebuild All started: Project: WiringColorSensor, Configuration: Debug ARM ------
1>  IPAddress.cpp
1>  LiquidCrystal.cpp
1>  Print.cpp
```

```
1>  Stepper.cpp
1>  Stream.cpp
1>  WString.cpp
1>  arduino.cpp
1>  BoardPins.cpp
1>  BcmI2cController.cpp
1>  BcmSpiController.cpp
1>  BtI2cController.cpp
1>  BtSpiController.cpp
1>  DmapSupport.cpp
1>  DmapErrors.cpp
1>  eeprom.cpp
1>  GpioController.cpp
1>  GpioInterrupt.cpp
1>  HardwareSerial.cpp
1>  I2c.cpp
1>  I2cController.cpp
1>  I2cTransaction.cpp
1>  NetworkSerial.cpp
1>  PCA9685Support.cpp
1>  PulseIn.cpp
1>  Servo.cpp
1>  Spi.cpp
1>  SpiController.cpp
1>  Adafruit_TCS34725.cpp
1>Adafruit_TCS34725.cpp(38): warning C4273: 'powf': inconsistent dll linkage
1>  C:\Program Files (x86)\Windows Kits\10\Include\10.0.10586.0\ucrt\math.h(715): note: see
previous definition of 'powf' (compiling source file Adafruit_TCS34725.cpp)
1>Adafruit_TCS34725.cpp(77): warning C4244: 'return': conversion from 'ULONG' to 'uint8_t',
possible loss of data
1>Adafruit_TCS34725.cpp(102): warning C4244: '=': conversion from 'ULONG' to 'uint16_t',
possible loss of data
1>Adafruit_TCS34725.cpp(103): warning C4244: '=': conversion from 'ULONG' to 'uint16_t',
possible loss of data
1>  StartupTask.cpp
1>  WiringColorSensor.ino
1>     Creating library C:\Users\Chuck\Documents\Visual Studio 2015\Projects\
WiringColorSensor\ARM\Debug\WiringColorSensor\WiringColorSensor.lib and object C:\Users\
Chuck\Documents\Visual Studio 2015\Projects\WiringColorSensor\ARM\Debug\WiringColorSensor\
WiringColorSensor.exp
1>  WiringColorSensor.vcxproj -> C:\Users\Chuck\Documents\Visual Studio 2015\Projects\
WiringColorSensor\ARM\Debug\WiringColorSensor\WiringColorSensor.dll
1>  WiringColorSensor.vcxproj -> C:\Users\Chuck\Documents\Visual Studio 2015\Projects\
WiringColorSensor\ARM\Debug\WiringColorSensor\WiringColorSensor.pdb (Partial PDB)
========== Rebuild All: 1 succeeded, 0 failed, 0 skipped ==========
```

You should not see any errors during the compilation. If you do and the errors are in your code, be sure to fix them until you get no errors. If you get strange errors, check the "Remove the Arduino IDE for Visual Studio Plugin" sidebar. Be sure to save your project files and make a backup if you must make changes to Visual Studio.

■ **Tip**   For more information about Arduino Wiring for UWP, including more sample projects, see https://
developer.microsoft.com/en-us/windows/iot/docs/arduinowiringprojectguide.

Deploy and Execute

Now it is time to deploy the application! Be sure to fix any compilation errors first. Like you have with other applications, you want to compile the application in debug first (but you can compile in release mode if you'd prefer) and you must turn on the debugger on your board.

If all of your wiring is ready, go ahead and power on your board. When ready, connect to the board to run the device portal, turn on the debugger, and then open the project properties to target the device and run with the remote debugger.

Recall from the Chapter 9, you must modify two settings; the **Remote machine** name and the **Authentication Mode**. Set the **Remote machine** name to the IP address of your device with the port specified by the remote debugger (e.g., 8116) when you started it from the device portal.

Now you can deploy your application. Go ahead and do that now but run it in debug so that you can see any error messages. Click the button a few times until you see the debug statements reporting the values read. If you encounter errors, stop the application and work through the problems.

The application is a startup application so you won't be able to start it by clicking the triangle icon in Device Manager, but you can click the **Add to Startup** link. This launches the application and in a moment you should see the triangle icon turn into a small square (toggles from start to stop) and the link will change to **Remove from Startup**. You cannot start the application with the triangle icon because this is a startup project. If you try to do this, you get a message that the application failed to start. Be sure to click the link instead. To stop the application, click the **Remove from Startup** link.

When the application starts successfully, you see two messages on the LCD, as shown in Figure 15-8. Here you see the message that the sensor was found and the application is ready for service.

Figure 15-8.   *Startup messages: WiringColorSensor project*

When you press the button, you see a message stating that the sensor is reading the data, as shown in Figure 15-9.

Figure 15-9.   *Reading sensor messages: WiringColorSensor project*

When the sensor has read data, it is displayed on the LCD, as shown in Figure 15-10.

Figure 15-10. *Displaying sensor data: WiringColorSensor project*

If you get something similar, congratulations! You've just made an Arduino Wiring solution run on Windows 10. How cool is that?

Go ahead and try it out with different objects. Just place an object over the sensor and press the button. You should see brightly colored objects produce RGB values of higher levels while dark objects (like a black spot) produce lower values.

Now it is time to look at the other two Arduino technologies starting with Windows Remote Arduino.

> ■ **Note** Since the following technologies have rather limited application in IoT solutions with any degree of complexity and given the features are a bit lower quality, I cover only the very basics of what is possible by way of simple examples. Also, this book is about Windows 10 IoT Core so a cursory examination of Arduino platform solutions is warranted.

Windows Remote Arduino

The Windows Remote Arduino is an interesting capability that permits you to control an Arduino remotely using Windows 10 (or Windows 8.1). There is a library (called Windows Remote Arduino) that you can include in your UWP application that allows you to write applications to connect to and control an Arduino remotely. Thus, you can write an application for your phone, IoT device, tablet, or desktop PC to interact with an Arduino.

The library supports connections via USB (directly connected to the Arduino), Bluetooth, and network (Ethernet). Sadly, there are some issues when using the library that limits you to either a network connection via Ethernet or USB connection. Clearly, a USB connection would have limited usability and while a Wi-Fi connection would be more convenient, it simply doesn't work very well.[6] Thus, the Ethernet network connection is what you use in this section to explore the capability.

> ■ **Note** Due to the limited capability in this feature, you can consider this section optional. You are welcome to explore the example further and experiment with the code yourself, but be warned there is little support (documentation) to help you, which is a bit disappointing and unusual for Microsoft.

The library and supporting code is open source and can be found on GitHub at `https://github.com/ms-iot/remote-wiring`. There are exactly two sample projects you can explore and they can be found at `https://github.com/ms-iot/windows-remote-arduino-samples`. However, the samples are a bit out of date and it takes some tweaking to get them to work (more than coverage in this book would warrant).

---

[6]I was able to get it to work, but the changes are extensive and it proved to be a bit unstable—maybe it was my network, maybe not. If you don't have an Ethernet shield, you can use a USB cable instead.

However, there is another option. Microsoft has created an application named the Windows Remote Arduino Experience (`https://www.microsoft.com/en-us/store/p/windows-remote-arduino-experience/9nblggh2041m`), which you can install from the Microsoft Store on your phone, tablet, or desktop PC. Given that the library code is a bit messy, you explore what is possible using this application.

The Windows Remote Arduino Experience application, hence Remote Arduino, is an application that connects to the Arduino that is loaded with a special sketch that uses the Firmata library (`https://www.arduino.cc/en/Reference/Firmata`). Fortunately, the latest release of the Arduino IDE has example Firmata sketches you can use with very little modification. So, don't worry, there isn't a lot of work to do here.

Unfortunately, while the Remote Arduino application supports Wi-Fi, Ethernet, and Bluetooth, I could not get the Bluetooth or Wi-Fi connections to work correctly without a lot of effort so I deem those not of sufficient quality for this book. However, the Ethernet connection works very well (as does the USB connect). Thus, you will use your Arduino with a network connection to make things more interesting.

The example project uses a simple (and very inexpensive)[7] analog sensor with only three legs (positive, signal, and negative), namely a TMP36 (`https://www.sparkfun.com/products/10988`), that you can wire to the Arduino header and use the Remote Arduino to read the value. If you want to implement the example, I've included the required hardware and a wiring guide for the Arduino.

Required Components

The following lists the components you need. You can get the TMP36 sensor from Adafruit (`www.adafruit.com`), SparkFun (`www.sparkfun.com`), or any electronics store that carries electronic components. The Arduino Ethernet shield is also available from Adafruit or SparkFun. Since this solution runs on your PC, you do not need any user interface added to the Arduino.

- Temperature sensor – TMP36

- Arduino (Uno or Mega-compatible)

- Breadboard (half or mini)

- (3) jumper wires (male-to-male)

- Arduino Ethernet (optional)

- Ethernet cable (optional)

- 5V power supply for Arduino (optional)

Figure 15-11 shows the Arduino Ethernet shield. Your shield should look very similar. However, I should warn you that there are a lot of Ethernet shields and modules available for the Arduino. Sadly, not all of them are compatible. In fact, most include their own bespoke (often open source) libraries. If you have a non-Arduino networking shield, you may need to modify this example substituting the appropriate code from the vendor's examples. In other words, it may not work so smoothly.

---

[7]It is not overly accurate but good enough for most uses.

Figure 15-11. *Arduino Ethernet shield*

The TMP36 temperature sensor is a small component with only three legs. It has a flat side that you can use to orient the pins. That is, with the flat side facing you, the pins are (from left to right), voltage in (5V), signal, and ground. Figure 15-12 shows a TMP36 sensor (oversized photo).

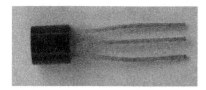

Figure 15-12. *TMP36 temperature sensor*

■ **Tip** If you want to implement this example but do not want to buy an Ethernet shield, you can connect your Arduino directly to your PC with the same USB cable you use to load the sketch.

Set up the Hardware

Since there are only a few connections, you will skip mapping the wires. There are only three required: one from the 5V power on the Arduino, one from GND on the Arduino, and another to one of the analog pins (A5) on the Arduino. These connect to the appropriate pin on the TMP36 plugged into a breadboard. Be sure to orient the sensor with the flat side facing you so you can get the wiring correct.

Arduino shields plug into the GPIO headers on the Arduino and all the pins pass through. More specifically, you can wire your Arduino through the Ethernet shield stacked on top of the board. Figure 15-13 shows the wiring needed for this example.

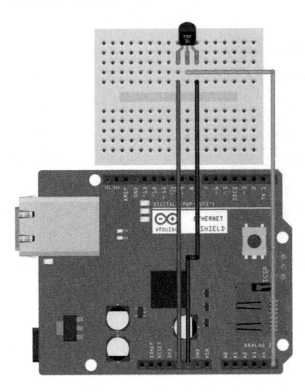

Figure 15-13. *Wiring for the Remote Arduino temperature project*

Now that you have the wiring done and the Arduino hardware is set up, let's now load the correct Firmata sketch on the Arduino.

COOL GADGET

One of the coolest gadgets for working with the Arduino is the Arduino mounting plate from Adafruit (www.adafruit.com/products/275). This small acrylic plate has space for a half-sized breadboard and an Arduino. It even has mounting holes for bolting the Arduino to the plate and small rubber feet to keep the plate off the work surface. Although you can make your own Arduino mounting plate from Lexan or Plexiglas (I have), the Adafruit product is just a notch better than what you can make yourself. For about $5.00 USD, you can keep your Arduino and breadboard together and avoid scratches on your table (from the sharp prongs on the bottom of the Arduino)—and, better still, avoid the nasty side effects of accidentally placing a powered Arduino on a conductive surface (never a good idea).

Prepare the Arduino

The sketch you will use is the StandardFirmataEthernet example sketch. To create the sketch, open the Arduino IDE and choose **File ➤ Examples ➤ Firmata ➤ StandardFirmataEthernet**. This opens a new IDE code window that you can use to compile and load the sketch onto the Arduino. Before you make changes, save the sketch as **ArduinoRemoteEthernet**.

Fortunately, the sketch has all the code you need to use an Arduino Ethernet shield. All you need is an IP address to assign to the Arduino. You could use DHCP, but you need to know the IP address to plug into the Remote Arduino application. Locate the following code in the sketch and change it to match your network configuration.

```
...
#define remote_ip IPAddress(10, 0, 1, 14)
// *** REMOTE HOST IS NOT YET WORKING ***
// replace with hostname of server you want to connect to, comment out if using 'remote_ip'
// #define remote_host "server.local"

// STEP 3 [REQUIRED unless using Arduin Yun]
// Replace with the port that your server is listening on
#define remote_port 5000

// STEP 4 [REQUIRED unless using Arduino Yun OR if not using DHCP]
// Replace with your board or ethernet shield's IP address
// Comment out if you want to use DHCP
#define local_ip IPAddress(10, 0, 1, 15)
...
```

Notice that I have marked only three lines in bold. You need only change these three values. Technically, you only need to change the local_ip value and not remote_ip, but you'll change both to be safe. Simply choose two IP addresses not used on your network. In the example, I choose 10.0.1.14 for the remote_ip and 10.0.1.15 for the local_ip. Again, you do not need the remote_ip so you can ignore it for now.

Finally, use the port 5000 instead of the default 3030. I found on my network that port 3030 was used for something else so it is best to change it.

■ **Note** If you elect to use a USB cable, you do not need to make the changes. Rather, simply load the StandardFirmata sketch and use it without any changes.

It is at this point where you need to use the appropriate USB cable to connect your Arduino to your PC. Check your Arduino to ensure that you have the right cable. Once connected, you can select the port in the Arduino IDE under **Tools ➤ Port**. Select it now. Next, select the correct board from the **Tools ➤ Board** menu. Now you're ready to compile and upload the sketch.

Begin by testing the compilation by clicking the small check mark button at the top of the IDE. Alternatively, you can click the **Sketch ➤ Verify/Compile** menu. There should not be any errors at this point. If there are, check your sketch for errors (recall that only three lines of code changed) and your Arduino IDE installation to ensure that it is set up correctly.

Once the sketch compiles correctly, upload it by clicking the small arrow button at the top of the IDE. Alternatively, you can click the **Sketch ➤ Upload** menu. The process only takes a moment. If you elect to use a power supply, you can disconnect your USB cable and then plug in the power supply. Your Arduino is now ready to be controlled remotely. Let's set up your PC.

Prepare Your PC

The only thing you need to do on your PC (phone, tablet, or desktop) is to install the Windows Remote Arduino Experience application. You can open the store application on your PC, search for **Remote Arduino Experience**, and then select the application. Install it as you would any other application from the Windows store. You may need to log in to the store if you haven't already done so. Once installed, you're ready to go.

Running the Remote Arduino Application

When your Arduino is ready—powered on and the appropriate cables (Ethernet) connected, you can launch the application. When it launches, you are asked to choose the connection type. The default is Bluetooth but you can change it by using the drop-down list, as shown in Figure 15-14. You need to select the Network option and enter the IP address you used in the sketch along with the default port of 5000. Note that if you chose to use a USB connection, simply select USB from the drop-down list and then click **Connect**.

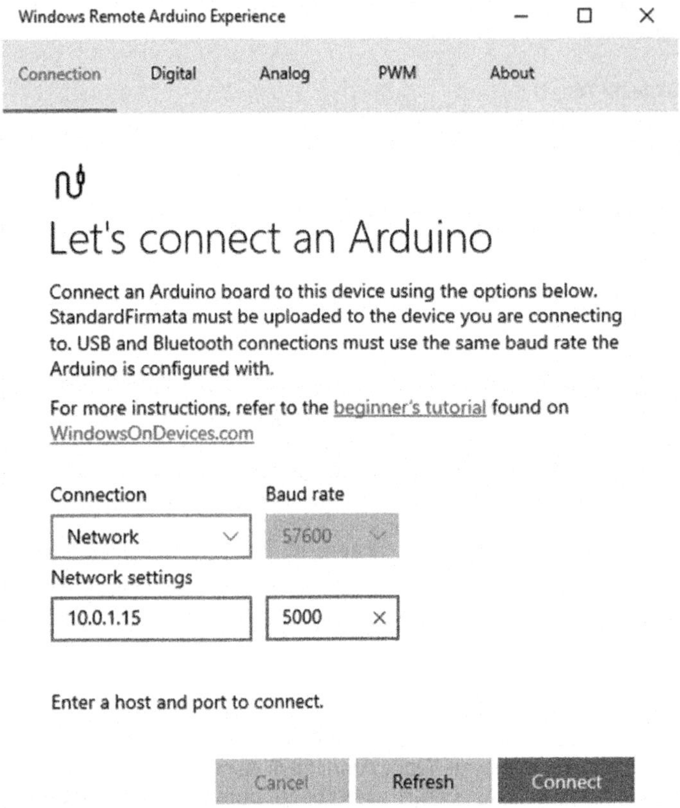

Figure 15-14. Connecting to the Arduino: Remote Arduino

Once the data is entered, click the **Connect** button. The connection should be nearly instantaneous. If it takes more than a few seconds, it will likely fail. If that happens, be sure to check your Arduino to make sure that it is correctly configured and connected to the network. You can always try pinging the IP address to ensure that it is reachable from your PC. If it is not, you may need to reconfigure your network or change the IP address to match the same domain (subnet) as your PC.

Once connected, you can explore all the pins on the Arduino. You see several tabs, as described next.

- *Digital*: Control the digital pins on the Arduino by turning them on or off, setting them to input or output.

- *Analog*: Read the values on analog pins.

- *PWN*: Control PWM to/from the PWM pins.

■ **Tip**   For a complete description of the pins on the Arduino, see the corresponding product page on
`arduino.cc`.

Since you are using an analog sensor, all you need to do is choose the analog tab and turn on the analog
pin (A5). Once you do that, the Remote Arduino application receives the values on that pin and displays
them. First, click the Input/Output slider and change it to Input. Once you do that, you will see the values
start appearing, as shown in Figure 15-15.

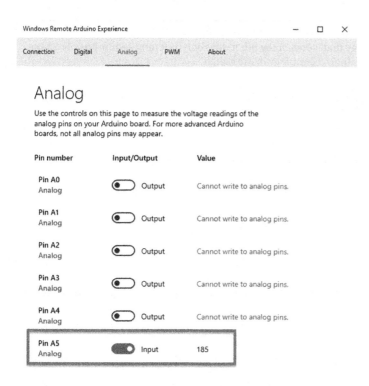

Figure 15-15.   *Reading Analog Pins: Remote Arduino*

Notice here you see a value of 185. Does that mean it is 185 degrees? No, it doesn't. Analog sensors
produce a value read on the pin from 0-1023 and you must interpret that value in order to make sense of the
data. In this case, the formula for the TMP36 (`https://learn.adafruit.com/tmp36-temperature-sensor/
using-a-temp-sensor`) is the following (courtesy of Adafruit). You first convert the value read (0-1023) into
volts and then you use that value to calculate the temperature in Celsius.

```
float voltage = reading * 5.0;
voltage /= 1024.0;
float temperature = (voltage - 0.5) * 100;
```

Let's test the formula. In Figure 15-15, the value is 185. Let's substitute that and see what you get.

```
voltage = 185 * 5.0 / 1024
voltage = 0.9033203125
temperature = (0.9033203125 - 0.5) * 100
temperature = 0.4033103125 * 100
temperature = 40.33
```

As it turns out, the temperature sensor was located under my monitor with the exhaust blowing directly on it. I measured that area of my desk with an infrared thermometer gun at about 42 degrees Celsius so this measurement is pretty close.

Clearly, this technology shows a lot of promise. If you have a need for this technology you may want to visit the GitHub repository and experiment with creating your own UWP Remote Arduino application. But be forewarned: because the code is outdated, it requires some tinkering to get it to work.

Now let's look at another technology that works in the other direction by letting Arduino applications connect to Windows 10 devices and use their technology in the sketch.

Windows Virtual Shields for Arduino

The Windows Virtual Shields for Arduino is another application that runs on your phone, tablet, or desktop PC but unlike the Remote Arduino that permits a Windows 10 application to control an Arduino, the Virtual Shields for Arduino application allows the Arduino to use the Windows 10 device as a sensor platform as if it were a shield attached to the Arduino.

Since the whole point is to leverage the sensors and hardware on Windows 10 in an Arduino sketch, it makes most sense to use a Windows 10 Phone or Tablet as the virtual shield. However, if you do not have a Windows 10 phone or tablet or your tablet or PC doesn't have sensors, you can still experiment with the technology by using your PC as a display device. Thus, you can write simple sketches that output data to Windows 10 via the Virtual Shields library.

■ **Note** Due to the limited capability in this feature, you can consider this section optional. You are welcome to explore the example further and experiment with the code yourself, but be warned there is little support (documentation) to help you, which is a bit disappointing and unusual for Microsoft.

The Arduino Virtual Shields library is open source and included in the Arduino IDE Library Manager of the latest versions of the IDE. So you can simply install it in the IDE and use the examples sketches without downloading anything. However, if you want to see the raw code, you can find it on GitHub at https://github.com/ms-iot/virtual-shields-arduino.

In order to complete the examples (or your own design), you must use the Microsoft application named Windows Virtual Shields for Arduino (https://www.microsoft.com/en-us/store/p/windows-virtual-shields-for-arduino/9nblgggz0mld), which you can install from the Microsoft Store on your phone, tablet, or desktop.

■ **Tip** If you want to see how the application works by examining the code, you can find it at https://github.com/ms-iot/virtual-shields-universal.

The Windows Virtual Shields for Arduino application, hence Virtual Shields, is an application that connects to the Arduino that is loaded with a special sketch that uses the Arduino Virtual Shields and JSON libraries. Fortunately, the latest release of the Arduino IDE has example sketches you can use with very little modification. So, don't worry, there isn't a lot of work to do here.

Unfortunately, while the Virtual Shields application supports Wi-Fi, Ethernet, and Bluetooth, unlike the Remote Arduino application, I could not get the Ethernet or Wi-Fi connection to work correctly without a lot of effort so I deem those not of sufficient quality for this book. However, the Bluetooth connection works (most of the time; more later) and the USB connection works well. Thus, you will use your Arduino with a network connection.

The example project uses a simple (and very inexpensive)[8] analog sensor with only three legs (positive, negative, and signal), namely a TMP36 (https://www.sparkfun.com/products/10988), that you can wire to the Arduino header, read the value, and display a message on the Virtual Shields application on your Windows 10 PC. To do so, you use a Bluetooth module from SparkFun. If you want to implement the example, I've included the required hardware and a wiring guide for the Arduino.

Required Components

The following lists the components that you need. You can get the TMP36 sensor from Adafruit (www.adafruit.com), SparkFun (www.sparkfun.com), or any electronics store that carries electronic components. The Bluetooth module is available from Adafruit or SparkFun. Since this solution runs on your PC, you do not need any user interface added to the Arduino.

- Temperature sensor – TMP36

- Arduino (Uno or Mega–compatible)

- Breadboard (half or mini)

- (7) jumper wires (male-to-male)

- SparkFun Bluetooth Mate Silver (https://www.sparkfun.com/products/12576)

- 5V power supply for Arduino (optional)

Figure 15-16 shows the SparkFun Bluetooth Mate Silver from SparkFun with a header soldered in place. You can use this module to connect your Arduino to your PC via Bluetooth. The module does not come with the header, so you have to solder one in place. You can use an angled header like shown here or a straight header. I found the angled header more convenient for breadboard use.

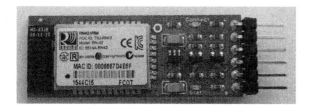

Figure 15-16. *SparkFun Bluetooth Mate Silver*

The TMP36 temperature sensor is the same one you used in the previous section. It is a small component with only three legs. It has a flat side that you can use to orient the pins. That is, with the flat side facing you, the pins are (from left to right), voltage in (5V), signal, and ground.

---

[8]It is not overly accurate but good enough for most uses.

■ **Tip**   If you want to implement this example but do not want to buy a Bluetooth module, you can connect your Arduino directly to your PC with the same USB cable that you use to load the sketch.

Set up the Hardware

Since there are only a few connections, you will skip mapping the wires. There are only seven required. One from the 5V power on the Arduino to the VCC connection on the Bluetooth module with a jumper to the TMP36, one from GND on the Arduino to the GND connection on the Bluetooth module with a jumper to the TMP36, one from the TX on the Bluetooth to the RX on the Arduino (digital pin 0), another from the RX on the Bluetooth module to TX on the Arduino (digital pin 1), and finally the signal pin on the TMP36 to one of the analog pins (A5) on the Arduino. Be sure to orient the sensor with the flat side facing you so you can get the wiring correct. Figure 15-17 shows the wiring needed for this example.

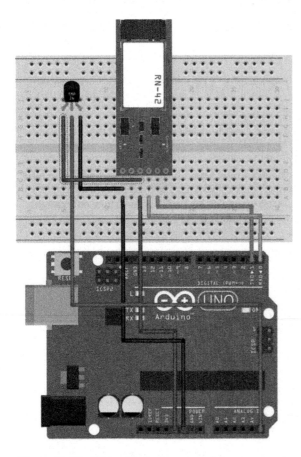

Figure 15-17. *Wiring for the Remote Arduino temperature project*

Now that you have the wiring done and the Arduino hardware is set up, let's now create a sketch on the Arduino.

Prepare the Arduino

To use the Arduino Virtual Shields library, you must download and install it along with the JSON library. Both of these can be installed via the Library Manager in the Arduino IDE. To install any library, open the IDE and (from any sketch window), click **Sketch ➤ Add Library ➤ Manage Libraries...** to open the Library Manager.

The Library Manager downloads the latest database. so it may take a few seconds before it is ready. Once the download is complete, click the search box and type **virtual shields**. This locates several potential matches. Scroll down until you find the library from Microsoft, as shown in Figure 15-18.

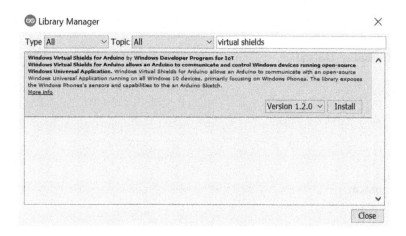

Figure 15-18. *Installing the Arduino Virtual Shields library: Arduino IDE*

To install the library, click the Install button and follow the prompts. When that is complete, install the Arduino JSON library from Benoit Blanchon. Open the Library Manager again (if you closed it) and search for **JSON**. Select the ArduinoJSON entry, as shown in Figure 15-19, and install that library. Once both are installed, you are ready to begin to write the sketch.

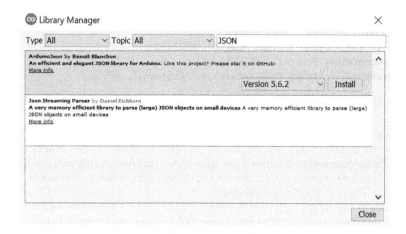

Figure 15-19. *Installing the ArduinoJSON Library: Arduino IDE*

The sketch that you will use as a base is the Basic-Text-Colored example sketch. To create the sketch, open the Arduino IDE and choose **File ➤ Examples ➤ Windows Virtual Shields for Arduino ➤ Basic-Text-Colored**. This opens a new IDE code window that you can use to compile and load the sketch onto the Arduino. Before you make changes, save the sketch as **ArduinoVirtualShield**.

Fortunately, the sketch has all the code that you need to use. All you need to do is add code to read the TMP36 sensor, convert it to temperature in Celsius, and print it. The Virtual Shields application on your PC is the user interface for the sketch. Fortunately, the code is really easy. All you need to do is modify the refresh() method, which is used to refresh the screen, and add a method to read the sensor. Name the new method readTemp(). The refresh() method is called whenever the Virtual Shields application connects or the user refreshes the screen. You will see how to do this in a later section. The following shows the code you need to add to this sketch.

```
...
// Read the TMP36 sensor and display the value
float readTemp()
{
  int TMP36 = 5;
  int rawtemp = 0;
  float volts = 0.0;
  float temp = 0.0;

  rawtemp = analogRead(TMP36);
  volts = (rawtemp * 5.0) / 1024.0;
  temp = (volts - 0.5) * 100;
  return temp;
}

// Callback for startup, reconnection, and when the pushing 'Refresh' button
void refresh(ShieldEvent* shieldEvent)
{
  float temperature = 0.0;

  // read the temperature
  temperature = readTemp();
  // display the temperature in Celsius
  screen.clear(ARGB(50,150,255));   // Show a blue screen
  screen.printAt(0, "Welcome to the Arduino Virtual Shield", WHITE);
  screen.printAt(1, "Remote Temperature Sensor!", WHITE);
  screen.printAt(2, "Temperature = " + String(temperature) + " Celsius.", WHITE);
}
...
```

Notice the new readTemp() method implements the formula described previously. To read the value from the sensor, you need only use the analogRead() method passing in the pin number (5). The changes to the refresh() method are wholesale changes. That is, just replace the code in the base sketch with the preceding code. Here you call the readTemp() method, clear the screen (changing it to a blue background), and then print a couple of messages. That's all the changes needed!

It is at this point where you need to use the appropriate USB cable to connect your Arduino to your PC. Check your Arduino to ensure that you have the right cable. Once connected, you can select the port in the Arduino IDE under **Tools ➤ Port**. Select it now. Next, select the correct board from the **Tools ➤ Board** menu.

You must disconnect the Bluetooth TX and RX wires (or just unplug the module from the breadboard) before uploading to avoid conflicts with the upload process because the TX and RX pins (wires) are used by the Arduino for loading sketches via USB as well as the module for Bluetooth connections. Thus, you need to unplug the module in order to load sketches on the Arduino. If you see errors when you upload, disconnect the Bluetooth module.

Begin by testing the compilation by clicking the small check mark button at the top of the IDE. Alternatively, you can click the **Sketch ➤ Verify/Compile** menu. Correct any errors presented. Once the sketch compiles correctly, upload it by clicking the small arrow button at the top of the IDE. Alternatively, you can click the **Sketch ➤ Upload** menu. The process only takes a moment. If you elect to use a power supply, you can disconnect your USB cable and then plug in the power supply. Your Arduino is now ready to be controlled remotely. Let's set up your PC.

Prepare Your PC

The only thing you need to do on your PC (phone, tablet, or desktop) is to install the Virtual Shields application. You can open the store application on your PC, search for **Virtual Shields**, and then select the application. Install it as you would any other application from the Windows store. You may need to login to the store if you haven't already done so. Once installed, you're ready to go.

Running the Virtual Shields Application

Before you can launch the application, you have to power on the Arduino and pair the Bluetooth module to your PC (phone, tablet, or desktop). Open the Bluetooth settings on your PC and wait for the device to show up. It should have a name similar to RNBT-4E6F shown in Figure 15-20. The second part of the name may differ, but it should start with RNBT. The passcode for the SparkFun Bluetooth module is 1234. Wait until the settings show the module is paired before launching the application.

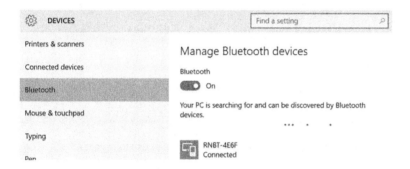

Figure 15-20. *Pairing the Bluetooth module: Windows 10 PC*

When the Bluetooth module is paired, you can launch the application. When you first start the application, it is at the output screen, as shown in Figure 15-21. Once you connect and return to this screen, the sketch fires the refresh() method, which reads the sensor and displays the data.

Figure 15-21. Output window: Virtual Shields

The icons or selections on the toolbar are shown in Figure 15-22 and described next.

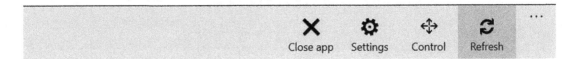

Figure 15-22. Toolbar: Virtual Shields

- *Close app*: Close the application.

- *Settings*: Connect to your device.

- *Control*: An interface with a set of preconfigured buttons that you can use to trigger events in your sketch (which you must write code to use).

- *Refresh*: Refresh the output screen by triggering the refresh() method in the Arduino sketch.

To change to the connection panel, click the small ellipse button at the bottom and then click **Settings**. This displays the connections dialog, as shown in Figure 15-23.

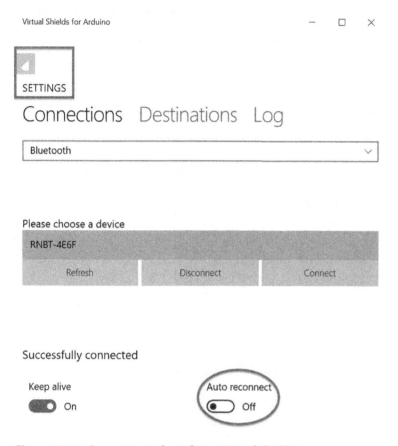

Figure 15-23. *Connecting to the Arduino: Virtual Shields*

Notice I have drawn a box and oval over two areas. The rectangular area is used to return to the main screen (also called *full screen* or simply the *output screen*), and the oval is a switch you may want to turn off. When it's on, the application attempts to connect to your device using the last known configuration. This can be maddeningly irritating if you change from USB to Bluetooth or have issues with connection. It is best to turn this off.

Also notice there are three areas: **Connections**, **Destinations**, and **Logs**. You only need the Connections panel where you select the communication type (Bluetooth is selected by default) and depending on the value you choose, the options section changes. For the Bluetooth option, you select the device from the list and then click **Connect**. It works the same way for USB but for network connections you are asked for a hostname (IP) and port.

▪ **Tip**   If you elected to not use the Bluetooth module, select the USB option, then select the COM port for the Arduino, and then click Connect.

If you do not see your Bluetooth device, you can click the **Refresh** button. This may happen if you started the application before pairing the device. When you're ready, select your device, and click the **Connect** button. This starts the handshake with the device. If you get errors, check your wiring to ensure that the Bluetooth module is connected properly. See the "What Can't I connect?" sidebar for tips on connecting to your device.

Once the device is connected, you can return to the output window by clicking the small icon in the upper-left area of the panel, as explained earlier. You may see a blank screen. If you do, click the Refresh button. Figure 15-24 shows an example of the output you may receive when running this example.

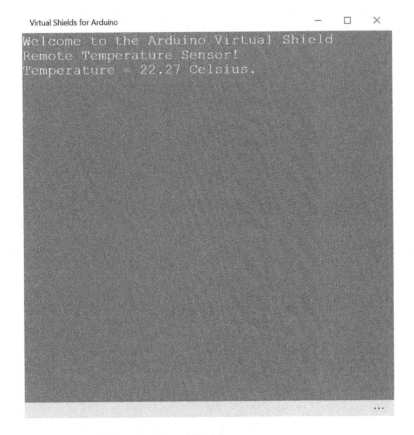

Figure 15-24. Full screen: Virtual Shields

If you see a similar message, congratulations! You've just successfully built an Arduino sketch that uses a Windows 10 device as a virtual shield. As you can see, this technology has a lot of promise and if you're like me, you can envision several ways it may be helpful. The limiting factor being whether you own a Windows 10 phone or similar device with sensors built in. Thus, this technology would be more interesting for those with similar devices. Of course, you can always use the older Windows tablet as a user interface for your Arduino sketch.

WHAT CAN'T I CONNECT?

You may encounter problems connecting to your Arduino either through Bluetooth (or if you are brave enough to try the network option) or even the USB connection. For whatever reason, the Virtual Shields application is not predictable and you may encounter failed connections or even lockups (the application freezes).

If you encounter connection issues with Bluetooth, be sure to turn off or relocate any other Bluetooth device in the area, such as your phone (unless you're running the application there), another computer, your wife's blood glucose monitor, your child's toy drone, and so forth. All of these can cause the Bluetooth module to behave erratically and can cause connection issues. I found the Bluetooth connection worked more reliably when there were no other Bluetooth devices nearby.

As for the lockups, that's something you're going to have to live with sadly. The application is a little buggy and can lockup when connecting or even after it has connected successfully and seldom during other times. The only fix for this is to restart the application. Fortunately, there is nothing you need to do on your Arduino. Simply close the application, restart it, and try your connection again.

I was unable to determine exactly what is causing the lockups, but I did discover I am not alone in experiencing such problems as others have reported similar problems. It is also clear that many others have had no problems at all. So be forewarned, this application may be another YMMV situation, which is yet again disappointing since the technology shows great promise for creating a hybrid Windows 10/Arduino IoT solution.

Summary

The Arduino community is a vast landscape of example code, documentation, and hardware designed for Arduino. So much so it would be a shame if you couldn't use some of that knowledge in your Windows 10 IoT Core projects. Fortunately, Microsoft has provided some technologies that can bridge the gap.

The Arduino Wiring—the most profound technology of the three discussed in this chapter—is arguably very useful. It permits you to reuse existing Arduino libraries and sketches with very little changes. This means you can reuse all of that knowledge to learn how to work with sensors and hardware. If you have worked in the Arduino world, you know how much this opens doors for reusing existing Arduino projects.

In this chapter, you explored the Arduino Wiring library by implementing a color sensing Arduino sketch running on your Raspberry Pi. You used the Adafruit Arduino library code without changes thereby demonstrating how you can reuse existing third-party Arduino code. You also added an LCD for the display, which demonstrated how you could reuse existing built-in Arduino code in your Windows 10 IoT Core application. Cool!

You also explored two other Arduino-related technologies: Remote Arduino and Virtual Shields. Sadly, these last two are interesting and show promise but given their low quality is not nearly as useful or important as the Arduino Wiring library.

In the next chapter, you take a look at how you can leverage the cloud for hosting data for your IoT solution using Microsoft Azure.

Azure IoT Solutions: Cloud Services

Now that you've seen a number of projects, ranging from very basic to advanced in difficulty, it is time to discuss how to make your IoT data viewable by others via the cloud. More specifically, you will get a small glimpse at what is possible with the Microsoft Azure cloud computing services and solutions.

I say a glimpse because it is not possible to cover all possible solutions available in Microsoft Azure for IoT solutions in a single chapter. Once again, this is a case where learning a little bit about something and seeing it in practice will help you get started. Like the other chapters where you've had a lightning tour, this chapter presents a few of the newer concepts and features of Microsoft Azure at a high level and in context of a sample project. You can therefore consider this chapter a bonus project chapter.

If you are just learning how to work with IoT solutions or have no plans to immediately host your solution in the cloud (or on Microsoft Azure), you can still learn quite a bit about the technology by following along and implementing the sample project.

Since the technologies presented are quite unique in implementation (but rather straightforward in concept), I keep the project hardware and programming to a minimal effort. In fact, you will reuse the weather sensor that you used Chapter 13. However, rather than saving the data in MySQL, you will send it to Microsoft Azure. You will also see how to present the data via a web site connected to the data in Microsoft Azure.

Let's take a brief, short tour of Microsoft Azure before examining the goals of the project and what you need to get started.

What Is Microsoft Azure?

That is a very interesting question. As it turns out, Microsoft Azure (hence Azure) is a total cloud platform for all manner of cloud solutions. In fact, Azure[1] is a growing collection of cloud products, such as virtual machines, databases, networking, storage, reporting, and more. Furthermore, Azure is Microsoft's platform and infrastructure for building, deploying, and managing applications and services, including platform as a service (PaaS) and infrastructure as a service (IaaS).

---

[1]Yes, the Azure web sites are all colored an interesting shade of blue.

© Charles Bell 2016

C. Bell, *Windows 10 for the Internet of Things*, DOI 10.1007/978-1-4842-2108-2_16

WHAT IS CLOUD COMPUTING?

The term *cloud computing* is sadly overused and has become a marketing term for some. True cloud computing solutions are services that are provided to subscribers (customers) via a combination of virtualization, grid computing (distributed processing and storage), and facilities to support virtualized hardware and software, such as IP addresses that are tied to the subscription rather than a physical device. Thus, you can use and discard resources on the fly to meet your needs.

These resources, services, and features are priced by usage patterns (called *subscription plans* or *tiers*), in which you can pay for as little or as much as you need. For example, if you need more processing power, you can move up to a subscription level that offers more CPU cores, more memory, and so forth. Thus, you only pay for what you need, which means that organizations can potentially save a great deal on infrastructure.

A classic example of this benefit is a case where an organization experiences a brief and intense level of work that requires additional resources to keep their products and services viable. Using the cloud, organizations can temporarily increase their infrastructure capability and once the peak has passed, scale things back to normal. Clearly, this is a lot better than having to rent or purchase a ton of hardware for that one event.

Sadly, there are some vendors that offer cloud solutions (typically worded as *cloud enabled* or simply *cloud*) that fall far short of being a complete solution. In most cases, they are nothing more than yesterday's Internet-based storage and visualization. Fortunately, Microsoft Azure is the real deal: a full cloud computing solution with an impressive array of features to support almost any cloud solution you can dream up.

If you would like to know more about cloud computing and its many facets, see `https://en.wikipedia.org/wiki/Cloud_computing`.

You can use Azure from a host of different operating systems so it is not tied exclusively to Microsoft Windows. Together with programming APIs, frameworks, and drivers in Visual Studio, you can build solutions to run as a web application in Azure. Better still, you can write applications for Windows, Android, and even iOS that integrate with Azure through JavaScript, Python, Node.js, and other Web development tools and languages.

Getting Started with Microsoft Azure

As part of your tour, you will see how to use Azure with IoT solutions by writing a sample project to send messages to and from Azure. Consider it a Hello, World for Microsoft Azure. Naturally, Azure has features that fully support IoT solutions and you will concentrate on those features in this chapter.

As you will see, there are tools built into the Azure portal and Visual Studio extensions that work with your source code project files to aid in setting up applications to interact with Azure features. In fact, you will see how to gather sensor data on your Raspberry Pi and send it to Azure to be stored, processed, and displayed and hosted as a web site. Yes, Azure completes your IoT solutions by providing the Internet-facing systems and features to share your data.

Although you will only scratch the surface of what is possible with Azure, the projects in this chapter give you a taste of what is possible for IoT solutions using Azure. Let's continue the lightning tour and see how to get started using Azure.

■ **Note** This lightning tour is only one path to learn how to use Azure. The presentation is tailored for using Azure with IoT solutions. As such, I cover only a small portion of the features of Azure. Also, there may be more than one way to achieve your goals using other features in Azure. Indeed, you could spend a lot of time learning everything there is to know about Azure.[2]

Sign Up for an Azure Account

The very first thing you need to do to use Azure is to sign up and create an account. While Azure is a paid service platform, Microsoft permits you to create a temporary free account that is active for 30 days. After the 30 days, you can continue to use Azure but any resources you created or used during your trial period may become billable. Fortunately, your account is not automatically converted to a paid account. You must specifically request the upgrade. However, if you build your IoT solutions using an Azure free account, you can continue to use them as paid resources once you upgrade your account.

The process to create an Azure account is quite easy but requires a valid Microsoft account to get started. If you do not have a Microsoft account, you need to create one by following the steps on https://signup.live.com. You do not have to have a Microsoft e-mail account—any valid e-mail account is OK to use for setting up your Microsoft account.

You also need to enter a credit card to complete the Azure account signup. Microsoft bills you $1 when you set up your account, but they refund you within a few days. This is Microsoft's way of verifying your account and is quite common for monthly subscription services that offer a trial period.

You may be wondering how Microsoft can offer a free account when all the resources are billable. They do this by crediting your account $200. This is normally more than enough to get you through the 30-day trial period. Having the credit card on file allows you to keep your resources going without interruption should you choose to continue your subscription.

Now, let's see how to create your Microsoft Azure account. Visit https://azure.microsoft.com/en-us/free/ and click **Start Now**. Be sure to read the conditions of signing up on the bottom of the page. As you will see, there are some services that may be free beyond the 30-day trial, including a feature called an IoT Hub.

The process starts with entering your information—including your name, e-mail, and phone number, and then clicking **Next**. You are then asked to verify the account via a text sent to your phone. Enter the phone number that you want to receive the text, and then click the **Text Me** button. You receive a code that you then enter in the text box. Click the **Verify Code** button. Once the code is verified, you must enter your credit card information and then click the **Next** button. Finally, read the agreement statement, check the box, and click the **Sign Up** button to complete the process.

Once the process is complete, you should be redirected to the portal at https://portal.azure.com. Figure 16-1 shows the initial portal screen that you see once logged in.

---

[2]While I consider myself a veteran of cloud computing, learning Azure has proven to be like peeling an onion. Every time I use a new feature, I find another cool thing I want to explore.

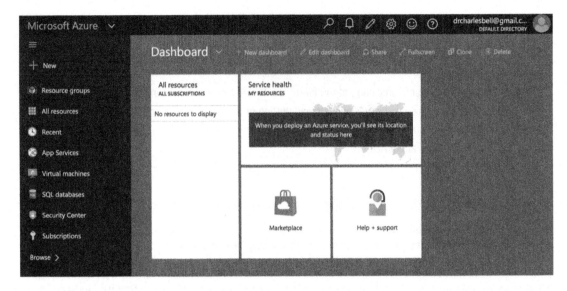

Figure 16-1. *Azure portal (initial login)*

While you see only a portion of the features of Azure in this chapter, there are a few features, terminology, and concepts that you should understand in order to complete the projects. The following lists the Azure features and concepts used in this chapter. I discuss the more complex topics in more detail in the following sections. You will see the others in action as you explore the projects.

- *Azure IoT Hub*: a relative recent addition to Azure, IoT Hubs allow you to create a service that allows you to send and store messages to the cloud as well as send messages from the cloud to a device. The IoT Hub can be connected to other Azure services to complete your IoT solution.

- *Azure IoT Suite*: a dedicated solution for IoT projects that provides preconfigured templates for creating two types of common IoT solutions: predictive maintenance and remote monitoring. It combines a number of existing Azure features with a streamlined setup process.

- *Device Explorer*: an application that can be used to monitor messages sent from and to devices connected to Azure

- *JavaScript Object Notation (JSON)*: an open-standard format that uses human-readable text to transmit data objects consisting of attribute–value pairs.[3]

- *Connected Services*: a Visual Studio feature that allows you to add connections to online of services, including the Azure IoT Hub. It provides huge time savings in adding Azure connectivity to your project.

---

[3]https://en.wikipedia.org/wiki/JSON

■ **Note** The Azure portal is a relatively new addition. The old management console (`https://manage.`
`windowsazure.com/`) is still operational, and depending on what resources you use, you may be redirected to
the management console.

Azure IoT Hub

The Azure IoT Hub is a service designed to allow bidirectional message passing between IoT devices and
the cloud (the IoT Hub or connected services). Thus, the IoT Hub can be considered the main Azure feature
for connecting your IoT solution to Azure. There are a number of features and benefits to using the IoT Hub,
including the following.

- Reliable device-to-cloud and cloud-to-device messaging

- Includes provisions for security credentials and access control

- Permits monitoring for device connectivity

- Microsoft provides a number of libraries to support a host of devices, languages, and
  platforms

Additionally, Microsoft has provided a feature in Visual Studio called Connected Services that, together
with the Azure IoT Hub Connected Services feature, simplifies application development. Furthermore, it
supports .NET and UWP applications.

For more information about the Azure IoT Hub, see `https://azure.microsoft.com/en-us/`
`documentation/services/iot-hub/`.

Azure IoT Suite

The Azure IoT Suite is a combination of existing Azure enterprise-grade features, IoT services, and
preconfigured solutions that allow you to get started quickly. The preconfigured solutions are complete
implementations of common IoT solutions. This allows you to get started experimenting with Azure and
linking your IoT solution to the cloud much more quickly and with far less effort than configuring the Azure
features yourself individually.

The list of capabilities of the preconfigure solutions in Azure is quite impressive. The following is a list of
the features and benefits of the preconfigured solutions.

- Easy-to-set up devices to collect data

- Built-in analytics to capture the data

- Sample web application to display the data in real time

- Sample alarms for watching the data for major events

- Visualize real-time and historical data

If that list isn't impressive enough, consider that you can take the preconfigured solution and modify it
as your needs change. That is, you can start out with the basic solution and then add more capability in the
form of faster processing, more devices, more storage, and so forth. It really is the best way to get your IoT
solution up and running in the cloud and even mature it as you continue to develop.

For more information about the IoT Suite, see `https://azure.microsoft.com/en-us/documentation/`
`suites/iot-suite/`.

Device Explorer

When developing Azure IoT solutions (or any Azure solution), developers need a tool to help monitor and verify the messages that their solutions are sending to and receiving from Azure. Device Explorer is an answer to fill that need. With Device Explorer, you can connect to your IoT Hub and monitor the messages send to and from Azure making it easier to diagnose problems during the development process.

The Device Explorer is a Windows application that you can run on your PC. You can download the Device Explorer and learn more about the application at `https://github.com/Azure/azure-iot-sdks/tree/master/tools/DeviceExplorer`. You use the Device Explorer in this chapter, so download and install it now.

Now that you have an Azure account and know a bit more about the technologies that you will use, let's now walk through creating a sample IoT application to demonstrate how to set up resources for storing IoT data in Azure.

Building IoT Solutions with Azure

Building IoT solutions that use Azure can be rather challenging. This is mostly due to the rather steep learning curve in understanding Azure and the Azure features needed. Rather than jump directly into a complex Azure-based IoT solution, let's set the stage for the chapter project, where you see a simple demonstration on how messages are exchanged with Azure. Essentially, you will implement the project in two stages: one to see how to write sample data to Azure, and another to see how to connect real sensors to store and view data in the cloud using the Azure IoT Suite.

The project in this section is designed to learn how to set up an IoT Hub and connect the application running on your device to send and receive messages. This teaches you how the data is sent to the cloud and prepares you for writing code to send live data to Azure.

Thus, the application is simplistic in the sense that you will not use a sensor; rather, you will use a user interface to get sample values and send those to the IoT Hub. Once again, this is so that you can learn the basics of working with the IoT Hub, and not make the project more difficult from the complexity of reading data from sensors and viewing them in the cloud.

The steps in this section require several steps, taking you from setting up your IoT Hub to configuring your PC, and finally, seeing the cloud messages in action, as shown next. I discuss each in more detail in the upcoming sections.

1. Set up an IoT Hub in Azure.

2. Set up your PC for working with the Azure IoT Hub.

3. Create the example application.

4. Use the Device Explorer to monitor messages sent from the device to the IoT Hub message queue.

5. Use the Device Explorer to send a message from the IoT Hub to the device.

These steps sound daunting[4] but if you follow along, you should be able to go through the steps in only a few minutes. Some of the steps you perform in this project can be reused in the chapter project.

---

[4]I've condensed the essentials to make it easy to follow.

Set up an IoT Hub

An IoT Hub is a feature in Azure for connecting devices to other features in Azure. Most notably, the IoT Hub provides a message queue for you to send messages to the cloud. These messages can then be consumed by other features (or other applications) in Azure or by another application connecting to Azure. Let's begin by logging into the Azure portal (`www.portal.azure.com`).

To add an IoT Hub, click **New** in the Azure portal. This opens a submenu that shows a list of categories. In the blade that opens, choose **Internet of Things**. Figure 16-2 shows the selections made.

Figure 16-2. Adding an IoT Hub

The Azure portal page is best viewed on a wide screen because it spreads things out quite a bit by opening sliding forms (called *blades*) for settings, selections of features, and so forth.

■ **Note** To make it easier to read the text, I use screen excerpts from the Azure portal.

On the next blade, you see the IoT specific features and products. Click the IoT Hub area. This opens a new blade, where you can fill in some information about the new IoT Hub, including the following.

- *Name*: Choose a name that is unique (Azure tells you if it is not). You have only a few characters to work with so keep it short.

- *Pricing Tier*: Use the free tier by clicking the arrow. Click **F1 Free** from the list and then click **Select**.

- *Number of IoT units*: Leave this set to the default.

- *Device-to-cloud Partitions*: Leave this set to the default. (Note: this equates to how many devices you can connect at one time.)

- *Resource Group*: Choose a name that is unique. You can use shared groups, but I recommend creating your own.

- *Subscription*: Leave this set to **Free Trial**.

- *Location*: Choose your location.

- *Pin to Dashboard*: Click this to place the IoT Hub on the dashboard for easy access.

Figure 16-3 shows the completed information for an IoT Hub I created when testing this project.

Figure 16-3. *Setting up an IoT Hub*

I highlight those areas you need to fill in when you create your own IoT Hub. Notice the names that I used. I simply used my initials and a short description. It's a bit cryptic, but remember, the names must be unique in the region where you are creating the resources. Add some numbers if you get a name violation to help make the names more unique.

■ **Note**   You can have only one free IoT Hub subscription per account.

The next step is the registration process. Azure merrily creates your IoT Hub for you in the background. An occasional pop-up message keeps you updated of the status. You can click the bell symbol in the top right of the portal to see your messages. I should note that, depending on resources being allocated; the registration could take some time. Now is a good time to refresh that beverage, pay some attention to the family, pet your dog, and so forth, and check back in about 15 to 30 minutes.[5]

Once created, you see a message stating, Deployments Complete. If at this point you do not see your IoT Hub in dashboard on the portal, you can try refreshing the screen or click **Resource Groups** to refresh the dashboard. Your IoT Hub should show up on your dashboard as a tile, as shown in Figure 16-4.

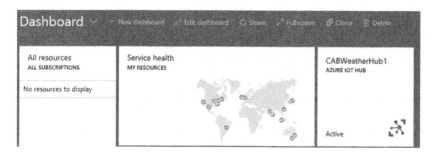

Figure 16-4. *New IoT Hub on Dashboard*

Click the tile for your IoT Hub. This takes you to the settings console, as shown in Figure 16-5. You can also see get to this by clicking **All Resources** and selecting your IoT Hub from the list.

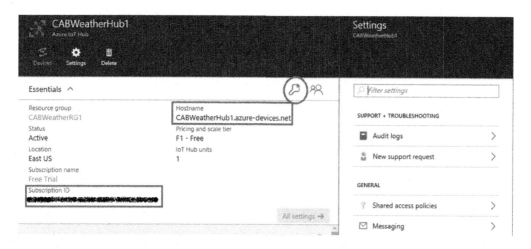

Figure 16-5. *Settings page for the IoT Hub*

---

[5]It usually only takes about 10 to 15 minutes, but since you're relaxing, take your time!

Notice the area marked Usage. You will use that in a moment to verify that your messages have been sent to the cloud. Notice also the subscription id and URI (hostname). Make a note of these strings, as you need them again in the project. I have redacted mine to protect my account.

While you are on this page, it is a good idea to click the small key icon to get the keys for the policy that you need to use. Click the key and then click **iothubowner** in the list. You see all the keys and permissions for this policy in the next blade. Copy the strings using the copy icon next to each and place these in a file somewhere. You will use the primary key and the connection string for the primary key, as annotated in Figure 16-6.

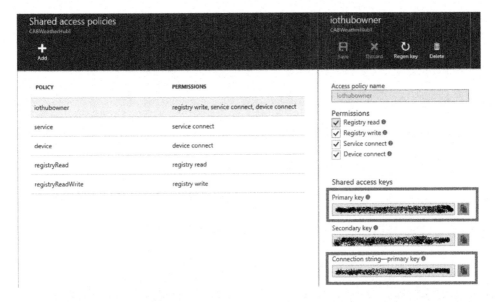

***Figure 16-6.** Policy settings and keys*

■ **Caution**   Your keys, Ids, and subscription-related tokens should be well protected. Someone with access to this data can create applications that can connect to your Azure account. Be sure to strip them or redact them before publishing any of your own code.

OK, the IoT Hub is ready for messages. Next, let's set up your PC.

Set up the PC

There are two things that you need to do on your PC to create the sample project: install the Device Explorer (if you haven't already) and install an add-in to Visual Studio called Connected Service for Azure IoT Hub. You will use the Device Explorer to test the sample application.

■ **Note**   Some older documentation speaks of something called the iothubexplorer (npmjs.com/package/ iothub-explorer/tutorial), which is a Node.js command-line tool. You can use that if you prefer command-line utilities, but you may need to install Node.js first.

You also need to install the Connected Service for IoT Hub. You can install this from within Visual Studio via the **Tools ➤ Extensions and Updates** dialog, as shown in Figure 16-7. The Connected Service for Azure IoT Hub is a utility to help you easily and quickly add Azure IoT Hub access to the project.

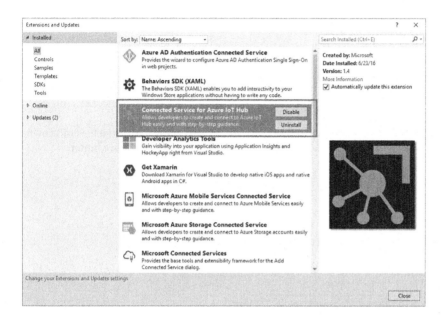

Figure 16-7. *Adding the Connected Services for Azure IoT Hub*

Now, let's create the sample Hello, World Azure style project.

Hello, World! Azure Style

In this section, you will see how to leverage the concepts and features you've seen in the previous sections to create a sample Azure IoT solution. In this case, you will create a simple application that uses a simulated sensor in the form of a user interface. You supply values, send them to the cloud, and see the last message received from the cloud.

The project covered all the basic things that you need to know to get an IoT solution talking to Azure. You will see the code for creating the connections to Azure as well as how to properly format and send messages using JSON strings.

As I mentioned previously, this project lays the groundwork for the chapter project. Indeed, much of what you develop in this project is used later. The only exception is the user interface. You use it as a means to test the Azure connection and message passing in this sample project, but you will not need it for the chapter project because it is a headless application.

Finally, this project runs on your PC, so there is no need for additional library references (but you can run it on your device if you want). Just make sure that you are connected to the Internet.[6]

Let's start with creating the project.

---

[6]It's easy to forget this step. Guess how I know.

New Project

You will use a C# project template for this project—the Blank App (Universal Windows) template. Use the name HelloAzure for the project name. You can save the project wherever you like or use the default location.

Azure IoT Hub Connected Service

You will use the connected service feature to add code to the project. The utility creates a new source file named AzureIoTHub.cs, which encapsulates what you need to communicate with the cloud. Let's run the connected service now.

Right-click the **References** item in Solution Explorer and choose **Add Connected Service**. Scroll down the dialog and select the **Azure IoT Hub** entry, as shown in Figure 16-8. Click the **Configure** button.

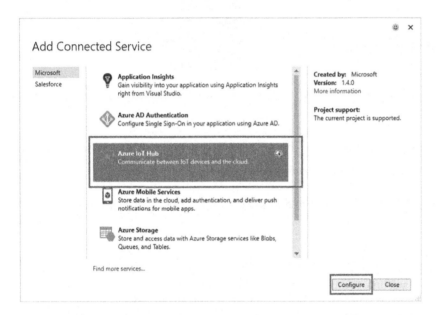

Figure 16-8. *Adding the Azure IoT Hub Connected Service*

You are asked to choose the security mode, as shown in Figure 16-9. Choose the hardcoded option. This is fine for small, experimental projects or projects for you own use. As you will see, you can choose a more restrictive method but this requires adding a bit more code in your project to use. Plus, it is still in an experimental stage. Click the **OK** button to continue.

Figure 16-9. Select Security Mode

On the configuration page, you are asked to log in to your Azure account. You should see your account in the upper-left corner of the dialog. If you do not see it, choose the account from the drop-down box. If you see a small yellow triangle, it means there was an issue logging in. To retry, click **Re-enter your credentials**. Once logged in, you see your IoT Hubs, as shown in Figure 16-10. If you do not see them, click **Refresh**.

Figure 16-10. Connecting to your Azure account

Next, select your IoT Hub and then click the **Add** button. A dialog will appear, asking you to create a new device. Simply click **New Device** and enter a name for your device. I used **HelloAzure**. Click the **OK** button to create the device and then click **OK** again to continue the registration. If you already created a device, you can select it from the list.

You see a dialog showing the progress of the registration. When done, you notice a number of new resources and service references added to your project, as well as a new code file named AzureIoTHub.cs, as shown in Figure 16-11. It even added support for JSON! This is the beauty of the connected service—all of this used to be a very manually intensive process, but it is now automated for you! Neat.

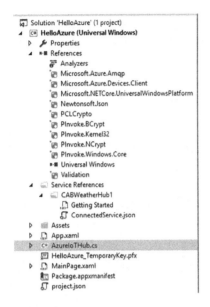

```
🔲 Solution 'HelloAzure' (1 project)
◢ C# HelloAzure (Universal Windows)
    ▷ 🔧 Properties
    ◢ ■■ References
        🔳 Analyzers
        🔳 Microsoft.Azure.Amqp
        🔳 Microsoft.Azure.Devices.Client
        🔳 Microsoft.NETCore.UniversalWindowsPlatform
        🔳 Newtonsoft.Json
        🔳 PCLCrypto
        🔳 PInvoke.BCrypt
        🔳 PInvoke.Kernel32
        🔳 PInvoke.NCrypt
        🔳 PInvoke.Windows.Core
        ■■ Universal Windows
        🔳 Validation
    ◢ 🔲 Service References
        ◢ 🔲 CABWeatherHub1
            🔲 Getting Started
            ⬚ ConnectedService.json
    ▷ 📁 Assets
    ▷ 📄 App.xaml
    ▷ C# AzureIoTHub.cs
        🔲 HelloAzure_TemporaryKey.pfx
    ▷ 📄 MainPage.xaml
        📄 Package.appxmanifest
        ⬚ project.json
```

Figure 16-11. *New files added to solution*

You also see a list of what was installed in the Output window, as shown next.

```
Adding Azure IoT Hub to the project.
Installing NuGet package 'Newtonsoft.Json' version 8.0.3.
Installing NuGet package 'Microsoft.Azure.Amqp' version 1.1.5.
Installing NuGet package 'Microsoft.Azure.Devices.Client' version 1.0.11.
Installing NuGet package 'PCLCrypto' version 2.0.147.
Installing NuGet package 'PInvoke.BCrypt' version 0.3.2.
Installing NuGet package 'PInvoke.Kernel32' version 0.3.2.
Installing NuGet package 'PInvoke.NCrypt' version 0.3.2.
Installing NuGet package 'PInvoke.Windows.Core' version 0.3.2.
Installing NuGet package 'Validation' version 2.2.8.
New service instance CABWeatherHub1 created
Successfully added Azure IoT Hub to the project.
```

Go ahead and explore that new code file. In it, you see code to connect to the IoT Hub and code to send and receive messages. We'll explore this code in more detail in a moment. There is also a link to the Getting Started page in Solution Explorer. Click that and learn a bit more about getting started writing C# Azure applications.

Now, let's build a user interface and then you'll add a class to simulate collecting data.

User Interface

Let's build a simple user interface to send messages to the cloud as well as display the last message received from the cloud. Click the MainPage.xaml file and add the code shown in Listing 16-1.

Listing 16-1. User Interface XAML Code

```
<Grid Background="{ThemeResource ApplicationPageBackgroundThemeBrush}">
        <StackPanel Width="400" Height="400">
            <TextBlock x:Name="title" Height="60" TextWrapping="NoWrap"
                    Text="Hello, World Azure Style!" FontSize="28" Foreground="Blue"
                    Margin="10" HorizontalAlignment="Center"/>
            <TextBlock Text="Temperature:" />
            <TextBox Name="InTemperature" Text="25.00" />
            <TextBlock Text="Pressure:" />
            <TextBox Name="InPressure" Text="10.00" />
            <Button x:Name="send_button" Content="Send Data" Width="75" ClickMode="Press"
                    Click="send_button_Click" Height="50" FontSize="24"
                    Margin="10" HorizontalAlignment="Center"/>
            <TextBlock Text="Message from Azure:" />
            <TextBlock Name="OutAzure" Text="" />
        </StackPanel>
    </Grid>
```

You create a number of labels as well as two text boxes to allow you to choose values for temperature and barometric pressure, and a button to send the data to Azure cloud. Figure 16-12 shows an example of what the user interface looks like when the application is running.

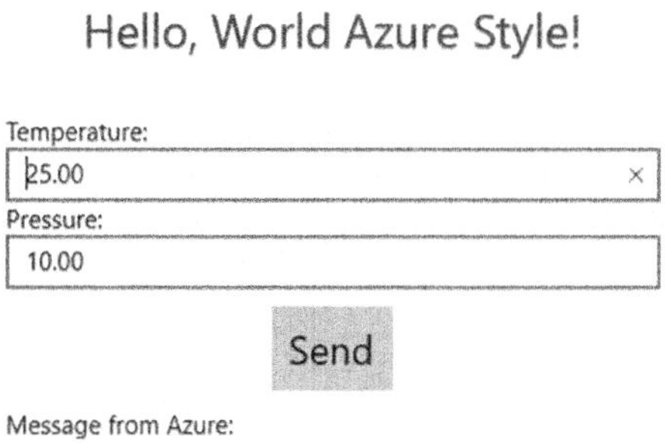

Figure 16-12. *Sample user interface*

There is just one more part you need to add to the code. You need a class to simulate encapsulation of capturing data from the sensor. Let's add a new class named `SensorData.cs`. You can add the new class by right clicking the project in the Solution Explorer and choose **Add ➤ Add New Item...** menu and then choose the **Class** entry in the list. Use the name **SensorData** and click the **Add** button.

Now, open that file and add the code to create a new setup attributes for the data values. Since you are simulating data from a sensor that reads temperature and barometric pressure, you create the two data attributes, as shown next.

```
namespace HelloAzure
{
    class SensorData
    {
        public double Temperature { get; set; }
        public double Pressure { get; set; }
    }
}
```

You use this class to fill in the data from the simulated sensor (the user interface). Return to the user interface MainPage.xaml.cs and add the code shown in Listing 16-2 to add the data from the user interface to the SensorData.

Listing 16-2. Code for Reading Sensor Data from User Interface

```
using System.Diagnostics;            // Add for debugging

namespace HelloAzure
{
    public sealed partial class MainPage : Page
    {
        public MainPage()
        {
            this.InitializeComponent();
        }

        private void send_button_Click(object sender, RoutedEventArgs e)
        {
            // Read data from the simulated sensor
            SensorData data = new SensorData();
            data.Temperature = Convert.ToDouble(this.InTemperature.Text);
            data.Pressure = Convert.ToDouble(this.InPressure.Text);
            Debug.WriteLine(String.Format("Data sent: {0}, {1}",
                data.Temperature, data.Pressure));
        }
    }
}
```

OK, now you're ready to complete the Azure segments of the code.

Communicating with Azure

Fortunately, most of the work is done for you. All you need to do is add code to send the data to the cloud and code to read data from the code periodically. Let's work on the code to send the data first.

To send a message to Azure, you must format it as a JSON object, which is really easy to do. In the AzureIoTHub.cs file, you have a class named SendDeviceToCloudMessageAsync(). All you need to do is modify that method to accept the data you "read" and format it into a JSON document and send it to Azure; most of the code is already there!

First, you must add two using clauses, as shown next.

```
using HelloAzure;       // Add for the SensorData class
using Newtonsoft.Json; // Add for JSON utilities
```

Next, you can modify the sample code and add code to serialize the data you "read" from the sensor. The following shows the modified method. I highlight the changes in bold.

```
public static async Task SendDeviceToCloudMessageAsync(SensorData data)
{
    var deviceClient = DeviceClient.CreateFromConnectionString(deviceConnectionString,
        TransportType.Amqp);
    var jsonMsg = JsonConvert.SerializeObject(data);
    var message = new Message(Encoding.ASCII.GetBytes(jsonMsg));
    await deviceClient.SendEventAsync(message);
}
```

Now, let's return to the send button click event and complete the code to send the data to Azure. However, since the code is a task, you need to change the signature of the button click method, as shown in Listing 16-3.

Listing 16-3. Updated Send Button Click Method Code

```
private async void send_button_Click(object sender, RoutedEventArgs e)
{
    // Read data from the simulated sensor
    SensorData data = new SensorData();
    data.Temperature = Convert.ToDouble(this.InTemperature.Text);
    data.Pressure = Convert.ToDouble(this.InPressure.Text);

    // Send data to the cloud
    await AzureIoTHub.SendDeviceToCloudMessageAsync(data);

    Debug.WriteLine(String.Format("Data sent: {0}, {1}",
        data.Temperature, data.Pressure));
}
```

OK, you're nearly there! The next thing you need to do is add code to read messages from Azure cloud periodically. For this, you use a timer event. You've already added this code to several other projects so I omit the details and simply present the completed code for the module. Modify the MainPage.xaml.cs file as shown in Listing 16-4.

Listing 16-4. Completed User Interface Code

```
using System;
using Windows.UI.Xaml;
using Windows.UI.Xaml.Controls;
using System.Diagnostics;            // Add for debugging
```

```
namespace HelloAzure
{
    public sealed partial class MainPage : Page
    {
        public MainPage()
        {
            this.InitializeComponent();
            DispatcherTimer timer = new DispatcherTimer();
            timer.Tick += timerTick;
            timer.Interval = TimeSpan.FromSeconds(3);
            timer.Start();
        }

        private async void send_button_Click(object sender, RoutedEventArgs e)
        {
            // Read data from the simulated sensor
            SensorData data = new SensorData();
            data.Temperature = Convert.ToDouble(this.InTemperature.Text);
            data.Pressure = Convert.ToDouble(this.InPressure.Text);

            // Send data to the cloud
            await AzureIoTHub.SendDeviceToCloudMessageAsync(data);

            Debug.WriteLine(String.Format("Data sent: {0}, {1}",
                data.Temperature, data.Pressure));
        }

        private async void timerTick(object sender, object e)
        {
            this.OutAzure.Text = await AzureIoTHub.ReceiveCloudToDeviceMessageAsync();
        }
    }
}
```

OK, you're now ready to test the application. Be sure that your code compiles without errors first.

Testing the Application

When ready, launch the Device Explorer and use the connection string you copied from the Azure IoT Hub Settings. Refer to Figure 16-6 if you forgot where this is. Paste the string into the IoT Hub Connection String text box on the **Configuration** tab and then click **Update**. This validates the connection string. Device Explorer fetches the correct keys for access, as shown at the bottom of the dialog. Figure 16-13 shows the completed process.

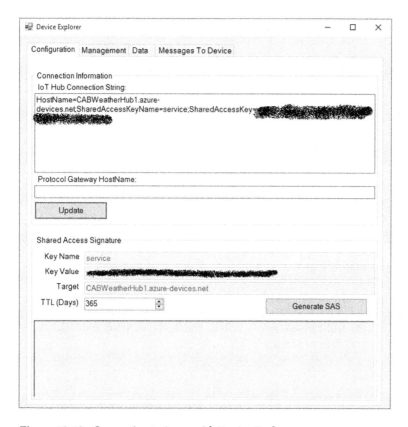

Figure 16-13. *Connecting to Azure with Device Explorer*

Next, you can click the **Data** tab to see messages from your device to the cloud. To do so, launch the application and click the **Send** button. Then, go back to the Device Explorer and click the **Monitor** button on the **Data** tab. Go ahead and click **Send** in your application a few times with different values. The values won't show up instantaneously, but they should show up within a few moments. You should see your events similar to Figure 16-14.

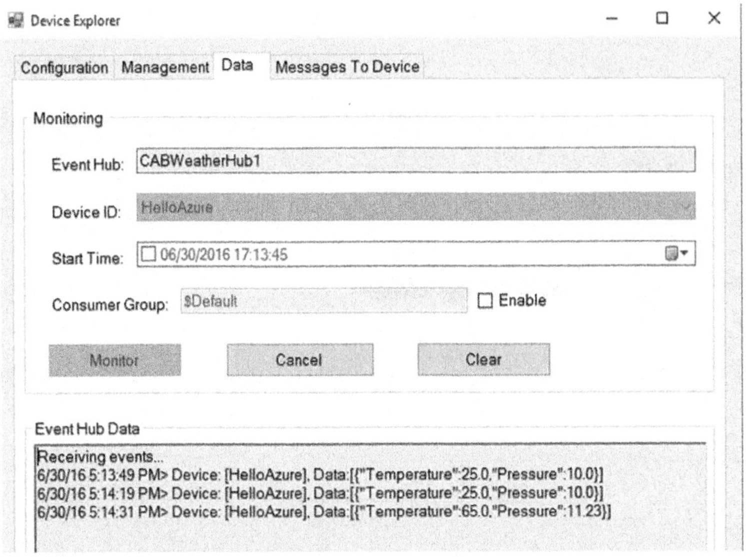

Figure 16-14. Monitoring messages in Device Explorer

OK, that was cool, eh? Now, let's send a message to the application from the cloud. You use the **Message To Device** tab in Device Explorer to send a message. Go ahead and click that tab now. In the **Message** text box, enter **Hello, from Azure!**, check the **Add Time Stamp** check box, and then click **Send**. You should see the output message in the Output area, as shown in Figure 16-15.

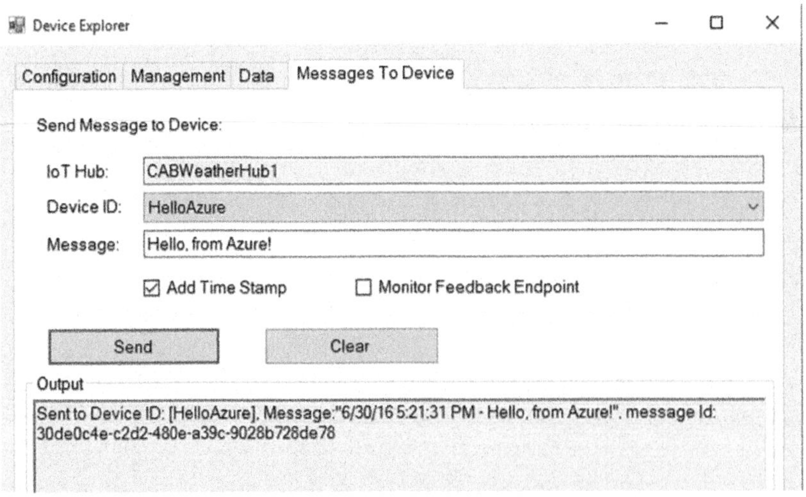

Figure 16-15. Sending a message to the cloud in Device Explorer

Now, go look at your application. Do you see the message? Figure 16-16 shows the result.

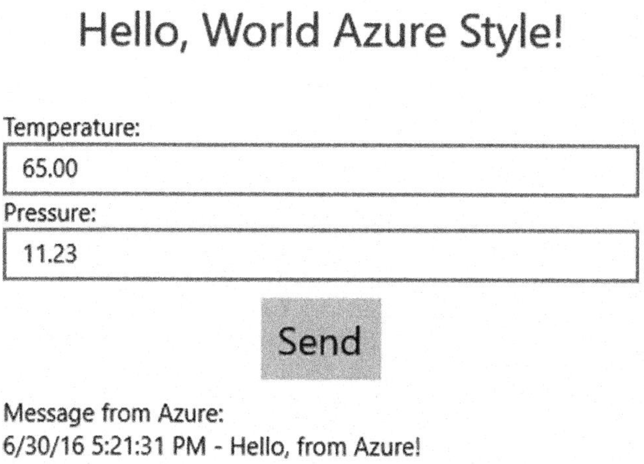

Hello, World Azure Style!
Temperature:
65.00
Pressure:
11.23
Send
Message from Azure:
6/30/16 5:21:31 PM - Hello, from Azure!

Figure 16-16. *Message received!*

So what just happened? Device Explorer formatted the message you wrote adding the timestamp and then sent it to the IoT Hub message queue. The application (in the timer tick event) connected to Azure and retrieved the message. How fantastic is that?

Now, let's go back to the Azure portal and look at the IoT Hub Settings page again. Figure 16-17 shows an excerpt from the page showing usage. Here you see that my IoT Hub sample logged 3 messages, which is exactly what I sent when I tested the application. You should see something similar in your IoT Hub usage data.

Usage
6/30/2016 UTC
CABWEATHERHUB1
0%
TOTAL
MESSAGES
3 / 8k
DEVICES
1

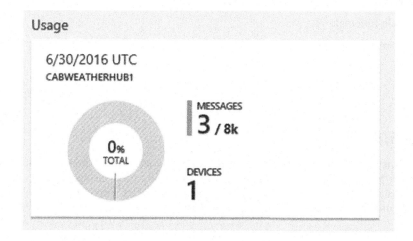

Figure 16-17. *Checking Usage in the Azure portal*

If you got all of this to work, take a victory lab around your house, apartment, or office. You've just created a sample project that stores data in the cloud.

In the next section, you see how to take this sample project and add real sensor data and view it in the cloud. Hold on, it's going to get a bit bumpy first. The technology and code is rather straightforward but a bit tedious to set up. However, I walk you through every step.

Overview

Now that you have seen how to get started writing IoT solutions with Azure, including how to set up an account, an IoT Hub, and a sample application that sends and receives messages from the cloud, let's now see how to send real data to Azure, process the data, and view it in a web site—all of which is hosted in Azure. To do this, you are going to discover several new tools and features. It's a bumpy ride through the many screens and a bit of a minefield with respect to terminology, but if you take your time and follow along, you can get this project running and see your IoT data in the cloud.

You want to make a version of the weather project from Chapter 13. Instead of writing the data to MySQL, you store it in an IoT Suite remote monitoring solution in Azure. While there's a lot to do, the IoT Suite makes it a lot easier than doing it yourself step-by-step in Azure. But first, let's get started with a review of the components and hardware that you need. I repeat some of the text from Chapter 13 for completeness.

Required Components

The following lists the components that you need. You can get the BMP280 sensor from Adafruit (`www.adafruit.com`) either in the Microsoft IoT Pack for Raspberry Pi or purchased separately, SparkFun (`www.sparkfun.com`), or any electronics store that carries electronic components. However, if you use a sensor made by someone other than Adafruit, you may need to alter the code to change the I2C address. Since this solution is a headless application, you do not need a monitor, keyboard, and mouse.

- Adafruit BMP280 I2C or SPI barometric pressure and altitude sensor
- Jumper wires: (4) male-to-female
- Breadboard (full size recommended but half size is OK)
- Raspberry Pi 2 or 3
- Power supply

Set up the Hardware

Although there are only four connections needed for this project, you will make a plan for how things should connect, which is good practice to hone. To connect the components to the Raspberry Pi, you need four pins for the BMP280 sensor, which requires only power, ground, and two pins for the I2C interface. Refer to Chapter 12 for how to wire the sensor. I include Figure 16-18 as a reminder of how things are connected.

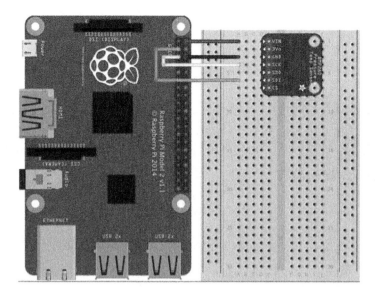

Figure 16-18. *Connections for the Weather Sensor project*

If you are following along with this chapter working on the project, go ahead and make the hardware connections now. Don't power on the board yet, but do double and triple check the connections.

Set up a New Azure Remote Monitoring Solution

Before you get into the code, let's set up a remote monitoring solution in Azure to host and present the data. The process is very straightforward (Microsoft has done an excellent job streamlining everything). Rather than connect to the Azure portal, you will use the Azure IoT Suite management web site at azureiotsuite. com/. You have to log in again using the same user account you used for the Azure portal (it's all the same technology, just a different console).

Once you log in, you are presented with two choices: to create a predictive maintenance solution or a remote monitoring solution. A predictive maintenance solution allows you to analyze data and detect situations where maintenance of your infrastructure, machinery, and so forth, is needed either through regular preventative maintenance or corrective maintenance. A remote monitoring solution allows you to connect your devices to see data collection in real time, set alarms for threshold and similar important events, and conduct analysis on the data. For more information about the preconfigured solutions, see https://github.com/Azure/azure-content/blob/master/articles/iot-suite/iot-suite-what-are-preconfigured-solutions.md.

Figure 16-19 shows an excerpt of the initial page that you see at the IoT Suite web site.

Figure 16-19. *Choose a preconfigured solution*

For this project, you set up a new remote monitoring solution. Click the **Select** button for the remote monitoring option. On the next screen, fill in the following text boxes.

- *Solution Name*: Choose a name for your remote monitoring solution. It must be unique among the other remote monitoring solutions in your region so it is a good idea to use your initials or perhaps a number in the name. The name is validated and if the name is available, a green check mark appears next to the name. For example, I used **CABWeatherHub1**.

- *Region*: Choose the closest region to your physical location. This isn't that critical, but it is convention and generally expected as a best practice. If the IoT Suite and all of its components are available in your region, a green check mark appears to the right. If you see a red X, choose another nearby region.

- *Subscription*: Choose the subscription you want to use. If you using the trial subscription select it; otherwise, Azure charges you for services setup.

Once you see three green check marks, as shown in Figure 16-20, you can click the **Create Solution** button to initiate the process.

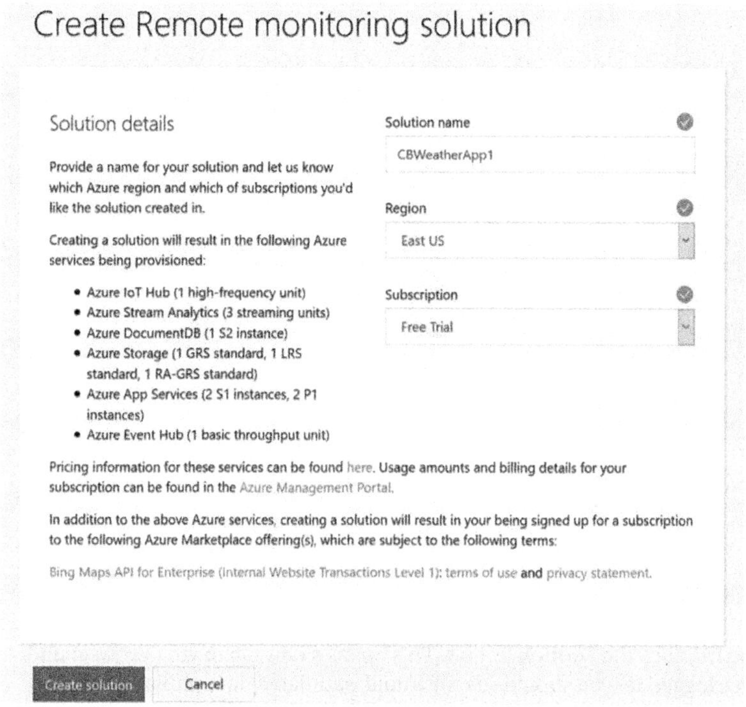

Create Remote monitoring solution

Solution details

Provide a name for your solution and let us know which Azure region and which of subscriptions you'd like the solution created in.

Creating a solution will result in the following Azure services being provisioned:

- Azure IoT Hub (1 high-frequency unit)
- Azure Stream Analytics (3 streaming units)
- Azure DocumentDB (1 S2 instance)
- Azure Storage (1 GRS standard, 1 LRS standard, 1 RA-GRS standard)
- Azure App Services (2 S1 instances, 2 P1 instances)
- Azure Event Hub (1 basic throughput unit)

Pricing information for these services can be found here. Usage amounts and billing details for your subscription can be found in the Azure Management Portal.

In addition to the above Azure services, creating a solution will result in your being signed up for a subscription to the following Azure Marketplace offering(s), which are subject to the following terms:

Bing Maps API for Enterprise (Internal Website Transactions Level 1): terms of use **and** privacy statement.

Solution name ✓

CBWeatherApp1

Region ✓

East US

Subscription ✓

Free Trial

Create solution   Cancel

Figure 16-20. *Setting up a Remote monitoring solution*

Notice the list of services and features that are set up for you. As you can see, there is an IoT Hub, 3 Stream Analytics jobs, a DocumentDB store, a standard storage service, 2 application services, and an Event Hub. Wow, that's a lot to configure. Again, you can do all of these steps yourself, but it is tedious and as I can attest, a little error-prone.

Once the process begins, it takes some time to execute. It could take as much as an hour to complete so take some time to pet your dog, catch up on your e-mail, and so forth. Once the process is complete, you see your new solution on the main page, similar to what's shown in Figure 16-21.

Figure 16-21. *Remote solution ready*

To see your solution's dashboard, click the **Launch** button. This opens a new tab in your browser and you may need to log in again. Once logged in, you should see the initial, simulated data similar to what is shown in Figure 16-22. You won't see all the features of the remote monitoring solution, but you see how to add a device and its data appears in the telemetry graph.

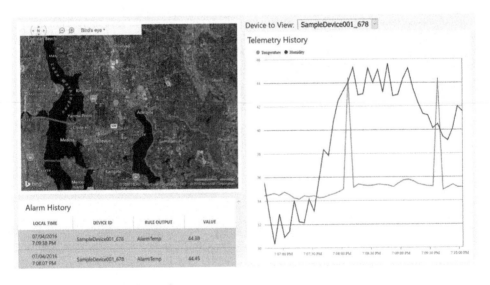

Figure 16-22. *Initial dashboard*

Wow, that's impressive. What you see here is a small map that places a dot at the location of each device and a graph that shows the data being sent to Azure from the devices in real time. You also see a list of alarms in the bottom-left.

The devices shown in the preconfigured solution are all simulated. That is, they aren't real and are generating fake data. This is to show you how the application works. However, the simulated devices consume Azure resources (they are billable) so you should feel free to poke around in the interface until the gee-gah[7] effect wears off and then disable the simulated devices to save resources (money).

To disable the devices, click Devices on the left side panel. You see a dialog with list of all devices defined, similar to what's shown in Figure 16-23. Notice the simulated devices.

Device List (4)

STATUS	DEVICE ID ▲	MANUFACTURER	MODEL NUMBER	SERIAL NUMBER	FIRMWARE	PLATFORM	PROCESSOR	INSTALLED RAM
● Running	SampleDevice001_678	Contoso Inc.	MD-14	SER14	1.14	Plat-14	i3-14	14 MB
● Running	SampleDevice002_678	Contoso Inc.	MD-0	SER0	1.0	Plat-0	i3-0	0 MB
● Running	SampleDevice003_678	Contoso Inc.	MD-11	SER11	1.11	Plat-11	i3-11	11 MB
● Running	SampleDevice004_678	Contoso Inc.	MD-4	SER4	1.4	Plat-4	i3-4	4 MB

(Left margin: FIND DEVICES)

***Figure 16-23.** Controlling devices*

Click one of the devices. This opens the device details pane on the right. To disable the device, choose the **Disable Device** link. Once disabled, you can delete them by clicking the device in the list and choosing **Remove Device...** from the panel on the right. Simply disabling them is enough for this project.

Next, you need to create a device that you can use from the IoT application. Click the **Devices** button on the left and then click the **Add a Device** at the bottom left of the panel. You are given a choice of creating a simulated or custom device. This opens a three-step dialog, which you can use to create your device.[8]

The first step is to choose a simulated or custom device. Choose the custom device option and click **Add New**. On the next page, choose the **Let me define my own device id** option and enter a name for your device. Once again, the name is validated and if it is available, a green check mark appears to the right. When ready, click the **Create** button. On the next page, you see the credentials for your device, which you can copy and save for use in the application. I have blanked out the credentials in the figure, but you should see them for your device. Figures 16-24 through 16-26 show examples creating a device with the dialog.

← ADD A DEVICE
STEP 1 of 3

Simulated Device

Software to simulate a device. Easily extensible for arbitrary events and commands; can run in a Windows Azure worker role. To create a simulated device, please follow the cooler sample instructions.

`Add New`

Custom Device

A physical hardware device.

`Add New`

***Figure 16-24.** Adding a new device (choose Custom Device)*

---

[7]A highly technical term to explain the blank looks, wry grins, and a host of assorted emotions displayed when experiencing cool technologies for the first time. Yes, it does wear off.

[8]Yes, this device appears in the list that the Azure Connected Services retrieves from Azure.

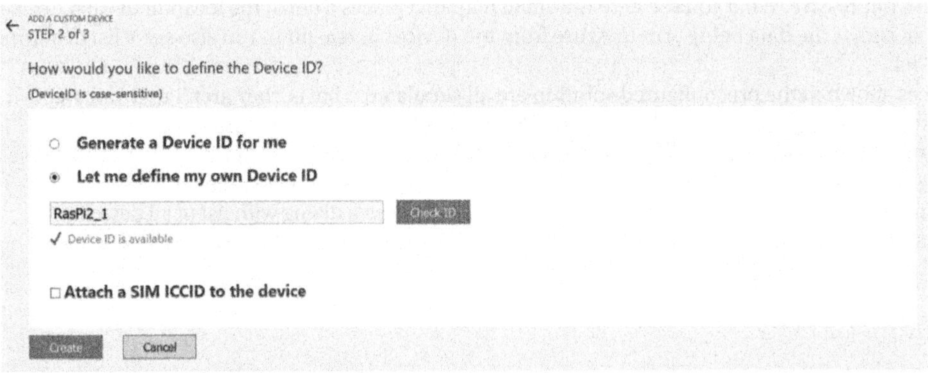

Figure 16-25. *Adding a new device (Device id)*

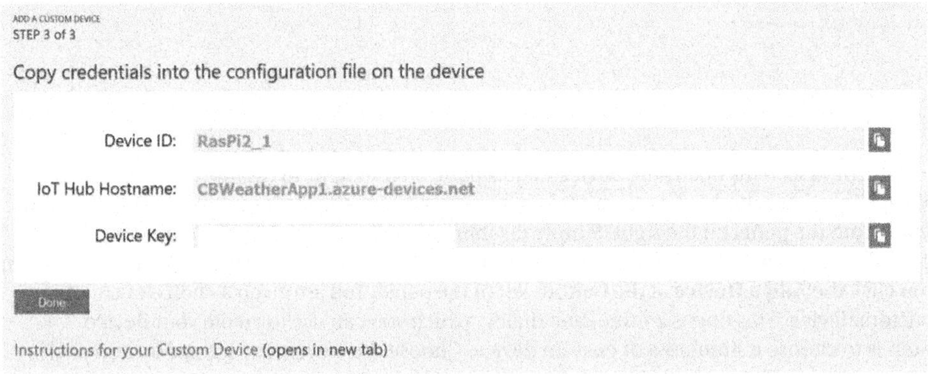

Figure 16-26. *Adding a new device (Credentials)*

When ready, click **Done**. After a few moments, you see your device in the list with a status of **Pending**. This changes once the device connects and starts sending data.

OK, you're almost done. There is just one more step you need to do to get your device fully operational with the remote monitoring solution. You must modify the metadata for the device. For this, you need to return to the Azure portal (`www.portal.azure.com`). Once you log in, you see all the services and features created for you during the remote monitoring setup. Figure 16-27 shows an example of my account. You see the IoT Hub, apps, and more created by the IoT Suite preconfigured remote monitoring solution.

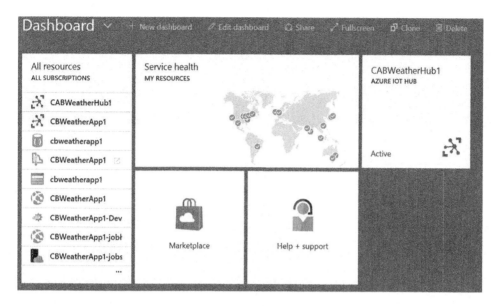

Figure 16-27. Returning to your Azure Portal to see the IoT Suite services

You must modify the device metadata in the Document DB resource in order to add the attributes that the remote monitoring solution expects, including the location and data elements. This seems to be a rough spot in the otherwise fantastic preconfigured solution. I expect that Microsoft will eventually include a better method of defining a custom device.

To modify the device metadata, click **All Resources** to show all of your Azure resources and then click the **Document DB** resource. Figure 16-28 shows these steps with my solution.

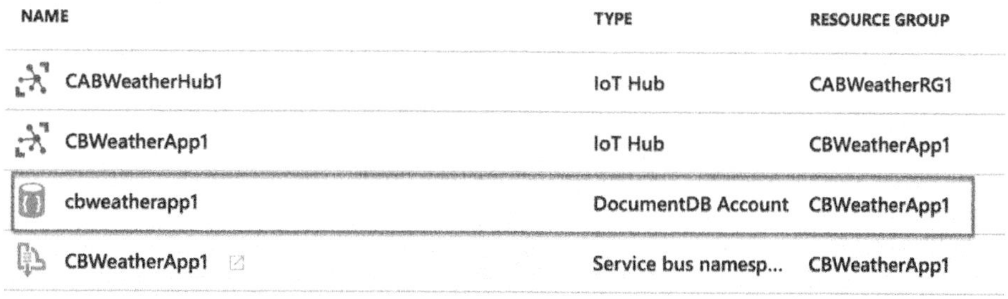

NAME	TYPE	RESOURCE GROUP
CABWeatherHub1	IoT Hub	CABWeatherRG1
CBWeatherApp1	IoT Hub	CBWeatherApp1
cbweatherapp1	DocumentDB Account	CBWeatherApp1
CBWeatherApp1	Service bus namesp...	CBWeatherApp1

Figure 16-28. Select DocumentDB Resource

Next, open the settings blade and then click **Document Explorer** at the top. This opens a blade that shows all the documents listed for this resource. These are files formatted as JSON documents. You see a number of files with strange names—one for each device. Click the file names until you find your new device (look for the device id). This opens the document and lets you edit the file.

Figure 16-29 shows an example of the Document Explorer blades for my solution.

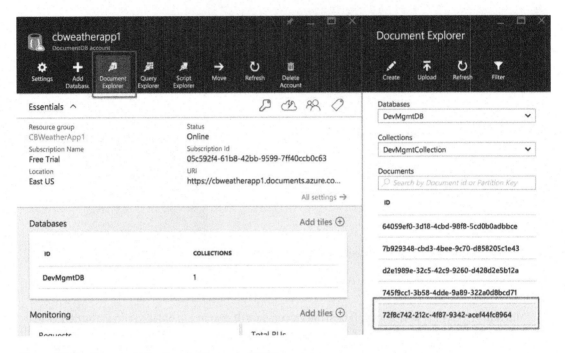

Figure 16-29. *Document Explorer in Document DB Resource*

You initially see only the device id and a few of the properties. What you need to do is add a few more properties, as shown in bold in Listing 16-5. Change your JSON document to include the new attributes (key/value pairs), as shown. Be sure to leave the existing attributes as they are and just add the missing ones. Use your own data, especially the latitude and longitude of your location so you can see your device on the map.

Listing 16-5. Modified JSON Document for the New Device

```
{
  "DeviceProperties": {
    "DeviceID": "RasPi2_1",
    "HubEnabledState": null,
    "CreatedTime": "2016-07-04T23:20:19.7587277Z",
    "DeviceState": "normal",
    "UpdatedTime": null,
    "Manufacturer": "Chuck",
    "ModelNumber": "test1",
    "SerialNumber": "N/A",
    "FirmwareVersion": "0",
    "Platform": "Win10IOTCore",
    "Processor": "ARM",
```

```
    "InstalledRAM": "1 MB",
    "Latitude": 37.5407,
    "Longitude": -77.436
  },
  "SystemProperties": {
    "ICCID": null
  },
  "Commands": [],
  "CommandHistory": [],
  "IsSimulatedDevice": false,
  "id": "72f8c742-212c-4f87-9342-acef44fc8964"
}
```

Once you finish editing the file, click Save at the top, as shown in Figure 16-30.

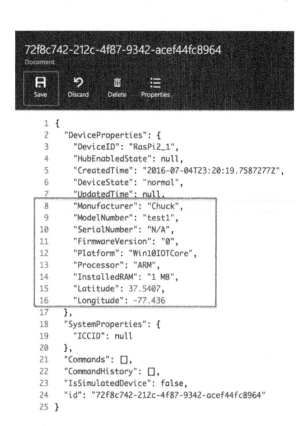

Figure 16-30. *Completing the JSON file for Device Settings*

OK, now you're ready to use the IoT Suite remote monitoring solution to see some data. Let's get started with writing the code for the project!

Write the Code

Now it's time to write the code for the project. You will use the same BMP280 sensor from Chapter 13 but instead of writing the data to MySQL, you will send it to the cloud. You will use the Glovebox.IoT.Devices and Units.NET libraries as well as the Azure IoT Hub connected service. You won't use a user interface, so this is a headless solution. Let's get started.

New Project

You will use a C# project template for this project—the Blank App (Universal Windows) template. Use the name AzureWeather for the project name. You can save the project wherever you like or use the default location. Once the project opens, double-click the MainPage.xaml.cs file. There are a number of namespaces you need to include. Go ahead and add those now, as shown next.

```
using System.Threading.Tasks;        // Add for Task.Run()
using System.Diagnostics;            // Add for debugging
using Glovebox.IoT.Devices.Sensors;  // Add for BMP280 (or BME280)
```

Here you added namespaces for threading so you can run a task (for the BMP280 library), diagnostics for writing debug statements to the log, and finally the sensor library from Glovebox. You'll see how to add the MySQL and Glovebox references in the next sections.

Next, you need to add some variables. You add a variable for the DispatcherTimer class, a variable for the MySQLConnection class, and a variable for the BMP280 class. The following shows the correct code for defining these variables. These are placed in the MainPage class.

```
private DispatcherTimer bmpTimer;    // Timer
private BMP280 tempAndPressure;      // Instance of BMP280 class
```

The code in the MainPage() function initializes the components, the BMP280 class and sets up the timer. In this case, you use a value of 5000, which is 5 seconds. You may want to consider making value greater if you plan to use the project for practical use cases. Listing 16-6 shows the complete MainPage() method.

Listing 16-6. MainPage Method

```
public MainPage()
{
    this.InitializeComponent();

    // Instantiate a new BMP280 class instance
    tempAndPressure = new BMP280();

    this.bmpTimer = new DispatcherTimer();
    this.bmpTimer.Interval = TimeSpan.FromMilliseconds(5000);
    this.bmpTimer.Tick += BmpTimer_Tick;
    this.bmpTimer.Start();
}
```

Adding References for the Hardware

You need to add a few references and run the Azure IoT Hub connected service. You need the Glovebox.IoT. Devices and Units.NET libraries as well as the Windows 10 IoT Extensions. While you have seen how to use each in this and previous chapters, the following paragraphs remind you of how to do each.

Glovebox.IoT.Devices

The Glovebox.IoT.Devices is a library that makes it easy to use I2C sensors like the BMP280. All you need to do is instantiate the class and call the methods to read the data. To add the Glovebox.IoT.Devices library, use the NuGet Package Manager from the **Tools ➤ NuGet Package Manager ➤ Manage NuGet Packages for Solution...** menu. Click the **Browse** tab and then type **glovebox** in the search box. Select the entry named Glovebox.IoT.Devices in the list. Tick the project name (solution) on the right and then click **Install**.

Units.NET

You also need to install the Units.NET library because the Glovebox.IoT.Devices library requires it. To add the Units.NET library, use the NuGet Package Manager from the **Tools ➤ NuGet Package Manager ➤ Manage NuGet Packages for Solution...** menu. Click the **Browse** tab and then type **units** in the search box. Select the entry named **Units.NET** in the list. Tick the project name (solution) and then click **Install**.

Windows 10 IoT Extensions

You must add the reference to the Windows 10 IoT Extensions from the project property page. You do this by right-clicking the **References** item in Solution Explorer.

Azure IoT Hub Connected Service

You will use the connected service like you did in the sample project. This creates a new source file named AzureIoTHub.cs, which encapsulates what you need to communicate with the cloud. Go ahead and run it now.

Right-click the **References** item in Solution Explorer and choose **Add Connected Service**. Scroll down in the dialog and select the Azure IoT Hub entry for the IoT Solution you created and click the **Configure** button. Remember to choose the hard-coded security option and then click **OK** button to continue. On the configuration page, you are asked to log in to your Azure account.

Next, select your IoT Hub and then click the **Add** button. You should see both the IoT Hub you created earlier as well as the IoT Hub for the remote monitoring solution. Figure 16-31 shows an example of what you will see when you are asked to select your IoT Hub.

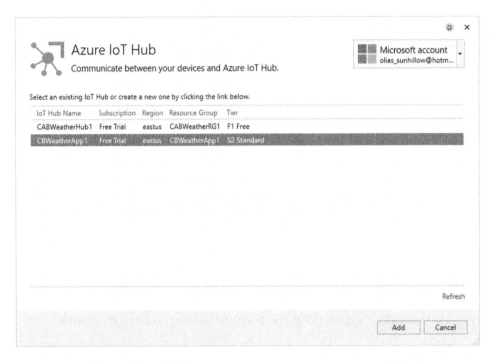

Figure 16-31. *Selecting the IoT Hub*

Next, a dialog appears, asking you to select your device, as shown in Figure 16-32. Choose the device that you created when you set up your remote monitoring solution. Click the **OK** button to select the device and click **OK** again to continue the registration.

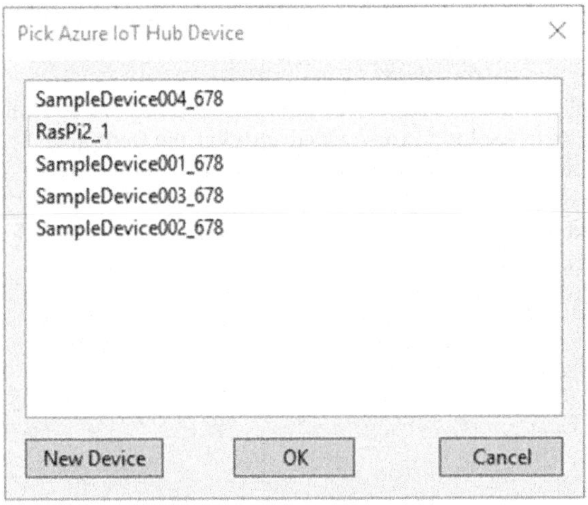

Figure 16-32. *Selecting your device*

Now you're ready to add the code to connect everything together. Let's begin by modifying the code you added using the IoT Hub connected service.

Create a Class for the Weather Data

Recall from the sample project that you need to add a new class to contain the data, which is updated by reading the data from the sensor and then sent to Azure using the IoT Hub class. Add a new class named WeatherData.cs. You can add the new class by right-clicking the project in the Solution Explorer. Select **Add ➤ Add New Item...** and then choose the **Class** entry in the list. Use the name **WeatherData** and click the **Add** button.

Now, open that file and add the code to create a new setup attributes for the data values. You need attributes for the device id, temperature, and barometric pressure, as shown next. Be sure to use the same name for device id that you used when you created your custom device.

```
namespace AzureWeather
{
    class WeatherData
    {
        public string DeviceId = "RasPi2_1";
        public double Temperature { get; set; }
        public double Pressure { get; set; }
    }
}
```

You use this class to fill in the data from the new methods to read the data in the MainPage.xaml.cs file.

Reading the Weather Data and Sending it to the Azure IoT Hub

Now let's add the code to read the weather data from the BMP280. You are using a timer to fire an event every 5 seconds to read (and save) the data. Thus, you need the event for the DispatcherTimer object defined earlier, named BmpTimer_Tick().

However, instead of putting the code to read the data in this event, you use another method and run it as a task (in a new thread). This is because the BMP280 object cannot run in the user interface thread. Thus, you create a new method named getData() to read the data.

The BMP280 class from the Glovebox.IoT.Devices library provides a variety of data from the sensor, including temperature in both Fahrenheit and Celsius, and barometric pressure in several units. You read all of these with the Temperature and Pressure attributes, as shown in Listing 16-7.

Listing 16-7. Code to Populate the WeatherData Class

```
private void BmpTimer_Tick(object sender, object e)
{
    var t = Task.Run(() => getData());
}
```

```
public async void getData()
{
    // Read data from the sensor
    WeatherData data = new WeatherData();
    data.Temperature = tempAndPressure.Temperature.DegreesCelsius;
    data.Pressure = tempAndPressure.Pressure.Bars;

    // Send data to the cloud
    await AzureIoTHub.SendDeviceToCloudMessageAsync(data);

    Debug.WriteLine(String.Format("Data sent: {0}, {1}",
        data.Temperature, data.Pressure));
}
```

You're almost done with the code. All that is left is modifying the code to write the weather data to Azure.

Writing the Data to the IoT Hub

Finally, you need to modify the AzureIoTHub.cs file to change the SendDeviceToCloudMessageAsync() method to use the new WeatherData class to format the data in JSON format and send it to Azure. This is very similar to the code you used in the sample project. Listing 16-8 shows the completed new method. You also need to add a couple of references for the weather class and the JSON class.

Listing 16-8. New SendDeviceToCloudMessageAsync Method in AzureIoTHub.cs

```
using AzureWeather;     // Add for the WeatherData class
using Newtonsoft.Json; // Add for JSON utilities
...
public static async Task SendDeviceToCloudMessageAsync(WeatherData data)
{
    var deviceClient = DeviceClient.CreateFromConnectionString(deviceConnectionString,
TransportType.Amqp);
    var jsonMsg = JsonConvert.SerializeObject(data);
    var message = new Message(Encoding.ASCII.GetBytes(jsonMsg));
    await deviceClient.SendEventAsync(message);
}
```

That's it, you're done! Now you can begin the deployment.

Deploy and Execute

Now it is time to deploy the application! Be sure to fix any compilation errors first. Like you have with other applications, you want to compile the application in debug first (but you can compile in release mode if you'd prefer) and you must turn on the debugger on your board. You do this with the device portal.

Go ahead and power on your board. When ready, connect to the board to run the device portal. Turn on the debugger, open the project properties to target the device, and run with the remote debugger.

Recall from the Chapter 9, you must modify two settings: the **Remote machine** name and the **Authentication Mode**. Set the **Remote machine** name to the IP address of your device with the port specified by the remote debugger when you started it from the device portal.

You may want to run the code in debug and watch the code run and fire events a few times. You can set a breakpoint in the SendDeviceToCloudMessageAsync() method and watch how the data is sent to Azure and displayed in the output window in Visual Studio.

Once you're ready to run the application normally, use the Device Portal to start the application. You can start or stop the application from the Apps tab on the Device Portal. The triangle icon can be used to start the application and the square icon shows a running application that you can stop.

When you run your application, since it is headless, you won't see anything happen. Let the application run for a few minutes and then return to your remote monitoring solution in the Azure IoT Suite. Launch your solution and then observe how the map changes. It should reorient to encompass all of your devices. That is, if you left the simulated devices running or did not remove them, you see dots located on the west coast of the United States. You should also see a dot for your device (RasPi2_1) located at the coordinates you put in the metadata.

To see the data for your device, click the dot and observe the data in the chart. Figure 16-33 shows an example of what you should see in the dashboard. This shows an interesting set of spikes in the data that I achieved by holding my finger on the sensor to increase the temperature. This does nothing for barometric pressure and since I ran the application indoors, the value remains constant.

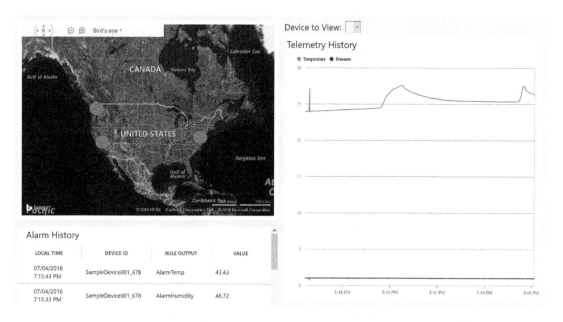

Figure 16-33. *Real-time data from the sensor in Azure*

If you see something similar to this, congratulations! You now know how to leverage Azure IoT Suite to create a remote monitoring solution. This gives you a firm foundation for starting to work with your own cloud-based IoT solutions.

■ **Note** The default setting for free IoT Hub resources, including the IoT Suite preconfigured solution is to retain data for only one day. To store the data longer, you need to upgrade your subscription to one of the paid levels.

Summary

Using cloud computing services, especially those that provide tools and features specifically for IoT solutions, such as Microsoft Azure, can be a steep learning curve. Fortunately, Microsoft has worked very hard to streamline the process to get you going quickly with the Azure IoT Suite preconfigured solutions. While the setup process is easy and the code to interact with the services is created for you (in template form) via the Azure IoT Hub Connected Service, there is nothing simple or basic about the features included in the Azure IoT Suite.

In fact, you can build all manner of cloud-based solutions and services with Microsoft Azure. In this regard, Azure is very powerful, which is exactly what you want when developing IoT Solutions that you want to put into service as either a reliable enterprise-grade product or even as a product you sell to others.

In this chapter, you close your journey of discovering how powerful Windows 10 IoT Core is and how you can use Windows 10 to build sophisticated IoT solutions from simple projects that you run for fun to full cloud-based solutions in Azure. Now it's time for you to engage your own imagination and put the tools and techniques you learned in this book toward building your own IoT solutions.

I hope you enjoyed the ride, and that reading about and working on the projects in this book were as much fun for you as I had writing them.

■ ■ ■

Where to Go from Here?

Now that you have had a thorough introduction to using Windows 10 IoT Core, which included the types of projects that you can create, and tutorials, as well as examples in three programming languages, it is time to consider what you can do beyond the pages of this book.

This chapter explores what you can do to continue the craft of building IoT solutions. Most people want to simply develop projects for themselves—for fun or to solve problems around the home or office. However, some of you may want to take your skills to the next level. Whatever the case, there are a few things that you should consider. The following sections look at sources for more example projects, discuss how to join the community of Windows 10 IoT enthusiasts through social media and other Internet resources, and explain how to become a contributing member of the growing throng of makers.

More Projects to Explore

If you want to work on more Windows 10 IoT projects, you will be happy to learn that there are many, perhaps hundreds, of examples that you can explore. The vast majority of examples are contributions from the community, ranging from a high-level overview to detailed instructions on how to complete the project. Sadly, most of the examples are presented with little or no instruction.[1] However, now that you have had detailed instructions on working with Windows 10 IoT projects and Visual Studio, you should be able to complete examples with little or no documentation.

There are two major repositories for Windows 10 IoT example projects: the Microsoft Windows 10 IoT web site (`https://developer.microsoft.com/en-us/windows/iot/samples`) and the Hackster.io general hardware community forum (`www.hackster.io/lists/windows-10-iot-core`).

Microsoft Windows 10 IoT Samples

The Microsoft sample page is really handy because you can filter the list of samples by tags. For example, you can narrow the list to only those samples that are written in C#, C++, headless, headed, or others. Many of the samples are presented in a number of languages, making them much easier to use. For example, you can visit the site and find a beginning project named Hello, World! implemented in C#, Python, or Node.js.

Fortunately, most of the projects on the site are documented and can be followed easily enough. That is, the documentation for some examples does not explain everything that you need to know to use the samples. In some cases, steps are missing or a familiarity with the tools and techniques are assumed. In a few cases (thankfully), the documentation is old and not applicable to the newer releases. Fortunately, this is rare and I have seen some improvements for the older samples.

---

[1]This is one of my major motivations for writing this book.

© Charles Bell 2016
C. Bell, *Windows 10 for the Internet of Things*, DOI 10.1007/978-1-4842-2108-2_17

That aside, the site is the best and most complete set of Windows 10 IoT samples available. There are very basic samples, samples that demonstrate how to work the with hardware, samples for use with the Arduino Wire library, and even some samples for creating utilities to run on your devices.

What makes this site even more useful is that the samples are stored in GitHub, allowing you to navigate directly to the source files and see the code in your browser, skipping the documentation or demonstration page. This makes it really nice if you just want to see how to implement some feature rather than walk through a long page of text. Of course, the best way to use the samples is to download the entire set of samples. Simply visit the GitHub site (`https://github.com/ms-iot/samples`) and click the **Download ZIP** button on the link. Once you download and unzip the file locally, you are able to see all the sample projects. How cool is that?

Figure 17-1 shows an excerpt from the site demonstrating the current list of tags available for searching the samples.

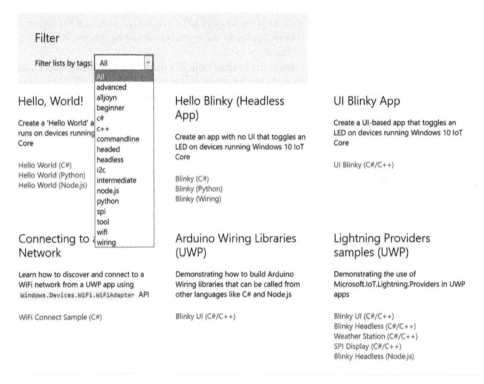

Figure 17-1. *Microsoft Windows 10 IoT Samples*

Most (if not all—I haven't checked all) of the samples are licensed under the MIT license, which makes it very convenient for everyone since the MIT license permits you to use and even publish the code. (See `https://opensource.org/licenses/MIT` for a sample of the MIT license.) This is really great because I have many times wanted to use a sample or demonstration of a project, only to discover the license doesn't permit it. Kudos to Microsoft for embracing the open source community.

Hackster.io

The Hackster site is a community dedicated to learning hardware. You can find all manner of hardware sample projects, including many of those for Windows 10 IoT, Raspberry Pi, Arduino, and more! There is a small but growing Windows 10 category. When you visit the parent site (`www.hackster.io`), you can choose the category by typing a subject. If you enter Windows 10, the site suggests the Windows IoT Core category. Just accept that and you will be presented with the list of samples for Windows 10 IoT.

On the category page, you see all of the projects you can explore. Each is marked with a relative level of difficultly, ranging from easy to advanced. You also see a count of the number of times the sample project was viewed and the number of thumbs-up ratings the project received from others in the community (you have to join Hackster to rate a project). Best of all, there is a comments section that you can use to encourage the designer or ask the designer for help with the project.

■ **Tip**   Use the golden rule when posting comments or questions in online forums. Resist the temptation to post opinions, fan flames of dissent or ridicule, and stick only to the facts.

What I like most about the Hackster site is that the samples are generally well documented and often include a number of photos of the project. Due to the unique structure of the site, the samples are organized in parts that make it easy to follow. For example, the web page for "Character LCD over I2C" (`www.hackster.io/dzerycz/character-lcd-over-i2c-ba8ee9`) contains an overview section that presents a short description of the project, associated tags, difficulty rating, publication date, and even the license. This makes reviewing the project very easy.

The Microsoft site has only a very terse introductory section that jumps directly into the documentation. For those that like to browse projects, the overview page is a great asset. Figure 17-2 shows the "Character LCD over I2C" overview. As you can see, it is a sample project for using a character-based LCD panel over the I2C bus.[2]

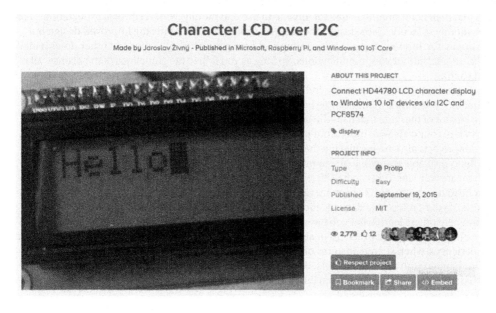

Figure 17-2. Hackster.io sample project

---

[2]This sample is an excellent example of a properly named project. Too often I find projects that are named differently or suggest a different implementation than what it really is. Be sure to use an accurate title when posting your own projects.

If you scroll down from the overview, you find sections that demonstrate how to connect the hardware (similar to how I introduce the projects in this book), a short walkthrough of the code, and descriptions and demonstrations on how to use the project.

Some samples include short videos to demonstrate or explain the project. At the end of the page, you find the comments section, which you can use to read what others have said about the project, as well as the questions that others have had regarding the project. If you get stuck on a sample, be sure to read all of the comments—there is a good chance that someone has already asked the question or solved the problem.

Now that you have seen a couple of resources for more sample projects, let's discuss how you can join the community and contribute to the growing repository of all things Windows 10 IoT.

Join the Community

Once you have mastered the sample projects in this book, as well as a few from other resources, it is time to take your hobby a bit further by joining the community of Windows 10 IoT developers and enthusiasts.

In this section, I discuss some of the reasons that you may want to share your knowledge. I also discuss etiquette for sharing and contributing, and suggest a few example communities that you may want to join or monitor. As you will see, not all of these are strictly Windows 10 (or even Microsoft) friendly, but they can be an excellent source for ideas.

Let's begin with why you would want to share.

Why Contribute?

As more and more free thinkers drive hobbies like IoT, the more prevalent the concept of sharing becomes. This is no accident. Many of the founders and pioneers of the IoT are open hardware and open source advocates. This applies not only to hardware and software but also to other intellectual products, such as the source code and documentation for IoT projects.

Many feel that their code should be free for anyone to use and modify, with reciprocal expectations. For example, if you modify someone else's design or code, you should share not only the improved design but also credit the originator. In some cases, this is as simple as listing the original author, but other times it may mean giving the original author your modifications. So long as you follow the guidelines of the license, all is fair and well in sharing.

However, depending on how the sample code was written (licensed), there may be some limitation to what can be shared. For example, it may not be possible to share code from a proprietary library. Although you may be the creator of the code that uses the library, you do not own the library and cannot share it. You most likely can share your code with others, but publication may be restricted.

Sharing your projects also means placing them someplace where others can find them. You may want to make them freely available to anyone or you may want to limit what people can do with your project. Fortunately, there are web sites that can handle either quite well.

So, why contribute your project? There are many reasons, including the fact that it can be a really good feeling to see one of your projects liked, used, and made by others. Perhaps the most important reason for contributing is to help others learn what you have, or better, learn how to avoid pitfalls or problems. In this way, we all benefit by learning best practices or simply better ways to implement our ideas. Finally, your own project and experiences, when shared, inspires others to create other projects or perhaps improve yours.

I've had this happen with my own projects. People have taken what I have made and improved it. Since they, in turn, shared their project, I was able to incorporate a lot of their improvements in my projects—making them even better than I envisioned.

Now, let's focus on how to go about sharing your projects.

How We Share

You may be wondering why anyone would want to give away for free something that they have worked on for hours. While it is true that the expectation is that you should share your cool projects with others, it isn't a hard-and-fast rule. In fact, there are some who have made their projects available for a fee as a precursor to selling the IoT solution in a commercial avenue. However, the vast majority of enthusiasts share their ideas and projects for free.

There are several communities where you can share your projects. You will see some of these in the next section. But first, there are some things that you must understand about sharing objects. Believe it or not, there is a set of rules—some written, some not—that you are expected to follow should you decide to embrace the IoT or any similar community. The following section lists some guidelines (rules) that you would do well to heed when sharing your ideas, projects, and commentary with the community.

Keep Your Designs Original

Nobody likes a copycat. You didn't like it when you were five years old and you won't appreciate it when you see something that you designed and shared for free being presented as the "design of the month" but credited to someone else.

Thus, you must do your homework to make sure that your design is unique. You don't have to purposefully alter your design so that it doesn't resemble someone else's, but you should do due diligence and at least search for similar projects. Remember, it is OK if you develop a similar project, but it generally is bad form (or perhaps a violation of the license) to simply reproduce something that someone else has published.

In the rare event that your project is nearly identical to another, so long as your work is your own, there shouldn't be a problem. In fact, this happened to me once. My response and that of the other designer were something like, "Cool project. Like minds, eh?" Once again, there is nothing wrong with this provided that you both acknowledge the resemblance and that there are no licensing issues.

If the other project is truly the same design but was licensed differently, you may have to negotiate with the other designer. This can happen when projects are licensed for ownership (e.g., commercial property), but it is rare given that most IoT sample repositories are sites where people share their projects for free.

Let's look at another non-source code example. What is the likelihood that a dozen different cases for a Raspberry Pi will be similar in size, have the same port openings, and perhaps even assemble the same way (snap together)? Very likely, yes? Does this mean that there is one original and 11 copies? No, certainly not. This is not what I am talking about.

What I mean by *unique* is that of those 12 cases, you should be able to identify some differences among them. Be that how they are constructed, if they are made from several parts, whether they have ventilation, and so forth. Even if all 12 designers started at the same time, there will be some minor differences. More importantly, each is its own work. That is, no one used the design of another to pass off as their own.

In a software project, the source code is most likely going to be a little different among the examples. While the degree of differences is something for lawyers to sort out, suffice it to say, if your code and another's are nearly identical, but created without knowledge of the other, it is OK to share your code— provided that there are no licensing conflicts.

Finally, when you share your project, and it is based on the work of another, you must annotate your code, documentation, and project web site, and give credit to the original designer. That is, you state unequivocally that your project is a derivation of the original. It is also good form to include a link to the original design along with a list of your modifications. Once again, this assumes that the license permits it.

Check the License

I have mentioned licensing under the aspect of downloading and using sample projects. Recall that most repositories require you to specify a license for your project. This permits the repository to host your project and communicate to everyone what your intentions are regarding ownership, permissions to use, and so forth.

As I stated previously, you need to check the license before using any design. If you plan to modify the design, you need to pay close attention to the license. The vast majority of licenses allow you to use the design and most allow you to modify it.

However, where some licenses differ is in regard to the ownership of the modifications. Some open source licenses, like GPL, permit modifications but require you to surrender those modifications to the original owner (the person or organization that created and licensed it) if you plan to distribute those changes. That is, you can modify it at will for personal use, but once you distribute those changes, you have to give them to the owner of the license.

I have only run into this a couple of times, but in those cases, the designer was prototyping designs for a commercial product. The license and indeed the text of the project made it clear she was looking for help with the design but that the design would not be made public. Watch out for this and tread lightly. Any work you do could be for the benefit of the owner and not yours to keep or profit.

■ **Tip**   When in doubt about a license, contact the originator and ask them directly.

Since most sample IoT projects are licensed for sharing and free modification, you normally don't have to worry too much. However, I recommend that you check the license before using any project, especially if you intend to share or publish your derivation.

Keep It Appropriate

Believe it or not, there are hobbyists and enthusiasts with impressively vivid imaginations who have come up with IoT projects that some may consider inappropriate or even obscene. No matter what your own views are, you should strive to tolerate the views of others. That doesn't mean you must compromise your own views—just be aware that yours may offend and strive to minimize the offense.

More specifically, don't upload projects with inappropriate themes to sites that are viewable by everyone. It is (or may be) fine to upload some project that promotes a theme, ideal, and so forth (provided there are no copyright violations), just don't upload projects or commentary that are clearly offensive or intended to cause harm.

For example, if you consider the fact that IoT projects are being used in schools to teach children the technology and techniques of working with hardware and designing software, you shouldn't upload projects with themes that parents may deem as inappropriate. The most obvious, of course, are offensive language, adult themes, and slanderous images.

You should check the usage and user agreement for the site that hosts your chosen repository as part of the post-no-post decision. Make sure that you read the section about what is and is not appropriate and adhere to that.

There is another angle to consider. You should avoid uploading sample projects that are or could potentially be illegal or unlawful. This may be difficult to discern considering the IoT community includes the entire globe. However, most sites have language to suggest what is and is not permitted. And some have language in the agreement that gives them (the site) the right to remove things that they deem inappropriate.

For example, I once saw a project for a *radio frequency identification* (RFID) reader that could be used to read RFIDs from a distance. This sounds harmless, but consider the number of things that use RFID, such as security badges, identification, and even credit cards. Clearly, reading RFID from things you own is fine (indeed, that's what the project demonstrated) but the project could be (and most likely has been) used for evil. Fortunately, others noticed this and the project site has been removed (the URL results in a 401 error).

So before you upload a design or sample project, make sure that you understand and agree to the terms of the user agreement, particularly what is and is not appropriate. Most times, a misunderstanding is not something that will get you into trouble, but if you do it more than once, it's likely that someone at the site will want to speak with you or restrict your access. Which brings me back to the opening of this section: be sure to respect the views of others—especially the intended audience of the site. If you disagree with the restrictions, find another site.

Annotate Your Work

I can tell if a sample IoT project is good or of high quality by the way that it is annotated and documented; that is, how well the designer described the project on its site. If I encounter a project that looks appealing only to discover that the designer didn't bother to describe how to connect the hardware, or explains the code in no more than seven words, or didn't provide any instructions, or worse, didn't present any photos of the actual implementation, I won't use it.

Thus, you should strive to provide as full a description as possible. You don't have to write a novel, novella, or a dissertation, but you should provide enough information to describe the intended use, the problem that it solves, as well as a set of instructions on how to write the source code, compile it, and deploy it.

The only exceptions are when you are still working on a project or you plan to make changes before finalizing it. Here, you should mark (annotate) the project with some verbiage about a work in progress, being experimental, and so forth. If your repository has a feature to mark the project as such, use that. This way, others will know that your project isn't quite ready for general adoption. One reason for doing this is to get feedback from others. I've done this myself with mixed results. Mostly, people are happy to comment that they like it, but don't comment or if they do, however encouraging, don't suggest any changes or improvements.

I would also suggest you provide some level of contact information so that others who have questions can contact you. Typically, viewers can easily contact you through the site, but you may want to provide other forms of contact (e.g., e-mail). You don't want to provide your home address and phone number, but an e-mail address is a nice way to make yourself open to the community.

For example, I have seen blogs, sample projects, and tutorials where people have posted their IRC handle, e-mail address, and even in one case their business phone number. Although I may not go quite that far, I suggest providing an e-mail address so that you can communicate with people who like your project. Plus, it's nice to connect 1:1 with someone to discuss your work!

Be a Good Citizen

Suppose you run across a sample project that not only isn't high quality but is also (in your opinion) designed or implemented incorrectly. Should you immediately comment and crush the designer's ego with a flippant remark about how dumb their code is? No, certainly not!

What I would do (most likely) is ignore the project altogether. I mean, why make things worse by pointing out the defects? I have found that the community at large (there are some exceptions) will likely do the same and not comment. Remember that one of the keys to determining whether a project is well designed (good) is the number of people who have used it. Typically, there is a counter you can check for this. If no one has liked it or even downloaded it, you can be sure it won't make it to the top of any search lists or as sample project of the month.

On the other hand, if you feel compelled to comment, be sure to either contact the designer privately or be as constructive as you possibly can. The goal should be to help the designer improve the project, not challenge their intellect (or pride).

When I do comment on projects that I find strange and perhaps flawed (and it is rare), I generally phrase my comments in the form of a question. A question normally doesn't put someone on the defensive, and if worded properly, it should also not offend.

For example, I may ask, "Have you found the code may hang if the user presses the button more than once?" This is a nice way of asking if the designer has tested his project under the conditions that you expect it to fail. This is good, constructive criticism in a very intellectual form. I am certain that if you think about what you are about to say, you can find other and perhaps more elegant ways of helping people improve their projects.

Now that we've seen why we share and how to share in a responsible manner, let's discover some of the communities that you may want to join or monitor.

Suggested Communities

There are quite a number of general Windows and Windows 10 web sites and online communities that you can visit and join. Most online communities have repositories that you can search for examples, tips, techniques, and complete projects that you can explore. Most also have one or more areas where members can comment, ask questions, or generally communicate with others on the forum. You typically must join to be able to post a reply or ask a question, but viewing is typically permitted by anyone.

The best way to use these resources is to visit them periodically. More specifically, you should read the articles (that interest you) and forum posts regularly. This allows you to keep up-to-date on current events, new techniques, and even new solutions to challenging problems.

Table 17-1 presents a brief list of online resources that you should consider visiting to keep abreast of the latest news about Windows 10, hardware, and IoT. For each forum or web site, I include the general topic, URL, and a brief description. However, you don't have to join any of these communities. And these are not the only forums that you can or should join. In some cases, you may want to simply monitor the site regularly.

Table 17-1. *Online Resources for Windows 10, IoT, and Hardware*

Topics	URL	Description
Windows 10	windows10forums.com	A general forum for Windows 10 organized into a number of categories ranging from hardware to debugging errors to installing software. An excellent source of up-to-the minute information from the user community.
Windows Development	https://msdn.microsoft.com/en-us/default.aspx	This site is the quintessential site for all things Windows programming. If you want a one-stop site for learning the ins and outs of Windows programming, there is none its equal. I use this site regularly as a non-member. Memberships are available for a fee with additional perks.
Windows 10 IoT Development	https://msdn.microsoft.com/en-us/internetofthings.aspx	The MSDN site for Windows 10 IoT Core.
Windows 10 IoT Core Forums	https://social.msdn.microsoft.com/forums/en-US/home?forum=WindowsIoT	The online forums for Windows 10 IoT Core (also part of MSDN).

(continued)

Table 17-1. (*continued*)

Topics	URL	Description
Windows 10 IoT Core	https://developer.microsoft.com/en-us/windows/iot	The starting point for Microsoft's Windows 10 IoT Core specific knowledge space. Includes links to online documentation, sample projects, and more.
Open Hardware	hackster.io	General how-to articles, including a number of Windows 10 IoT Core example projects.
General Electronics	adafruit.com	An online electronics store offering a library of how-to articles covering interesting hardware, including Raspberry Pi, Arduino, and more.
General Electronics	sparkfun.com	An online electronics store offering a library of how-to articles covering interesting hardware, including Raspberry Pi, Arduino, and more.
Raspberry Pi	https://thepihut.com/	An online electronics store offering products for the Raspberry Pi. They also have a number of blogs and tutorials on using the Raspberry Pi in its native operating system.
Raspberry Pi	raspberrypi.org	An online community devoted to the Raspberry Pi.
Arduino	arduino.cc	An online community devoted to the Arduino microcontroller family.

Notice that there are web sites for Windows 10,[3] Windows 10 IoT, Raspberry Pi, general hardware, and similar resources. Although many do not specifically address or cover Windows 10, most have a growing repository of knowledge that is often surprisingly helpful with Windows 10 IoT Core.

For example, I try to connect to the Raspberry Pi sites to keep tabs on what is going on there. I can often get ideas for IoT projects or simply ideas for features by seeing projects implemented for the Raspberry Pi and its native operating system. I find it is often the case that while the source code can be quite different, most of the hardware (connections, etc.) apply without modification, which makes sense since the hardware is not tied to the operating system running on the device (but the libraries that drive the hardware does).

---

[3]Some are even a bit hostile toward Windows users. Should you encounter this absurd bias, simply refrain from replying in kind and ignore the negative remarks. I've found most who post derogatory commentary often know little of which they speak, basing their answers on innuendo, assumptions, and general inaccuracies.

■ **Tip**   Don't discount the power of a keyword search using your favorite online search tool. I have often found obscure gems of information that aren't posted on the more popular sites. In most cases, they are well-written blogs. I often start with a keyword search before I visit the sites listed in Table 17-1. If nothing else, it confirms whether what I am researching is unique or ubiquitous. You should also consider using key phrases from error messages as search terms to get help for specific errors.

There are also a few excellent periodicals that you should consider reading. I exclude the typical Windows periodicals—not because they aren't helpful but because they are far too general for IoT-related research. Rather, the following are periodicals that I've found to be very helpful in my IoT research or any electronics or hobby project.

- *The MagPi Magazine.* A monthly magazine devoted to all things Raspberry Pi. Includes many articles on sample projects, news about the Raspberry Pi, peripherals, and general hardware reviews. (`www.raspberrypi.org/magpi`)

- *Make.* A bimonthly magazine devoted to the broad realm of the maker community presenting sample projects, tutorials, hardware, tools, drones, robots, and more! It truly is a one-stop periodical for all things hacking, tweaking, and general DIY for hobbyists, enthusiasts, and professionals alike. (`http://makezine.com`)

Now that you've seen a number of online communities (but in no way a complete list, as more are added seemingly weekly), as well as a couple of periodicals that you can buy, let's discuss the next step that you can take once you have become a productive contributing member of IoT and IoT-related online communities—becoming a maker.

Become a Maker

The next progression in your growth from novice to enthusiast (and even professional) is to practice your craft regularly and share your knowledge with the world. One excellent outlet for this desire is to become a maker. Makers are widely regarded as experts in their areas of interest. The best part is that a maker's interests can vary greatly from one to another. That is, becoming a maker isn't about learning a specific set of techniques (academic or otherwise). It really is about practicing a craft.

What's a Maker?

Sadly, there is no single definition of what a maker is or should be. This is because a maker is someone with highly creative skills who desires to tinker and work on projects that can range from mechanical sculptures that spit fire to electronic gismos, to new ways to recycle materials to build things on the cheap.

Indeed, a maker is simply an artisan, a craftsman, a hobbyist, or an enthusiast who desires to create things. Hence, *maker*. There are many kinds of makers. However, what unites them is the willingness to share their techniques and skills with others. As a result, to truly become a maker, you must participate in the community of makers.

Share Your Ideas

You can become a maker and not contribute at all to the online communities. And while that's perfectly fine, if you really want to help make the community stronger, you should become more involved. The best way to do this is to join one or more of the online forums and start contributing.

That doesn't mean you have to start off by posting some fantastically cool and successful IoT project complete with award-winning documentation and bulletproof code that computer science professors will someday teach young minds to mimic your brilliance. Sure, you may scoff (or even laugh) at that, but I have encountered people who are afraid of posting anything lest they be proven wrong or ridiculed for inaccuracies.

The best way to avoid that is to start out slow by first asking questions of your own. You can also start by throwing out a few positive comments for those ideas and projects you like. As you become more involved with the topics and techniques, your knowledge will expand to the point where you can start answering others' questions.

I encourage you to consider this level of involvement if you want to become more involved with Windows 10 IoT Core. I believe that interacting with the community and gaining a reputation for being helpful and sharing ideas is one of the things that separates a hobbyist from an enthusiast.

Thus, if you want to become a maker, you should share your ideas with others.

Attend an Event

Another way that you can become more involved with the maker community is to attend a Maker Faire (www.makerfaire.com). These events are held all over the world. The events allow makers to showcase their creations, teach others, and generally celebrate all things maker. See the Maker Faire web site for events in your area.

If you live in or near a larger city, you may find that there are local user groups for Raspberry Pi, MySQL, Arduino, and Windows 10. There are even Maker user groups. Try searching for events and groups in your area to find out where and when they meet. Most groups have an open door policy and invite one and all to attend their meetings. Some, however, do have dues for charter members, but this often comes with additional perks such as access to tools, labs, and discounts on bulk purchases.

Once you have attended an event (even one that has nothing to do with Windows 10 IoT Core), such as a Raspberry Pi meetup, you'll be hooked. You can learn quite a lot from others this way. And who knows—perhaps one day you will present at an event. Speaking from experience, it can be very satisfying sharing your knowledge with others in an open forum like a conference, user group, meetup, or Maker Faire. That's when you know you've obtained the reputation, skills, and knowledge that a typical maker possesses.

That doesn't mean that by the time you read this chapter you're an expert at all things IoT, or that you even profess to be an expert, but you should be well on your way all the same.

Summary

Taking their IoT skills to the next level is what most people are ultimately inspired to achieve. However, even if you do not want to become a traveling maker, teaching the world, you can learn quite a lot by simply joining the Windows, Windows 10, IoT, and open hardware online communities.

In this chapter, I presented suggestions on how best to interact with online communities, how to join an online community, and how to take your skills to the highest level of enthusiasm and become a maker. This chapter therefore rounds out our journey down the Windows 10 IoT road. I hope that I have inspired you to continue practicing your skills and to join the community of likeminded enthusiasts.

I sincerely hope that this book has opened many doors for you concerning Windows 10, IoT, and anything open hardware. May your IoT projects all be successful, but when they aren't, may you learn something in the process. Just remember to share your experiences—good or bad—with the world and to give back a little of what you can take from the efforts of others.

Appendix

This appendix presents a list of the hardware required to complete each chapter, along with a consolidated list for the book. The appendix concludes with suggestions regarding acquiring the hardware needed. While these lists are included in each chapter and discussed in greater detail, listing them here helps to see all of the hardware used in the book as a set and helps when planning to purchase the components that you do not already own.

Hardware by Chapter

This section presents a list of the hardware for each chapter that contains a project or sample that you can try on your own. We begin with a look at the base components that apply to all of the chapters with projects.

Base Components for All Chapters

The following lists the hardware needed for all the projects in the book. That is, you should acquire these items to complete the projects in this book.

- Raspberry Pi 2 or 3, MinnowBoard Max–compatible, or DragonBoard 410C
- Wi-Fi dongle (Raspberry-compatible) or Ethernet cable
- 5V 2A power supply
- Micro-SD card (minimum of 8GB)
- breadboard (half or full sized)
- jumper wires: male-to-female
- jumper wires: male-to-male

Some projects that are *headed*, meaning they have a user interface that you can interact with that runs on your device.

- HDMI monitor
- USB keyboard (Raspberry-compatible)
- USB mouse (Raspberry-compatible)

© Charles Bell 2016
C. Bell, *Windows 10 for the Internet of Things*, DOI 10.1007/978-1-4842-2108-2

The following are accessories that you may want to acquire that could make the projects a bit more fun or perhaps enhance your experience.

- Enclosure for your device
- Motorola Lapdock (to provide a monitor, keyboard, and mouse)
- USB hub

Required Hardware by Chapter

This section presents the required hardware list for each chapter that has a project. I include a short synopsis of the main project for each chapter. Those chapters without projects are omitted.

■ **Note**   Each chapter is annotated as *headed* (requires a monitor, keyboard, and mouse) or as *headless* (no user interface hardware is needed).

Chapter 3: Headless

This chapter explores the origins of the Raspberry Pi, including a tour of the hardware and a short primer on how to use its native operating system. The chapter demonstrates how easy it is to write programs to control hardware on the Raspberry Pi using a Python script. The project solution is a simple script to turn an LED on and off.

- (1) 560 ohm 5% 1/4W resistor
- (1) 10mm red LED
- (1) breadboard
- (2) jumper wires: male-to-female

Chapter 5: Headed

This chapter provides a crash course in Visual C++ that covers the basics of syntax and constructs of a Visual C++ application, including a walk-through of building a real C++ application that blinks an LED. The project covers XAML, including how to wire events to controls, and even a little about how to use the dispatcher timer.

- (1) 560 ohm 5% 1/4W resistor
- (1) 10mm red LED
- (1) breadboard
- (2) jumper wires: male-to-female

Chapter 6: Headed

This chapter provides a crash course in C# that covers the basics of syntax and constructs of a Visual C# application, including a walk-through of building a real C# application that blinks an LED. The project covers XAML including how to wire events to controls, and even a little about how to use the dispatcher timer.

- (1) 560 ohm 5% 1/4W resistor
- (1) 10mm red LED
- (1) breadboard
- (2) jumper wires: male-to-female

Chapter 7: Headless

This chapter provides a crash course in Python that covers the basics of syntax and constructs of a Python application, including a walk-through of building a real Python application that blinks an LED. The project covers how to work with headless applications, including how to manage a startup background application.

- (1) 560 ohm 5% 1/4W resistor
- (1) 10mm red LED
- (1) breadboard
- (2) jumper wires: male-to-female

Chapter 9: Headless

This chapter introduces the Adafruit Microsoft IoT Pack for Raspberry Pi. You look at a project that shows how to read sensor, thereby making the transition from experiments to actual, usable projects. This chapter shows that the kit is a viable option for getting the right amount of hardware to start building IoT solutions.

- (1) pushbutton (breadboard pin spacing)
- (2) red LEDs
- (2) yellow LEDs (or blue is OK)
- (1) green LED
- (5) 150 ohm resistors (or equivalent to match LEDs)
- (7) jumper wires: male-to-female

Chapter 10: Headless

This chapter presents how to use an ADC, how to connect and set up an SPI device, how to read a potentiometer, and finally how to use the debug feature to write out statements to the output window. While the project itself is rather simplistic, the emphasis and therefore the learning part of the project lie in discovering how to write code for all of these features.

- (1) 10K ohm potentiometer (breadboard pin spacing)
- (2) red LEDs
- (2) yellow LEDs (or blue is OK)
- (1) green LED
- (5) 150 ohm resistors (or appropriate for your LEDs)
- (1) MCP3008 ADC chip
- (19) jumper wires: (8) male-to-male, (11) male-to-female

Chapter 11: Headed

This chapter presents how to use a special library, Microsoft IoT Lightning Providers, to get access to PWM and SPI interfaces. The project introduces how to use the new library to interface with the ADC via an SPI interface, how to read values from an LDR, and how to use PWM to control the brightness of an LED. While the project itself is rather simplistic, the code is another example of more complex programming for IoT solutions.

- (1) LED (any color)
- (1) 10K ohm resistor
- (1) 150 ohm resistors (or appropriate for your LEDs)
- (1) light-dependent resistor (photocell)
- (1) MCP3008 ADC chip
- (13) jumper wires: (5) male-to-male, (8) male-to-female

Chapter 12: Headed

This chapter presents a depth of complexity that demonstrates how to combine a number of advanced tools and techniques from using a code library written by someone else to building a C++ headed application to incorporating a C# and C++ project in the same solution to read weather sensors (a very popular choice for IoT project). Combining all of these together makes this project the most complex in the book. It provides the best example of the power of Visual Studio and UWP IoT applications.

- (1) Adafruit BMP280 I2C or SPI barometric pressure and altitude sensor
- (4) jumper wires: male-to-female

Chapter 13: Headless

This chapter introduces the concept of storing IoT data in a database. It introduces MySQL presenting a short tutorial on how to connect your device to MySQL for storing data. The project shows how to add a database component to a C# project as a means to use MySQL to store data and thus write more IoT projects that can persist data for later retrieval and analysis using proven technologies.

- (1) Adafruit BMP280 I2C or SPI barometric pressure and altitude sensor
- (4) jumper wires: male-to-female

Chapter 14: Headless

This chapter takes a minor off ramp from our regularly scheduled Windows 10 IoT project highway to examine how to control hardware remotely via the Internet. The project for this chapter demonstrates how to build a nifty out of office sign with a mechanical flag and LEDs controlled from a web page. This represents the fundamental building blocks for other remote controlled IoT projects.

- (1) SparkFun Servo Trigger (sparkfun.com/products/13118)
- (1) servo (adafruit.com/products/169) or (sparkfun.com/products/9065)
- (4) red LEDs
- (1) green LED

- (5) 150 ohm resistors

- (1) solder breadboard (optional)

- (11) jumper wires: (10) male-to-female, (1) male-to-male

Chapter 15: Headless

This chapter is another off ramp that examines how to leverage the vast and growing repository of Arduino sketches and examples for use with Windows 10 IoT Core. The chapter demonstrates several small projects to use special libraries in Visual Studio to write Arduino-style sketches on our Raspberry Pi as well as how to control an Arduino remotely from Windows 10 UWP applications.

- (1) Arduino Uno

- (1) USB cable (for programming Arduino)

- (1) Adafruit RGB color sensor with IR filter and white LED–TCS34725 sensor

- (1) 1602 (16 characters on 2 lines) LCD module

- (1) potentiometer (breadboard-friendly)

- (1) pushbutton

- (1) 10K ohm resistor

- (1) 220 ohm resistor

- (12) jumper wires: male-to-female

- (12) jumper wires: male-to-male

- (1) Raspberry Pi 2 or 3

There are also some optional components that you may want to have if you want to implement all of the examples in the chapter. These include the following. See the chapter for more details about the optional components.

- (1) Temperature sensor: TMP36

- (1) Arduino Ethernet shield

- (1) SparkFun Bluetooth Mate Silver (www.sparkfun.com/products/12576)

- (1) Ethernet cable

- (1) 5V power supply for Arduino

Chapter 16: Headless

This chapter introduces an optional foray into the world of cloud computing by demonstrating how to incorporate an Azure Remote Monitoring solution to present data collected from a weather sensor. The chapter shows how easy it is to expand your IoT solution from a simple, local network solution to a worldwide cloud-based solution.

- (1) Adafruit BMP280 I2C or SPI barometric pressure and altitude sensor

- (4) jumper wires: male-to-female

Consolidated Hardware List

This section presents a table that lists the hardware needed to complete the core projects in this book. This includes all of the projects through Chapter 13, which presents a solid overview of how to start building IoT projects and should be sufficient to satisfy (and challenge) beginners and intermediate readers alike. Chapters 14, 15, and 16 are optional for most readers, due to the added complexity and additional hardware requirements.

The following tables present the required hardware for projects through Chapter 13 and the required hardware for the optional projects. If the component can be found in a kit, the tables list the source for purchasing the components as a kit. However, most components can be purchased from an online or retail electronics store, such as Adafruit or SparkFun. In some cases, the component is from a specific vendor.

Once again, it is assumed you will have purchased a board (e.g., Raspberry Pi), a power adapter, and so forth, as described in the "Base Components for All Chapters" section.

Table A-1. *Required Hardware (through Chapter 13)*

Component	Quantity	Sources
Breadboard (half or full)	1	Adafruit Microsoft IoT Pack for Raspberry Pi
Jumper wires: male-to-female	1 set	Adafruit Microsoft IoT Pack for Raspberry Pi
Jumper wires: male-to-male	1 set	Adafruit Microsoft IoT Pack for Raspberry Pi
LEDs (red)	4	Adafruit Microsoft IoT Pack for Raspberry Pi
LEDs (green)	2	Adafruit Microsoft IoT Pack for Raspberry Pi
LEDs (yellow)	2	Adafruit Microsoft IoT Pack for Raspberry Pi
10K ohm resistor	1	Adafruit, SparkFun
560 ohm resistors	1	Adafruit Microsoft IoT Pack for Raspberry Pi
150 ohm resistors	5	Adafruit, SparkFun
Adafruit BMP280 I2C or SPI barometric pressure and altitude sensor	1	Adafruit Microsoft IoT Pack for Raspberry Pi
MCP3008 ADC chip	1	Adafruit Microsoft IoT Pack for Raspberry Pi
Light-dependent resistor (photocell)	1	Adafruit Microsoft IoT Pack for Raspberry Pi
10K ohm potentiometer (breadboard pin spacing)	1	Adafruit Microsoft IoT Pack for Raspberry Pi
Pushbutton (breadboard pin spacing)	1	Adafruit Microsoft IoT Pack for Raspberry Pi

▓ **Note** Chapters 14, 15, and 16 also require most of the components listed in Table A-1.

Table A-2. *Required Hardware (Chapters 14-16)*

Component	Quantity	Sources
RGB color sensor with IR filter and white LED– TCS34725	1	Adafruit Microsoft IoT Pack for Raspberry Pi
Arduino Uno	1	Adafruit, SparkFun
USB cable (for programming the Arduino)	1	Adafruit, SparkFun
SparkFun servo trigger	1	sparkfun.com/products/13118
Micro hobby servo	1	adafruit.com/products/169 or sparkfun.com/products/9065
(1) Solder breadboard (optional)	1	Adafruit, SparkFun

Suggestions for Purchasing the Hardware

In Chapter 9, you discover the main source for the components as the Adafruit Microsoft IoT Pack for Raspberry Pi, which contains all of the hardware you will need for the projects through Chapter 13 as well as Chapter 16. Only Chapters 14 and 15 require hardware components not found in the kit. However, there is an alternative, as described in the sidebar in Chapter 9. I repeat the discussion on both options next for reference.

Adafruit Microsoft IoT Pack for Raspberry Pi

The Microsoft IoT Pack for Raspberry Pi 3 comes in two varieties: one with the Raspberry Pi (www.adafruit.com/products/2733) and one without the Raspberry Pi (www.adafruit.com/products/2702) for those who already own a Raspberry Pi 2 or 3.

The kit with the Raspberry Pi costs about $114.95 and the kit without the Raspberry Pi costs about $75.00. Clearly, if you already have a Raspberry Pi, you can save some money there. In fact, for those who want to use a different low-cost computer board, you can buy the kit without the Raspberry Pi—except for the micro-SD card with Windows 10 and possibly the power supply, all of the other components will work with other boards. The kit comes with a number of handy components, including prototyping tools and a few sensors.

■ **Note** You will not need every component included in the kit, but the extra parts will come in handy when developing your own IoT projects.

There are three categories of components: electronic components included in the kit, accessories for the Raspberry Pi, and sensors.

The electronic components provided in the kit include the following.

- (2) breadboard trim potentiometer
- (5) 10K 5% 1/4W resistor
- (5) 560 ohm 5% 1/4W resistor
- (2) diffused 10mm blue LED
- (2) diffused 10mm red LED
- (2) diffused 10mm green LED
- (1) electrolytic capacitor: 1.0uF
- (3) 12mm tactile switches

The list of accessories in the kit is long. The following includes all of the accessories included in the kit. I describe some of these in more detail.

- *Adafruit Raspberry Pi B+ case– smoke base / clear top*: An excellent case to protect your Pi from accidents.

- *Full-size breadboard*: Plenty of space to spread out your circuits.

- *Premium male-to-male jumper wires, 20 × 6 inches (150mm)*: Jumps from one port to another on the breadboard. They're extra long and come molded in a ribbon so you can peel off only those you need.

- *Premium female-to-male extension jumper wires, 20 × 6 inches*: Jumps from male GPIO pins to the breadboard ports. They also come molded in a ribbon.

- *Miniature Wi-Fi module*: A Raspberry Pi–approved Wi-Fi dongle (not needed for the Raspberry Pi 3).

- *5V 2A Switching power supply with a 6-foot micro-USB cable*: Meets the Raspberry Pi requirements for power.

- *MCP3008 – 8 channel 10-bit ADC with SPI interface*: A breakout board you can use to expand the number of SPI interface channels for larger IoT projects

- *Ethernet cable, 5-foot*: A nice touch considering the kit has a Wi-Fi dongle. Good to have a backup plan!

- *8GB class 10 SD/MicroSD memory card*: Windows 10 IoT core preloaded!

The sensors included with the kit are an unexpected surprise. They provide what you need to create some interesting IoT solutions. Best of all, they are packaged as breakout boards making them easy to wire into our circuits. The following lists the sensors included in the kit.

- *(1) photocell*: A simple component to measure light.

- *Assembled Adafruit BME280 temperature, pressure, and humidity sensor*: Measures temperature, barometric pressure, and humidity.

- *Assembled TCS34725 RGB color sensor*: Measures color. Comes with an infrared filter and white LED.

Adafruit Parts Pal

If you are planning to use a board other than the Raspberry Pi, are on a more limited hobby budget, or want only the bare essentials, there are alternatives to the Microsoft IoT Pack from Adafruit. In fact, Adafruit sells another kit that includes almost everything you need for the projects in this book. It doesn't come with sensors, but all of the basic bits and bobs are in there, and you can always buy the sensors separately.

The Adafruit Parts Pal comes packaged in a small plastic case with a host of electronic components (www.adafruit.com/products/2975). The kit includes the following components prototyping tools, LEDs, capacitors, resistors, some basic sensors, and more. In fact, there are more components in this kit than the Windows IoT Pack for the Raspberry Pi 3. Better still, the kit costs only $19.95, which means it's a good deal (and the case is a great bonus). Although you may not need all of the parts in this kit, it offers a great start for building a supply of electronics parts for future projects.

- (1) storage box with latch

- (1) half-size breadboard

- (20) jumper wires: male-to-male, 3 inches (75mm)

- (10) jumper wires: male-to-male, 6 inches (150mm)
- (5) 5mm diffused green LEDs
- (5) 5mm diffused red LEDs
- (1) 10mm diffused common-anode RGB LED
- (10) 1.0uF ceramic capacitors
- (10) 0.1uF ceramic capacitors
- (10) 0.01uF ceramic capacitors
- (5) 10uF 50V electrolytic capacitors
- (5) 100uF 16V electrolytic capacitors
- (10) 560 ohm 5% axial resistors
- (10) 1K ohm 5% axial resistors
- (10) 10K ohm 5% axial resistors
- (10) 47K ohm 5% axial resistors
- (5) 1N4001 diodes
- (5) 1N4148 signal diodes
- (5) NPN transistor PN2222 TO-92
- (5) PNP transistor PN2907 TO-92
- (2) 5V 1.5A linear voltage regulator, 7805 TO-220
- (1) 3.3V 800mA linear voltage regulator, LD1117-3.3 TO-220
- (1) TLC555 wide-voltage range, low-power 555 timer
- (1) photocell
- (1) thermistor (breadboard version)
- (1) vibration sensor switch
- (1) 10K breadboard trim potentiometer
- (1) 1K breadboard trim potentiometer
- (1) Piezo buzzer
- (5) 6mm tactile switches
- (3) SPDT slide switches
- (1) 40-pin break-away male header strip
- (1) 40-pin female header strip

The only thing that I feel is missing are the male-to-female jumpers, but you can buy them separately (www.adafruit.com/product/1954). For only $1.95 more, they're worth adding to your order!

Index

© Charles Bell 2016

C. Bell, *Windows 10 for the Internet of Things*, DOI 10.1007/978-1-4842-2108-2

Get the eBook for only $5!

Why limit yourself?

Now you can take the weightless companion with you wherever you go and access your content on your PC, phone, tablet, or reader.

Since you've purchased this print book, we're happy to offer you the eBook in all 3 formats for just $5.

Convenient and fully searchable, the PDF version enables you to easily find and copy code—or perform examples by quickly toggling between instructions and applications. The MOBI format is ideal for your Kindle, while the ePUB can be utilized on a variety of mobile devices.

To learn more, go to www.apress.com/companion or contact support@apress.com.

Apress®
THE EXPERT'S VOICE™

CPSIA information can be obtained
at www.ICGtesting.com
Printed in the USA
LVOW02s1438280217
525683LV00004B/245/P